Automotive Science and Mathematics

Allan Bonnick

LONDON AND NEW YORK

First published by Butterworth-Heinemann

This edition published 2011 by Routledge
2 Park Square, Milton Park, Abingdon, Oxon OX14 4RN
711 Third Avenue, New York, NY 10017, USA

First edition 2008

British Library Cataloguing in Publication Data
A catalogue record for this book is available from the British Library

Library of Congress Cataloging-in-Publication Data
A catalog record for this book is available from the Library of Congress

ISBN: 978-0-7506-8522-1

Printed and bound in Great Britain by the MPG Books Group

Contents

Preface

One of the main aims of this book is to provide a course of study of science and mathematics that constantly demonstrates the links between these disciplines and the everyday work of technicians in the automotive field.

The subject matter has been chosen to provide full cover for the Science and Mathematics of the BTEC and IMI National Certificates and Diplomas and the related Technical Certificates and NVQs up to and including Level 3.

The needs of students in the 14 to 19 age group who may be following a scheme of vocational studies have been borne in mind during the writing of the book. It is hoped that these students and their teachers will find the links between theory and practice that are demonstrated in the text to be helpful in strengthening students' desire to continue with their education.

The topics start at a fairly basic level and the coverage should provide the necessary skill for trainees and students to demonstrate competence in key skills.

The coverage of some topics, such as vehicle dynamics and heat engines (thermodynamics), is at the advanced end of National Level 3 and will be found helpful by HNC/HND and Foundation Degree students.

Answers are provided to assist those who may be studying privately and a set of solutions is available on the Elsevier website for lecturers, teachers and other training providers.

Units and symbols

Formulae and the associated symbols are frequently used to describe the relationship between various factors; for example, power $P = 2 \times \pi \times T \times N$.

In this formula, P stands for power in watts, T is torque in newton metres and N is the number of revolutions per second. In formulae, the multiplication signs are normally omitted and the above equation is written as $P = 2\pi TN$.

In order to simplify matters, many countries have adopted the international system of units which is known as the Système Internationale, normally referred to as SI units. This system is used in this book because it is the system that is widely used in engineering and science.

SI units

Quantity	Symbol	SI unit
Mass	m	Kilogram (kg)
Length	l	Metre (m)
Time	t	Second (s)
Velocity	U, v	m/s
Acceleration	a	m/s^2
Electric current	I	Ampere (A)
Temperature	K	Kelvin

Derived SI units

Quantity	Symbol	Base SI units	Derived unit
Area	A	$m \times m$	m^2
Volume	V	$m \times m \times m$	m^3
Density	ρ	kg/m^3	kg/m^3

Derived SI units

Quantity	Symbol	Base SI units	Derived unit
Force and weight	F, W	$kg\,m/s^2$	N (newton)
Pressure	P	$kg/m\,s^2$	N/m^2 (pascal)
Energy	E or U	$kg\,m^2/s^2$	Nm = J (joule)
Power	P	$kg\,m^2/s^3$	J/s = W (watt)
Frequency	f	s^{-1}	Hz (hertz)
Electric charge	Q	$A \times s$	C (coulomb)
Electric potential difference	V	$kg\,m^2/As^3$	V (volt)
Electrical resistance	R		Ω (ohm)

Prefixes

Prefix	Symbol	Multiply by
Tera	T	10^{12}
Giga	G	10^9
Mega	M	10^6
Kilo	k	10^3
Milli	m	10^{-3}
Micro	μ	10^{-6}
Nano	n	10^{-9}

Glossary

Adiabatic
An ideal process in which there is no interchange of heat between the working substance and its surroundings. The adiabatic index is a basic feature in the consideration of operating cycles for internal combustion engines.

Air–fuel ratio
The amount of air required for combustion of a given mass of fuel. It is expressed as a ratio such as 14.7:1. This means that 14.7 grams of air are required for each 1 gram of fuel.

Alternating current (a.c.)
A voltage supply that varies its polarity, positive and negative, in a regular pattern.

Analogue
A varying quantity such as voltage output from an alternator.

Brake power
The actual power that is available at the output shaft of an engine. It is the power that is measured on a dynamometer called a brake. Brake power is quoted in kW but the term brake horse power (bhp) is often used; $1\,\text{bhp} = 0.746\,\text{kW}$.

Capacitance
The property of a capacitor to store an electric charge when its plates are at different electrical potentials.

Centre of gravity
The point at which the entire weight of the object is assumed to be concentrated.

Darlington pair
A circuit containing two transistors that are coupled to give increased current gain. It is used for switching in automotive circuits where high current is required.

Differential lock
A device that temporarily disables transmission differential gears in order to improve traction in difficult driving conditions.

Elasticity
The property of a material to return to its original shape when it is stretched or otherwise deformed by the action of forces.

Equilibrium
A state of balance. The equilibrant of a system of forces is the single force that will produce balance in the system; it is equal and opposite to the resultant.

Forward bias
When the polarity of the emf (voltage) applied to a p–n junction diode is such that current begins to flow, the diode is said to be forward biased.

Gradeability
The maximum angle of a gradient that a vehicle is capable of climbing.

HEGO
Heated exhaust gas oxygen sensor. These are exhaust gas oxygen sensors that are equipped with a heating element that reduces the time that it takes for the sensor to reach a satisfactory operating temperature.

Inertia
Inertia is the resistance that a body offers to starting from rest or to change of velocity when it is moving. The mass of a body is a measure of its inertia.

Joule
The joule is a unit of energy equal to 1 N m.

Kelvin
The Kelvin temperature is used in most engineering calculations. It starts from a very low temperature that is equivalent to −273°C. Temperature in K = temperature in C + 273.

Kinetic energy
The kinetic energy of a body is the energy that it possesses by virtue of its velocity.

Lambda
Lambda λ is symbol from the Greek alphabet that is used to denote the chemically correct air–fuel ratio for an internal combustion engine. The exhaust gas oxygen sensor is often referred to as a lambda sensor because it is used to detect the percentage of oxygen in the exhaust gas.

LED
Light-emitting diodes are diodes that emit light when current is passed through them. The colour of the light emitted is dependent on the semiconductor material that is used in their construction.

Load transfer
Load transfer is the apparent transfer of load from front to rear of a vehicle that occurs when the vehicle is braked or accelerated. Load transfer from side to side also occurs as a result of centrifugal force when the vehicle is cornering.

Mass
The mass of an object is the quantity of matter that it contains. Mass is measured in grams (g) or kilograms (kg).

Newton
The newton is the unit of force in the SI units system. It is the force that will produce an acceleration of $1\,m/s^2$ when it is applied to a mass of 1 kg that is free to move. 1 newton is approximately equal to 0.225 lbf.

NO_x
The different forms of oxides of nitrogen.

Ogive
An ogive is the bell-shaped curve that derives from a cumulative frequency graph.

Oxygen sensor
Exhaust gas oxygen sensors (EGOs) detect the percentage of oxygen in the exhaust gas of an engine.

Pollutant
Pollutants are the gases and other substances that arise from the operation of motor vehicles. In connection with engine emissions CO_2, NO_x and CO are among the substances that harm the atmosphere and the environment in general.

Quartiles
A system used in statistics to divide a set of data into four equal parts. The parts are called first quartile, second quartile etc.

Resultant
The resultant of a number of forces acting at a point is the single force that would replace these forces and produce the same effect.

Roll centre
The roll centre height of a vehicle is the distance from the ground to the point at which the vehicle body will tend to roll when subjected to a side force, such as the centrifugal force produced by turning a corner. The height of the roll centre is determined by the type of suspension system. The front roll centre and the rear roll centre are normally at different heights and the imaginary line drawn between the two roll centres is known as the roll axis.

Semiconductor
Semiconductors are materials that have a higher resistivity than a conductor but a lower resistivity

than a resistor. Semiconductor materials are the basis of transistors, diodes, etc. and are widely used in the construction of integrated circuits.

Selective catalyst reduction
A system used on some diesel engine vehicles to reduce emissions of NO_x. A liquid such as urea is injected into the exhaust stream where it works with a catalyst so that NO_x is converted into N_2 (nitrogen) gas and H_2O (water vapour).

Specific fuel consumption
The mass (weight) of fuel that each kW of power of an engine consumes in 1 hour under test bed conditions. SFC is measured in kg/kWh and it is a measure of an engine's efficiency in converting fuel into power.

Tonne
The tonne is a unit of mass that is used for large quantities. 1 tonne = 1000 kg.

Traction control
A computer-controlled system that co-ordinates anti-lock braking, differential lock and engine management functions to provide optimum vehicle control under a range of driving conditions.

Xenon
Xenon is one of the noble gases. It is used in small quantities in the manufacture of headlamp bulbs to give an increased amount of light.

Young's modulus
Young's modulus of elasticity is an important elastic constant used in describing the properties of elastic materials. Young's modulus is denoted by the symbol E; it is calculated by dividing stress by the strain produced.

Zirconium
Zirconium is a metallic element used in the construction of voltaic-type exhaust gas oxygen sensors.

This page intentionally left blank

1
Arithmetic

1.1 Terminology of number systems

Prime
Prime numbers are numbers that are divisible by themselves and by 1.

Examples of prime numbers are: 1, 3, 5, 7, 11, 13, etc.

Integer
An integer is a whole number, as opposed to a fraction or a decimal.

Digit
The symbols 1, 2, 3, 4, 5, 6, 7, 8, 9, that are used to represent numbers, are called digits.

Rational
A rational number is any number that can be written as a vulgar fraction of the form a/b where a and b are whole numbers.

Irrational
An irrational number is one that cannot be written as a vulgar fraction. If an irrational number is expressed as a decimal it would be of infinite length. Examples of irrational numbers that appear often in mechanical calculations are π and $\sqrt{2}$.

Real number
A real number is any rational or irrational number.

Vulgar fraction
A fraction has two parts: a numerator and a denominator, e.g. $\frac{1}{2}$.

In this example the numerator is 1 and the denominator is 2.

Improper fraction
An improper fraction is one where the numerator is larger than the denominator, for example 3/2.

Ordinal number
An ordinal number is one that shows a position in a sequence, e.g. 1st, 2nd, 3rd.

1.2 The decimal system

In the decimal system a positional notation is used; each digit is multiplied by a power of 10 depending on its position in the number. For example:

Example 1.1
$567 = 5 \times 10^2 + 6 \times 10^1 + 7 \times 10^0$, or $5 \times 100 + 6 \times 10 + 7 \times 1$. That is, 5 hundreds, 6 tens and 7 ones.

The decimal point is used to indicate the position in a number, after which the digits represent fractional parts of the number.

For example, 567.423 means 567 plus $4/10 + 2/100 + 3/1000$.

Addition and subtraction of decimals

When adding or subtracting decimals the numbers must be placed so that the decimal points are exactly underneath one another; in this way the figures are kept in the correct places.

Example 1.2

(a) Simplify $79.37 + 21.305 + 10.91$

```
 79.37
 21.305
 10.91
-------
111.585
-------
```

(b) Simplify $65.42 - 23.12$

```
65.42
23.12
-----
42.30
-----
```

Multiplication and division – decimals

Multiplying and dividing by powers of 10

In decimal numbers, multiplication by 10 is performed by moving the decimal point one place to the right.

For example, $29.67 \times 10 = 296.7$

To multiply by 100 the decimal point is moved two places to the right; to multiply by 1000 the decimal place is moved three places to the right, and so on for higher powers of 10.

To divide decimal numbers by 10 the decimal point is moved one place to the left, and to divide by 100 the decimal point is moved two places to the left.

Example 1.3

$132.4 \div 10 = 13.24$, and $132.4 \div 100 = 1.324$

Long multiplication

The procedure for multiplying decimals is shown in the following example.

Example 1.4

Calculate the value of 27.96×56.24

Step 1

Count the number of figures after the decimal points. In this case there are four figures after the decimal points.

Step 2

Disregard the decimal points and multiply 2796 by 5604.

Step 3

Perform the multiplication

```
    2796
    5624
--------
   11184
   55920
 1677600
13980000
--------
15724704
--------
```

Step 4

There now needs to be as many numbers after the decimal point as there were in the original two numbers, in this case four. Count four figures in from the right-hand end of the product and place the decimal point in that position. The result is 1572.4704.

Long division

Example 1.5

Divide 5040 by 45.

```
                Dividend
Divisor----45) 5034 (112----Quotient
```

Set the problem out in the conventional way. The three elements are called dividend (the number being divided), divisor (the number that is being divided into the dividend), and quotient (answer), as shown here.

Step 1

```
45) 5040 (1
    45        How many times does 45 divide into 50?
     5          It goes once.
              Put a 1 in the quotient. 1 × 45 = 45.
              Write the 45 directly below the 50.
              Subtract the 45 from the 50 and write the
                remainder directly below the 5 of the 45.
```

Step 2

```
45) 5040 (11
    45
    ──
     54       Bring down the 4 from 5040. Divide 45
     45         into 54. It goes once.
     ──       1 × 45 = 45. Put another 1 in the
     90         quotient.
              Write the 45 directly below the 54 to
                give the remainder of 9.
              Bring down the 0 from the last place
                of 5040.
```

Step 3

```
45) 5040 (112
    45
    ──
     54       45 divides into 90 exactly twice.
     45         2 × 45 = 90.
     ──       Write the 2 in the quotient.
      90      Write the 90 under the other 90.
      90        There is no remainder.
      ──      The answer is: 5040 ÷ 45 = 112.
       0
```

Division – decimals

Example 1.6

Calculate the value of $26.68 \div 4.6$

Step 1

Convert the divisor 4.6 into a whole number by multiplying it by 10; this now becomes 46.

To compensate for this the dividend 26.68 must also be multiplied by 10, making 266.80. Then divide 266.80 by 46.

Step 2

Proceed with the division:

```
46) 266.8 (5.8
     230       Produced by 5 × 46
     368       ~You have reached the decimal point.
                  Place a decimal
                  point in the quotient, next to the 5.
     368       Divide 46 into 368 = 8
     000       8×46=368 to subtract. Zero remainder.
```

There is no remainder, so $26.68 \div 4.6 = 5.8$.

When there is a remainder

Example 1.7

Divide 29.6 by 5.2

Multiply 5.2 by 10 to make it a whole number and multiply 29.6 by 10 to compensate. The long division then proceeds as in Example 1.5.

```
52) 296 (5.692307
    260
    ───
     360
     312
     ───
      480
      468
      ───
       120
       104
       ───
        160
        156
        ───
         400
         364
         ───
          36   This is the remainder
          ──
```

The calculation could continue but in many problems the degree of accuracy required does not warrant further division.

The procedure for making approximations and deciding how many decimal places to show in an answer is shown in Section 1.3.

1.3 Degrees of accuracy

Rounding numbers

At a particular petrol station in Britain in 2005 the price of petrol was 93.6 pence/litre. Taking 1 gallon to be equal to 4.5461 litres, the price per gallon of the petrol would be $93.6 \times 4.5461/100 =$ £4.2551496. This number is of little value to most motorists because the smallest unit of currency is one penny. For practical purposes the price per gallon would be rounded to the nearest penny. In this case the price per gallon would be shown as £4.26.

This type of rounding of numbers is performed for a variety of reasons and the following rules exist to cover the procedure.

Decimal places

It is common practice to round numbers to a specified number of decimal places.

Example 1.8

For example, 4.53846 is equal to 4.54 to two decimal places, and 4.539 to three decimal places.

The general rule for rounding to a specified number of decimal places is:

1. place a vertical line after the digit at the required number of decimal places; delete the digits after the vertical line.

If the first digit after the vertical line was 5 or more, round up the number by adding 1 in the last decimal place.

Example 1.9

4.53846 to 2 decimal places

> 4.53|846
>
> 4.54 to 2 decimal places.

4.53846 to 3 decimal places

> 4.538|46
>
> 4.538 to 3 decimal places.

Significant figures

In many cases numbers are quoted as being accurate to certain number of significant figures. For example, vehicles are often identified by engine size in litres. An example is one model in the 2005 Ford Fiesta range that has an engine size of $1388\,\text{cm}^3$. This is equivalent to 1.388 litres because $1000\,\text{cm}^3 = 1$ litre, and the car is known as the 1.4 litre Fiesta. This is a figure that is given as accurate to two significant figures because it is the size in litres that is significant.

The rule for rounding to a specified number of significant figures is:

> If the next digit is 5 or greater the last digit of the rounded number is increased by 1. A nought in the middle of the number is counted as significant. For example, the number 6.074 correct to three significant figures is 6.07.

1.4 Accuracy in calculation

When working in significant figures the answer should not contain more significant figures than the smallest number of significant figures used in the original data.

When working in decimal places the usual practice is to use one more decimal place in the answer than was used in the original question. In both cases the degree of accuracy used should be quoted.

1.5 Powers and roots and standard form

A quantity such as 3×3 can be written as 3^2; this is called 3 squared, or 3 to the power of 2. The small figure 2 to the right is called the **index** and it tells us how many times 3 is multiplied by itself. The number that is being multiplied, in this case 3, is called the **base**. When the index is a fraction it is also known as a root; for example $9^{1/2}$ means 'square root of 9', and this written as $\sqrt{9}$.

General rules for indices

In the following the symbol 'a' is used to represent any number; a^n means 'a' multiplied by itself n times. In this case 'a' is the base and n is the index.

The rules are as follows.

1. To multiply together powers of the same base, add the indices.
2. To divide powers of the same base, subtract the indices.
3. To raise one power to another, multiply the indices.

Example 1.10

(i) Simplify $a^2 \times a^2$

Using rule 1 in the above list, $a^2 \times a^2 = a^{(2+2)} = a^4$, which means 'a' multiplied by itself four times.

(ii) Simplify $a^3 \div a^2$

Using rule 2 above, $a^3 \div a^2 = a^{(3-2)} = a^1$, and this is written as a.

(iii) Simplify $(a^2)^2$

Using rule 3 above, $(a^2)^2 = a^{(2\times2)} = a^4$

Negative indices

When the index is negative as in a^{-1} it means $\frac{1}{a}$, or the reciprocal of a.

For example $2^{-1} = \frac{1}{2}$

Decimal indices

Decimal indices such as $8^{0.4}$ occur in some calculations. These may be calculated as follows.

1. Press AC
2. Enter 8
3. Press the shift key
4. Press x^y
5. Enter 0.4
6. Press =

The result is $8^{0.4} = 2.297$

Fractional indices

The square root of 25 is normally written as $^2\sqrt{25}$ and it means the number which, when multiplied by itself twice, gives 25. In this case it is 5.

Another way of writing the square root of 25 is $25^{½}$.

In general $^n\sqrt{a} = a^{1/n}$. The fractional index represents the root, e.g. square root and cube root, and the denominator of the index represents the root to be taken. For example $^3\sqrt{a} = a^{1/3}$.

1.6 Standard form

Writing numbers in standard form is often used to avoid mistakes in reading very large or very small numbers. For example, 20 000 000 may be written as 2×10^7, which means 2 multiplied by 10 seven times, and 0.0005 can be written as 5×10^{-4}, which means 5 divided by 10 000.

In effect, when the index is positive (+) it shows the number of places that the decimal point must be moved to the right.

Example 1.11

$2.186 \times 10^3 = 2186$. The decimal point is moved three places to the right.

When the index is negative (−) it shows the number of places that the decimal point must be moved to the left.

Example 1.12

$3.24 \times 10^{-2} = 0.0324$. The decimal point is moved two places left.

Multiplying and dividing numbers in standard form

When numbers are written in standard form the rules of indices can be used to facilitate multiplication and division.

Example 1.13
Simplify: $(25 \times 10^3) \times (3 \times 10^2)$

$$(25 \times 10^3) \times (3 \times 10^2) = 25 \times 3(10^3 \times 10^2)$$
$$= 75 \times 10^{(3+2)} = 7.5 \times 10^6$$

Simplify: $\frac{(8 \times 10^4)}{(2 \times 10^2)}$

$$\frac{(8 \times 10^4)}{(2 \times 10^2)} = \frac{8}{2} \times 10^{4-2} = 4 \times 10^2$$

1.7 Factors

Numbers such as 24 can be made up from 6×4. So 6 and 4 are **factors** of 24.

Example 1.14
$88 = 2 \times 44 = 11 \times 8$. The factors here are 2 and 44, or 11 and 8.

Highest common factor (HCF)

Take two numbers such as 40 and 60.

The factors of 40 can be 5 and 8, or 10 and 4, or 2 and 20.

The factors of 60 can be 5 and 12, 15 and 4, 30 and 2, 20 and 3.

One number, 20, is the largest factor that is common to 40 and 60.

The highest common factor (HCF) of a set of numbers is the greatest number that is a factor of all of the numbers in the set.

Multiples

88 is a multiple of 2, 44, 11, or 8.

Lowest common multiple (LCM)

The lowest common multiple of a set of numbers is the smallest number into which each of the given numbers (factors) will divide exactly without leaving a remainder.

For example, the LCM of 3 and 4 is 12 because that is the smallest number into which both 3 and 4 will divide without leaving a remainder.

1.8 Fractions

Addition and subtraction

When adding or subtracting fractions they must first be changed to give a common denominator.

For example $1/3 + 5/6$.

This can be changed to $2/6 + 5/6$.

This gives a common denominator of 6, which is the smallest (lowest) number into which 3 and 6 will divide without leaving a remainder.

A procedure that is used for addition and subtraction of fractions is shown here:

Example 1.15
Simplify $1/3 + 5/6$

The lowest common denominator $= 6$

So $1/3 = 2/6$. The 2/6 is added to 5/6 to give $7/6 = 1 + 1/6$.

This is normally written

$$1/3 + 5/6 = \frac{1 \times 2 + 5 \times 1}{6} = \frac{2+5}{6} = \frac{7}{6} = 1\frac{1}{6}$$

Subtracting fractions

Use the same procedure as for addition.

Example 1.16
Simplify $5/6 - 1/2$

The common denominator is 6

$$5/6 - 1/2 = \frac{5-3}{6} = \frac{2}{6} = 1/3$$

Fractions and whole numbers

Mixed numbers such as $2\frac{1}{2}$ and $3\frac{1}{4}$ can be expressed as improper fractions.

Example 1.17
Convert the following mixed numbers to improper fractions.

$2\frac{1}{2}$, $3\frac{1}{4}$

$$2\tfrac{1}{2} = \frac{(2 \times 2) + 1}{2} = \frac{5}{2}$$

$$3\tfrac{1}{4} = \frac{(3 \times 4) + 1}{4} = \frac{13}{4}$$

The procedure is: multiply the whole number part by the denominator of the fractional part and then add the number obtained to the numerator of the fractional part.

The procedures for adding, subtracting, multiplying and dividing fractions can then be used.

Example 1.18

The specified oil capacity of a certain engine is $4^1/_4$ litres. Owing to an error during an oil change, $4^5/_8$ litres are put into the engine when it is refilled.

Determine the amount of oil that must be drained out in order to restore the correct oil level on the dip-stick.

Solution

The amount of oil to be drained out $= 4^5/_8 - 4^1/_4$

$$4^5/_8 = \frac{37}{8}$$

$$4^1/_4 = \frac{17}{4} = \frac{34}{8}$$

$$= \frac{37-34}{8} = \frac{3}{8} \text{ litre}$$

Combined addition and subtraction

The procedure here is the same as for addition or subtraction, paying regard to the plus and minus signs.

Example 1.19

Simplify $^1/_2 + ^1/_4 - ^1/_8$

The common denominator is 8

Answer: $^1/_2 + ^1/_4 - ^1/_8 = \dfrac{4+2-1}{8} = ^5/_8$

Multiplication and division of fractions

Cancelling is the process of simplifying fractions when multiplying and dividing.

Example 1.20

Simplify $\dfrac{7 \times 56}{8}$

This reduces to $7 \times 7 = 49$ because $\frac{56}{8} = 7$

Eliminating the 8 in this way is called cancelling.

Multiplying fractions

The procedure here is to multiply the numerators together and then multiply the denominators together.

Example 1.21

$$\frac{5}{3} \times \frac{4}{7} = \frac{5 \times 4}{3 \times 7} = \frac{20}{21}$$

Dividing fractions

To divide by a fraction the rule is to invert it and then multiply.

Example 1.22

Evaluate $\dfrac{1}{2} \div \dfrac{1}{4}$

$$\frac{1}{2} \div \frac{1}{4} = \frac{1}{2} \times \frac{\mathbf{4}}{\mathbf{1}} = \frac{4}{2} = 2$$

Note: The $\dfrac{1}{4}$ has been inverted to give $\dfrac{4}{1}$

Order of performing operations in problems involving fractions

In order not to get confused when dealing with fractions it is important to perform operations in the correct sequence. The rule is as follows.

1. **Brackets** – deal with brackets first.
2. **Divide** and **multiply** and deal with **of**.
3. **Add** and **subtract**.

Some people find the acronym **BODMAS** (brackets, of, divide, multiply, add, subtract) useful as a reminder of the order in which operations on fractions are performed.

Example 1.23

Simplify $\frac{1}{16}+\left(\frac{1}{2}\div\frac{1}{6}\right)$

$$\frac{1}{16}+\left(\frac{1}{2}\div\frac{1}{6}\right)$$

Work out the bracket first

$$\left(\frac{1}{2}\div\frac{1}{6}\right)=\frac{1}{2}\times\frac{6}{1}=3$$

Then do the addition

$$\frac{1}{16}+3=3\frac{1}{16}$$

Placing fractions in order

Example 1.24

Place the following fractions in order, smallest first: 5/8, 3/4, 6/10, 2/5, 2/3.

As most work is now done in decimal it is best to convert the fractions to decimals.

First, divide the bottom into the top of the fraction and then express the result to three decimal places.

$5/8 = 0.625$, $3/4 = 0.750$, $6/10 = 0.600$, $2/5 = 0.400$, $2/3 = 0.667$

Placed in order, smallest first; 2/5, 6/10, 5/8, 2/3, 3/4

1.9 Ratio and proportion. Percentages

A ratio is a comparison between similar quantities. In general a ratio of m to n may be written as m : n or as a fraction m/n.

In order to state a ratio between two quantities they must be in the same units.

Example 1.25

In an engine, 500 ml of gas is compressed into a space of 50 ml.

State this as a ratio in the form m : n and m/n.

In the form m:n the ratio is 500 : 50, which can be reduced to 10 : 1.

As a fraction $m/n = 500/50 = 10/1$.

Examples of ratios in vehicle technology

Example 1.26 Aspect ratio of tyres

The ratio of the height of a tyre to the width of the tyre is known as the aspect ratio, or the height-to-width ratio.

In the tyre section shown in Figure 1.1 the height of the tyre h = 200 mm and the width w = 250 mm.

The aspect ratio is $\frac{h}{w}=\frac{200}{250}=0.8$; this is usually expressed as a percentage which, in this case, is 80%.

The tyre size that is moulded into a tyre wall gives the aspect ratio. For example on a tyre marking of 185/70 – 14, the 185 is the tyre width in millimetres, 70 is the aspect ratio, and the 14 is the rim diameter in inches.

Example 1.27 Gear ratios – gearbox

Gearbox ratios are expressed in terms of engine speed in rev/min compared with the output speed of the gearbox shaft in rev/min.

If an engine is running at 3000 rev/min and the gear that is engaged gives a gearbox output shaft speed of 1000 rev/min, the ratio is 6000 : 1000 and this is given as 6 : 1.

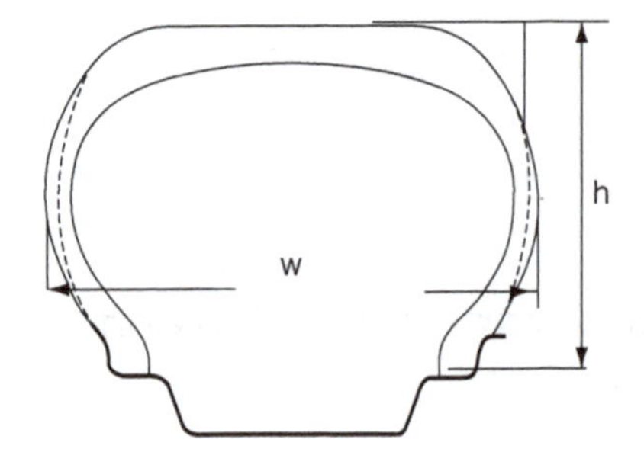

Fig. 1.1 Aspect ratio of a tyre (Example 1.26)

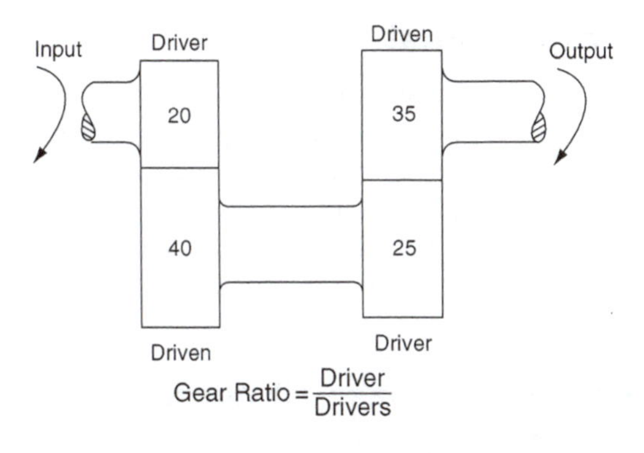

Fig. 1.2 Gearbox ratio (Example 1.27)

Calculation of gearbox ratios:

Gearbox ratio

$$= \frac{\text{number of teeth on the driven gears multiplied together}}{\text{number of teeth on the driving gears multiplied together}}$$

Figure 1.2 shows the gear train for a gearbox in second gear. The driven gears have 40 teeth and 35 teeth respectively, and the driving gears have 20 and 25 teeth respectively.

The formula for calculating the gear ratio is normally shortened to give

$$\text{gearbox ratio} = \frac{\text{driven}}{\text{driver}} \times \frac{\text{driven}}{\text{driver}}$$

In this example the gear ratio $= \frac{40}{20} \times \frac{35}{25} = \frac{1400}{500} = 2.8:1$

Example 1.28 Air–fuel ratio

The air-to-fuel ratio of the mixture that is supplied to an engine is the ratio between the mass of air supplied and the mass of that is mixed with it.

For proper combustion of petrol a mass of 14.7 kg of air is required for 1 kg of fuel.

This gives an air-to-fuel ratio = 14.7 : 1.

Example 1.29 Compression ratio

The compression ratio of a piston engine is the ratio between the total volume inside the cylinder when the piston is at bottom dead centre and the volume inside the cylinder when the piston is at top dead centre. These two volumes are shown in Figure 1.3.

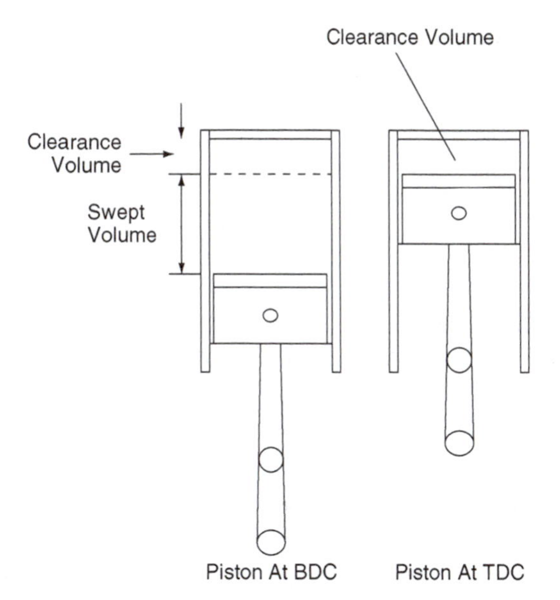

Fig. 1.3 Compression ratio (Example 1.29)

Calculate the compression ratio of an engine that has a swept volume of $450\,\text{cm}^3$ and a clearance volume of $50\,\text{cm}^3$.

Solution

Compression ratio

$$= \frac{\text{swept volume} + \text{clearance volume}}{\text{clearance volume}}$$

$$= \frac{450+50}{50} = \frac{500}{50} = 10:1$$

$$= \frac{500}{50}$$

$$= 10:1$$

Percentages

Percentages are used when making comparisons between quantities.

Example 1.30

A person may earn £200 per week while a friend earns £220 per week.

The friend earns £20 per week more.

This may be expressed as a percentage as follows.

Percentage extra $= \frac{20}{200} \times 100 = 10\%$

Example 1.31

An engine that develops 120 kW is tuned to raise the power output to 145 kW. Calculate the percentage increase in power.

Solution

The power increase $= 145 - 120 = 25\,\text{kW}$

$$\begin{aligned}\text{Percentage increase} &= \frac{\text{increase}}{\text{original}} \times \frac{100}{1}\\ &= \frac{145-120}{120} \times 100\\ &= \frac{25 \times 100}{120}\\ &= 20.8\%\end{aligned}$$

1.10 The binary system

Arithmetic operations in systems that use computer logic are normally performed in binary numbers because the logic circuits for binary numbers are simpler than for other number systems. In binary numbers, each digit is multiplied by the appropriate power of 2.

For example, binary $101 = 1 \times 2^2 + 0 \times 2^1 + 1 \times 2^0 = 5_{10}$

Most significant bit (MSB)

In computing language each digit of a binary number is called a bit. The digit at the left-hand end of the number is the most significant bit (MSB); the digit on the right-hand end is the least significant bit (LSB).

Hexadecimal

Hexadecimal numbers have a base of 16.

Letters are used to denote digits greater than 9.

For example $A = 10_{10}$, $B = 11_{10}$, $C = 12_{10}$, $D = 13_{10}$, $E = 14_{10}$, $F = 15_{10}$

The number $A2F_{16} = 10 \times 16^2 + 2 \times 16^1 + 15 \times 16^0 = 2560 + 32 + 15 = 2607_{10}$

Converting base 10 numbers to binary

The following example shows a method that requires continuous division by 2, as shown in Example 1.32.

The layout and positioning of the remainder part of the division are important.

Example 1.32

Convert 53_{10} to binary.

2 /53
2 /26 – 2 into 53 goes 26 times, remainder 1. (LSB)
2 /13 – 2 into 26 goes 13 times, remainder 0
2 /6 – 2 into 13 goes 6 times, remainder 1
2 /3 – 2 into 6 goes 3 times, remainder 0
2 /1 – 2 into 3 goes once, remainder 1
0 – 2 does not go into zero so the remainder is 1. (MSB)
This final digit is the most significant bit.

The binary number reads as: starting at the bottom 110101.

The binary equivalent of $53_{10} = 110101_2$.

Uses of binary numbers in vehicle systems

The electronic control unit (ECU) in a vehicle system is a computer. Most computers operate internally on binary numbers. In a computer, the 0s and 1s that make up a binary number, or code, represent voltages. For example, in a system known

Table 1.1 Some ASCII codes

Character	ASCII code
A	1000001
k	1101011
8	0111000

as TTL (transistor–transistor logic) a voltage of 0 volts to 0.8 volts acts as a binary 0 and a voltage in the range 2 volts to 5 volts acts as a binary 1. The information and instructions that are transmitted around a computer-controlled system on a vehicle are made up from binary numbers, or codes. Each 0 or 1 in the binary code represents a voltage level, or state of a transistor switch – either on or off.

Binary codes

The ASCII code (American Standard Code for Information Interchange) is widely used in computing and related activities. Letters and numbers are represented by a 7-bit code. Some examples are shown in Table 1.1.

1.11 Directed numbers

Directed numbers are numbers that have a plus or minus sign attached to them, for example $+4$ and -3.

A typical example of the use of a directed number is in the description of low temperatures such as those below freezing point.

The strength of antifreeze solutions is a guide to the protection against frost a solution will give. For example, a solution of 1 part of antifreeze to 3 parts of water gives protection against ice damage down to a temperature of −25°C.

Directed numbers occur in many applications, and in some cases it may be helpful to think of directed numbers in terms of movement on a number-line, as shown in Figure 1.4.

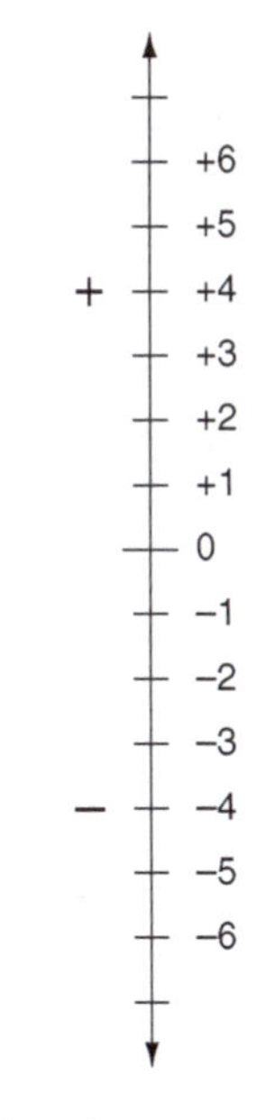

Fig. 1.4 Directed numbers

For example, $-2-4$ means start at -2 on the number-line and then move down another 4 to give -6.

Rules for dealing with directed numbers

Addition of directed numbers, where all the signs are the same.

1. To add together numbers whose signs are the same, add the numbers together.
2. The sign of the sum of the numbers is the same as the sign of each of the numbers.

Example

Find the value of: $+5+3+8$.

Using the above rule, add the numbers together $= 16$; the sign of each number is $+$. this means that the result is $+16$. The plus sign is normally omitted and the absence of a sign in front of the number shows that it is a plus number.

Example

Find the value of: $-3-2-4$.

Add the numbers together to give 9; the sign is minus, so $-3-2-4=-9$.

Addition of directed numbers where the signs are different

To add together two directed numbers whose signs are different, subtract the smaller number from the larger one, and give the result the sign of the larger number.

To find the value of: $-2+4$.

On the number scale, start at -2 and from this point move four steps in the plus direction, which takes you to $+2$.

$$-2+4=2$$

Subtraction of directed numbers

To subtract a directed number use the normal method of subtraction and place the number to be subtracted beneath the other number; then change the sign of the number that is being subtracted and add the numbers together. In other words, *change the sign of the bottom line and add.*

Find the value of: $-3-(+2)$

The bottom line is $+2$; change its sign and it becomes -2, and $-3-2=-5$

Multiplication of directed numbers

Example

$+2+2+2=+6$

But, 3×2 is the same as $+2+2+2=+6$

Example

$(-4)+(-4)+(-4)=-12$

But $(-4)+(-4)+(-4)$ is the same as $3\times(-4)=-12$

Division of directed numbers

Using the procedure as for multiplication $3\times(-4)=-12$

The justification for this is:

$$\frac{-12}{3}=-4$$

and

$$\frac{-12}{-4}=3$$

When dividing: like signs produce a plus (+)

Unlike signs produce a minus (−).

When multiplying and dividing directed numbers

$$(+)\times(+)=(+) \qquad (+)\div(+)=(+)$$
$$(-)\times(-)=(+) \qquad (-)\div(-)=(+)$$
$$(-)\times(+)=(-) \qquad (-)\div(+)=(-)$$
$$(+)\times(-)=(-) \qquad (+)\div(-)=(-)$$

1.12 Summary of main points

1.13 Exercises

1.1 Figure 1.5 shows the spring and slipper assembly that is used pre-load with respect to the rack and pinion gears of a steering system. In order to provide the correct pre-loading force, the gasket B and two shims C must have a total thickness of 0.89 mm. If the gasket is 0.14 mm in thickness and one of the shims has a thickness of 0.25 mm, determine the thickness of the other shim.

1.2 The compression height of a piston is the distance from the centre of the gudgeon pin to the highest part of the piston crown. Figure 1.6 shows a piston where the compression height is the dimension H. Calculate the compression height of this piston.

1.3 Figure 1.7(a) shows the position of thrust washers that are used to absorb end thrust on a crankshaft. Figure 1.7(b) shows the procedure for checking the end float on the crankshaft that is required to ensure proper lubrication of these thrust washers.

The end float on a certain crankshaft must be 0.15 mm but when measured it is found to be 0.35 mm. In order to rectify this excess end float, oversize thrust washers are available. How much thicker must each thrust washer be to ensure that the end float of the crankshaft is 0.15 mm? (Note: one thrust washer for each side of the main bearing.)

1.4 Figure 1.8 shows a piston and connecting rod assembly. In this assembly the gudgeon pin centre is offset to one side of the piston. Given the two dimensions shown in Figure 1.7, calculate the amount of gudgeon pin offset.

1.5 Figure 1.9 shows a cross-section of a cylinder head and combustion chamber. The cylinder head face is to be machined to restore its flatness. The dimension X must not be less than 25.2 mm. When measured prior to machining, the dimension X is found to be 26.1 mm. Determine the maximum depth of material that can be machined off the cylinder head face when attempting to restore its flatness.

1.6 Figure 1.10 shows part of the head of a poppet valve. Determine the dimension X.

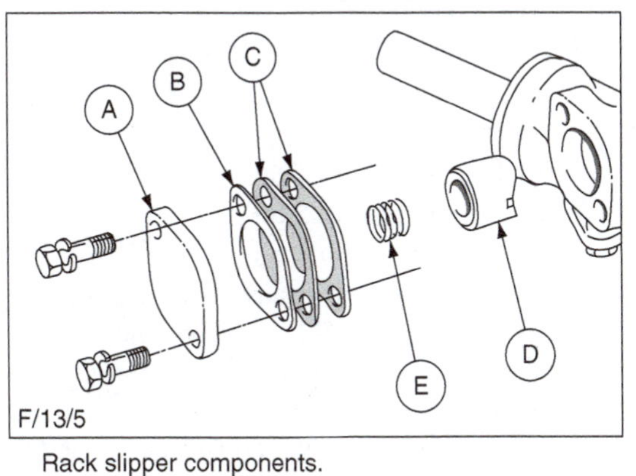

Rack slipper components.
A - Cover plate B - Gasket C - Shims
D - Slipper E - Spring

Fig. 1.5 (Exercise 1.1)

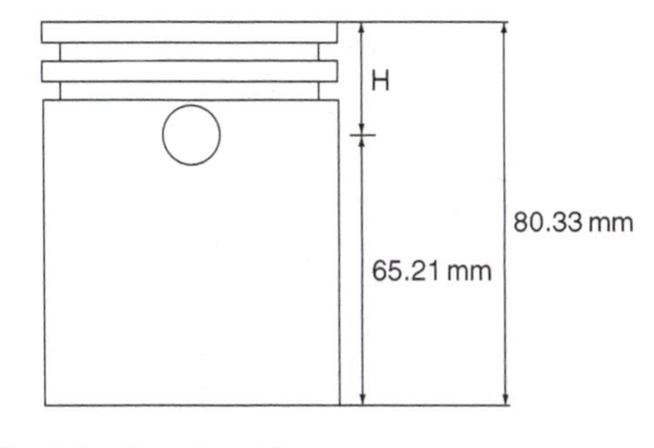

Fig. 1.6 (Exercise 1.2)

(a)

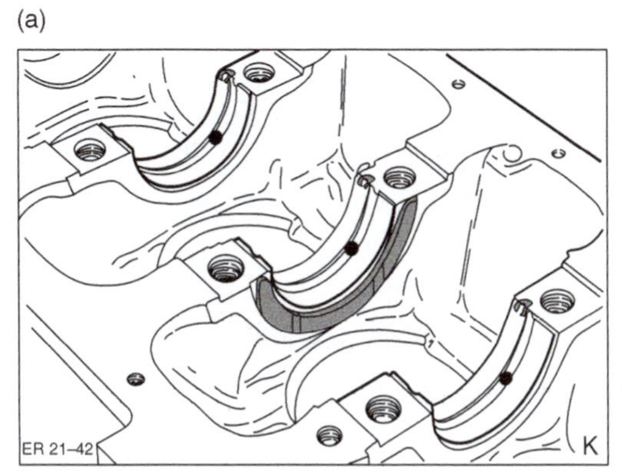

Installation of bearing shells and thrust half washers.

(b)

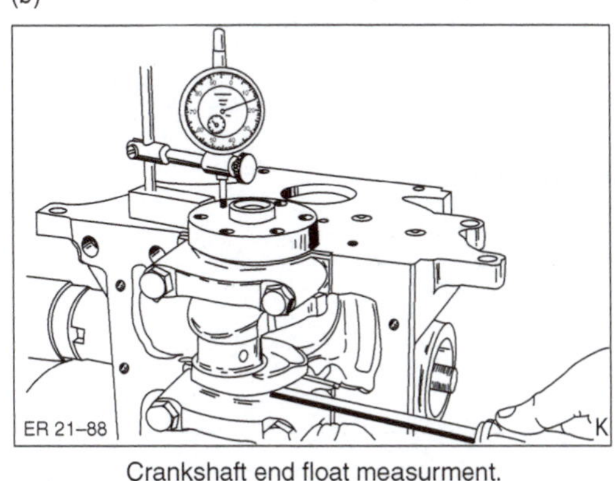

Crankshaft end float measurment.

Fig. 1.7 (Exercise 1.3)

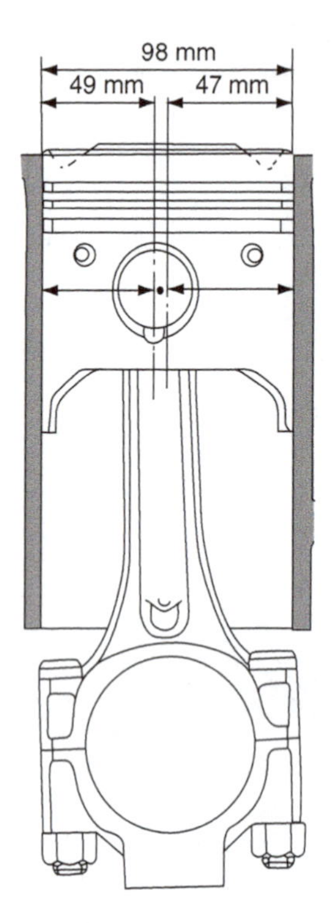

Fig. 1.8 (Exercise 1.4)

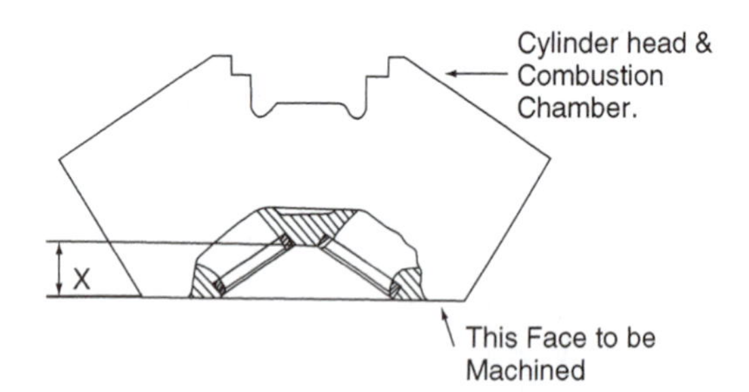

Fig. 1.9 (Exercise 1.5)

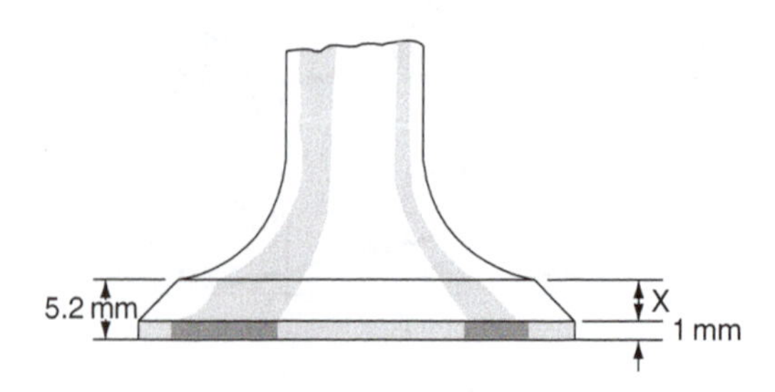

Fig. 1.10 (Exercise 1.6)

1.7 The overall gear ratio for a vehicle in a certain gear is determined by multiplying the final drive ratio by the gearbox ratio. Given that a vehicle has a second gear ratio of 1.531 : 1 and a final drive ratio of 4.3 : 1, calculate the overall gear ratio. Give the answer correct to decimal places, without using a calculator.

1.8 Determine the value of each of the following.

(a) $18.6 \div 6.2$, (b) $778 \div 389$, (c) $172.8 \div 1.2$, (d) $31.42 \div 0.40$

1.9 A vehicle travels 140 km and uses 8.67 litres of fuel. Calculate the average fuel consumption in km/litre.

1.10 The swept volume of an engine is calculated by multiplying the cross-sectional area of the cylinder by the length of the stroke and the number of cylinders.

Calculate the total swept volume of a 4-cylinder engine that has a stroke of 90 mm and a bore whose cross-sectional area is $55.6\,\text{cm}^2$. Give the answer in cm^3 and litres, correct to two decimal places.

1.11 Evaluate the following, correct to two decimal places.

(a) $\dfrac{76.25 \times 3}{38.125}$, (b) $37 \div 2.62$,

(c) $2.179 \div 3.142$

1.12 Write down the following numbers correct to the number of significant figures stated.

(a) 37.8651	to 4 S.F.
(b) 48.703	to 4 S.F.
(c) 39 486 621	to 5 S.F.
(d) 0.00765	to 2 S.F.

1.13 A technician's job is advertised at a rate of pay of £387 for a 39-hour week. calculate the hourly pay rate and give the answer correct to the nearest penny.

1.14 A certain garage charges £42 per hour for technicians' work on vans.

A particular repair takes 6 hours and 20 minutes and the materials used have a total cost to the customer of £102.58.

Calculate the total charge to the customer before VAT is added.

1.15 A vehicle travels 10.36 km on 1 litre of fuel. How much fuel will be used on a journey of 186 km at the same rate of fuel consumption?

1.16 The fuel tank of a certain vehicle weighs 30 kg when empty. The capacity of the fuel tank is 220 litres. Calculate the weight of a tankful of fuel when it is filled with diesel fuel that has a density of 0.83 kg/litre.

1.17 Simplify the following by giving each answer as a single power.
(a) $a^2 \times a^3 \times a$, (b) $x^3 \div x^2$, (c) $(5^3)^2$, (d) $(t \times t^2)^2$, (e) $10^3 \times 10^2 \div 10^4$

1.18 Determine the value of each of the following.
(a) $27^{1/3}$, (b) $81^{1/2}$, (c) $9^{0.5} \times 9^{1.5}$, (d) $10\,000^{0.25}$

1.19 Find the value of each of the following, giving the answer correct to three decimal places.
(a) 8^{-1}, (b) 2^{-3}, (c) $64^{1/6}$, (d) $25^{-1/2}$

1.20 Find the HCF of the following numbers.
(a) 3 and 6, (b) 9, 27 and 81, (c) 57 and 19, (d) 8, 16 and 56

1.21 Find the HCF of the following.
(a) 13, 52, 65; (b) 18, 54, 216; (c) 28, 196, 224

1.22 During a check of the antifreeze strength of the coolant in five vehicles it is found that the following quantities of antifreeze are required to bring the coolant up to the required strength for protection against frost.
Vehicle 1 – $1/3$ litre
Vehicle 2 – $1/2$ litre
Vehicle 3 – $5/8$ litre
Vehicle 4 – $1/2$ litre
Vehicle 5 – $3/8$ litre.
Determine the total amount of antifreeze required.

1.23 Simplify $4\tfrac{1}{2} + 3\tfrac{3}{4} + 17/8$

1.24 Simplify the following.
(a) $1/2 - 1/5$ (b) $3/4 - 5/8$ (c) $5/8 - 1/4$ (d) $7/8 - 1/4$

1.25 Simplify the following.
(a) $\frac{1}{2} + \frac{1}{3} + \frac{4}{5} - \frac{1}{6}$ (b) $2\frac{1}{3} - 1\frac{1}{4} + \frac{5}{8}$ (c) $\frac{1}{32} - \frac{1}{16} + \frac{3}{8} + 9$

1.26 Simplify the following.
(a) $\frac{1}{81} \div \frac{1}{9}$, (b) $\frac{4}{5} \times \frac{5}{8} \div \frac{7}{8} \times \frac{8}{9}$, (c) $\frac{2}{7} \times \frac{7}{8} \times \frac{1}{16}$

1.27 Simplify the following.
(a) $\left(\frac{1}{3} \div \frac{1}{4}\right) - \frac{1}{6}$, (b) $\left(\frac{1}{81} + \frac{1}{9}\right) \times \frac{12}{4}$, (c) $\left(\frac{4}{3} \div \frac{1}{6}\right) + \frac{3}{8} + \frac{1}{32}$

1.28 Write these numbers in order, largest first.
0.2 0.21 0.0201 0.201 0.2001

1.29 Place these fractions in order, smallest first:
$\frac{3}{8}$ $\frac{4}{5}$ $\frac{5}{6}$ $\frac{1}{2}$

1.30 Convert 55_{10} to binary

1.31 Convert $A2F_{16}$ to a number to base 10.

2
Statistics – An introduction

2.1 Definition

Statistics is a branch of mathematics that is used to help people to understand the information that is contained in large amounts of data such as numbers.

Example 2.1
During the 1990s attempts were made to plan ahead so that sufficient training places would be made available to provide the trained workforce that the motor industry would need up to the year 2000. The Department of Education and Science provided figures that showed the number of school leavers who were expected to be looking for job opportunities in each of the years shown in Table 2.1.

From this table it is fairly easy to see that the largest number of school leavers occurs in the year 2000. However, it is not quite so easy to gain an overall impression of the way that the number of school leavers changes over the following years. This aspect of this group of figures can more readily be seen when they are presented in the form of a chart, or graph, as shown in Figure 2.1.

From this picture of the numbers of school leavers it is easy to gain an overall impression of the numbers in one year compared with another. This is just a simple explanation of one use of statistical methods.

Table 2.1 Numbers of school leavers looking for work 1994 to 2000 (Example 2.1)

ȳ Year	No. school leavers (000s)
1994	571
1995	605
1996	622
1997	625
1998	623
1999	628
2000	629

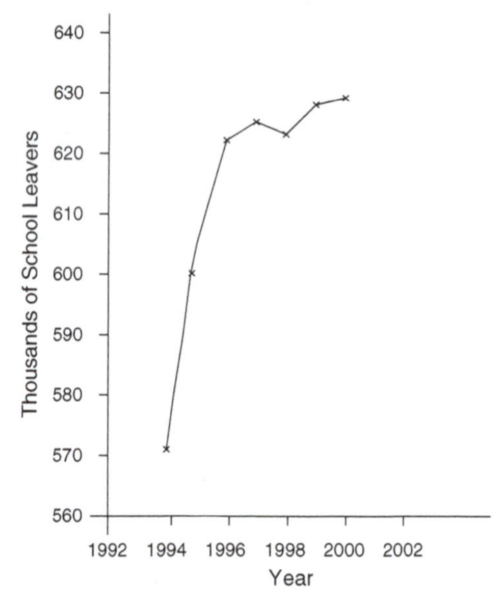

Fig. 2.1 Numbers of school leavers in the period 1992–2000 (Example 2.1)

2.2 Collecting and sorting raw data

The numbers and other facts that are collected for various purposes are called data. In the form that data is collected they are known as raw data. Raw data is often presented as a list and one of the first tasks in processing raw data is to sort a list into some required order.

Example 2.2

Table 2.2 shows the examination marks achieved by a group of 30 technicians.

To study these examination scores and similar lists of figures, the figures may be grouped into **classes**, or categories. In this particular case it has been decided to make classes of 10 marks per class. There are two scores between 20 and 30, so the two dashes are placed in the tally section of the tally chart and the process is repeated for the other scores as shown in the **tally chart** in Table 2.3.

In order to simplify the counting the fifth tally mark is crossed through the other four and counting then proceeds in fives ~~////~~.

The frequency is the number of times that a particular entry occurs; for example, scores in the 31 to 41 class occur four times, i.e. with a frequency of 4.

Table 2.2 Examination scores achieved by a group of 30 technicians (Example 2.2)

87	43	40
63	97	38
51	49	61
50	72	57
23	43	29
38	42	88
52	40	52
65	59	47
58	70	45
47	71	49

Table 2.3 Tally chart (Example 2.2)

Score	Tally	Frequency
20–30	//	2
31–41	////	4
42–52	~~////~~ ~~////~~ //	12
53–64	~~////~~	5
65–76	////	4
77–88	//	2
89–99	/	1
	Total	30

2.3 Making sense of data

Once data has been collected it must be organised in such a way to enable trends and patterns to be seen. The variables that make up the data can be discrete, or continuous.

Discrete variables

An example of a discrete variable is the number of people working in a garage business. This number may be 1, 2, 3 and so on; it cannot of course be 1.3 or 2.6 etc. The number of people working in a garage is a **discrete variable**.

Continuous variables

An example of a continuous variable is the diameter of a crankpin. The nominal size of a particular crankpin is 50 mm. Depending on the accuracy of measurement the actual diameter could be 64.95 mm, or 50.05 mm. In this case the crankpin diameter is a **continuous variable**.

In general there are two methods of dealing with data.

1. **Descriptive statistics**. Here the data is organised into tables from which graphs, charts and other pictorial presentations can be made up.
2. **Statistical inference**. A typical example here is the use of statistics in quality control. A recent case concerned the apparent large number of failures that occurred when anti-lock

braking systems were fitted to large vehicles. The question was raised 'Are the number of failures being reported unusually high and, if they are, what may be the causes?' A large survey was conducted and the results were analysed to see what patterns existed in the data; the results were presented in a report that used both descriptive and analytical statistics. One result was that failure rates were higher than experts expected and on further examination a number of contributory factors to an apparent high failure rate were found to play a part.

2.4 Descriptive statistics – pictographs

Pie charts and bar charts are examples of pictographs.

Pie charts

Pie charts can be used to highlight certain features of data contained in a report. For example, a report by the Office of Fair Trading on garage repair and servicing that was published in the year 2000 contained information which included that there were 16 000 independent garages, 6500 franchised ones, and 3800 Immediate Fit centres in the United Kingdom. This information can be made to have a greater impact by displaying it in the form of a pie chart, as shown in Figure 2.2.

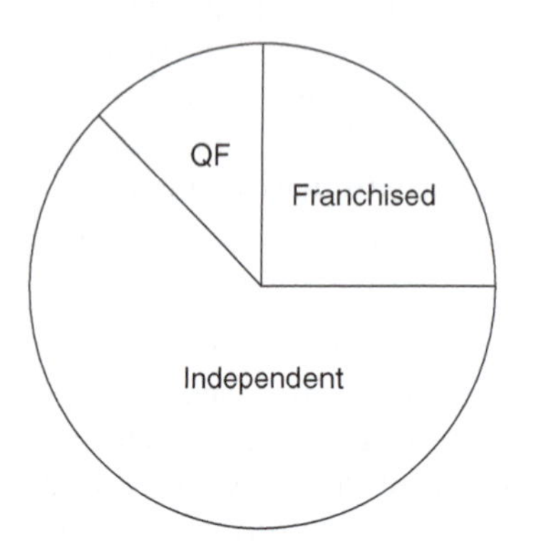

Fig. 2.2 Pie chart showing composition of the vehicle repair and servicing industry

Constructing pictographs

Example 2.3 The pie chart

In the example of the composition of the garage industry in Britain the total number of garages is $16\,000 + 6500 + 3800 = 26\,300$. The proportions that each sector represents as a fraction of the total are:

Independents	$16\,000/26\,300 = 0.61$
Franchised	$6\,500/26\,300 = 0.25$
Immediate Fit	$3\,800/26\,300 = 0.14$

Each sector occupies an equivalent proportion of the 360° of the circle that forms the basis of the pie chart. In this case, each garage sector occupies the following angle, to the nearest degree.

Independents	$0.61 \times 360 = 220°$
Franchised	$0.25 \times 360 = 90°$
Immediate Fit	$0.14 \times 360 = 50°$

From this, with the aid of compasses and a protractor, the pie chart can be constructed.

Example 2.4

In Figure 2.3, which shows worldwide diesel engine production, the vertical scale represents the number of engines produced. The width of each

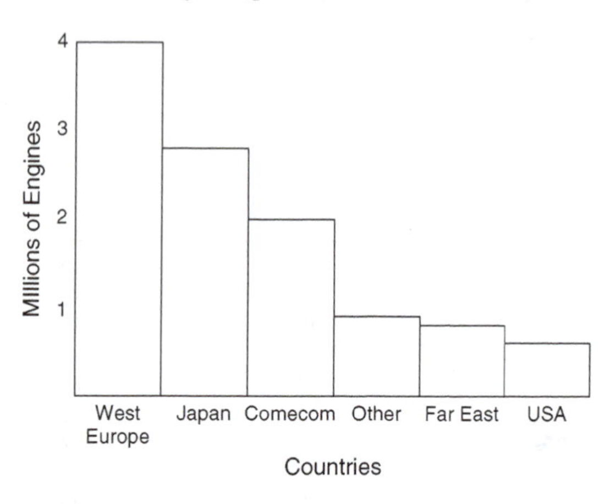

Fig. 2.3 Bar chart showing worldwide distribution of diesel engine production in 1987 (Example 2.4)

bar is the same and a label is added to show which country is being represented on the diagram; the height of the bar represents the number of diesel engines produced in that particular country.

2.5 Interpreting data. Statistical inference

Data and the related statistics are normally collected for some specific purpose. The process of collection of data normally starts by someone asking a question; data is then collected and analysed using statistical methods, and, finally, the results are interpreted. This section provides an introduction to methods of analysing data.

Frequency and tally charts

Frequency

This is the number of times that an event occurs in a given period. For example, a breakdown centre may receive 35 calls for assistance in one hour on a Saturday night. The frequency of calls is 35.

Frequency distributions

When summarising data, use may be made of **classes**, or **groups**. The number of items of data that fall into a class is called the **class frequency**. When data is presented in groups, as in Table 2.4, the data is referred to as **grouped data**. Although grouping data may destroy some of the detail in the original data an advantage is gained because the grouped data gives a very clear picture of patterns, which are an important part of statistical analysis.

When data has been collected a table is then constructed that shows data by classes together with the corresponding class frequency. The table thus constructed is a **frequency table** and it displays a **frequency distribution**.

Table 2.4 Heights of applicants for Auto Technician programme (Example 2.5)

Height (inches)	No. applicants
60–62	5
63–65	18
66–68	42
69–71	27
72–74	8
Total	100

Example 2.5

Table 2.4 shows a frequency distribution for the height in inches of 100 applicants for places on an Automotive Technician training scheme.

Class interval – class limits

In the first column and second row of Table 2.4, the figures denoted by 60–62 constitute a class interval. The number 60 is the **lower class limit** and the number 62 is the **upper class limit**.

Class boundaries

The heights were measured to the nearest inch, which means that the class 60–62 can include heights of 59.5 to 62.5 inches. These figures are known as class boundaries. 59.5 is the **lower class boundary** and 62.5 is the **upper class boundary**.

Class width

The class width is the difference between two successive lower class boundaries, or two successive upper class boundaries. In this example the figure is 62.5 minus 59.5 = 3.

The tally chart and frequency distribution

One of the reasons for collecting data is to enable users to look for patterns that may indicate some trend, or relationship between various parts of the data. The tally chart is a chart that assists the user to assemble data, and it also shows patterns in frequency distribution.

By adding a column to Table 2.4 to make the tally chart of Table 2.5 it is possible to look for

Table 2.5 Tally chart, showing patterns of frequency distribution

Height (inches)	Tally chart	No. applicants
60–62	/////	5
63–65	//////////////////	18
66–68	//	42
69–71	///////////////////////////	27
72–74	////////	8
Total		100

patterns in the data. In this case there is a clear tendency for the scores to cluster around the 66 to 68 mark, with rather more scores above this level than below it.

2.6 Importance of the shape of a frequency distribution

Certain patterns in frequency charts and graphs indicate that some mathematical relationship exists between some aspects of the data being examined. Various methods are used to display data so that the patterns may be observed. Among the methods of displaying frequencies are the:

- histogram
- frequency polygon
- cumulative frequency plot.

The histogram

The histogram is a form of bar graph that is used show patterns in frequency distributions. The normal procedure is to make each bar of equal width. The height of the bar then represents frequency.

A histogram for the data in Table 2.4 will have the width of each rectangle equal to the class width of 3. The centre of each rectangle is placed at the class mark, 61, 64, 67, etc., as shown in Figure 2.4.

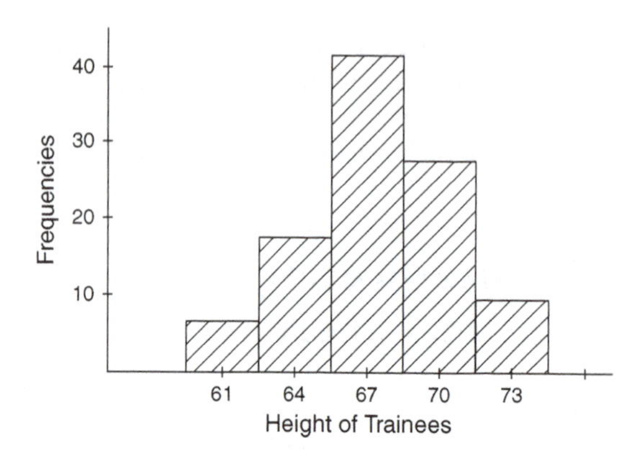

Fig. 2.4 The histogram

Provided that the base lengths of each rectangle are equal, the height of each rectangle is proportional to the class frequency.

It is clear that the pattern of the data is similar to that displayed in the tally chart; however, the histogram displays the pattern in a more informative way, because the information is presented in a more compact fashion.

The frequency polygon

If the mid-points of the top of each vertical bar of the histogram are joined together by straight lines a frequency polygon of the type shown in Figure 2.5 is produced. This type of figure also

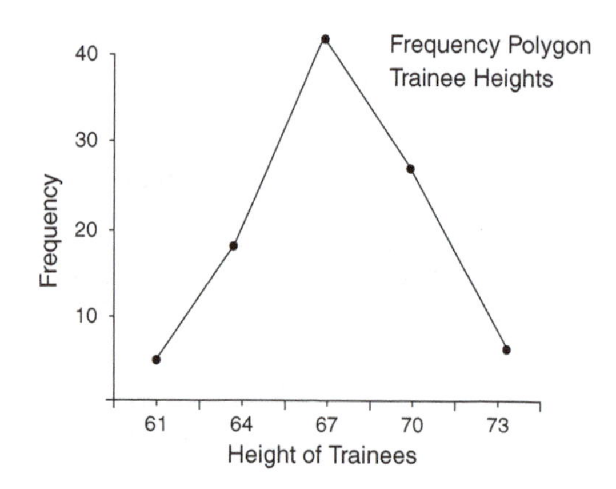

Fig. 2.5 Frequency polygon

provides an informative picture of the frequency distribution.

Relative frequency

Relative frequency is the frequency of a class divided by the total frequency of all classes, and it is normally expressed as a percentage. For example, the table of heights of applicants for automotive training can be adapted to show relative frequencies.

The percentage (relative) frequency is obtained by dividing each separate class frequency by the total number all frequencies. In this case the total number of frequencies of all classes is 100; the relative frequency of the class 66–68 is 42 so the relative frequency is $42 \times 100/100 = 42\%$. The relative frequency plot is identical to the frequency plot except that the scale on the vertical axis is a percentage.

Cumulative frequency

Cumulative frequency plots are another way of displaying frequencies.

Cumulative frequency is the total frequency of all values less than the upper class boundary of a given class interval.

A table displaying cumulative frequencies is called a **cumulative frequency distribution**. Table 2.6 can be converted into a cumulative frequency distribution by the following method: take the class interval 69–71; here the upper class boundary is 71.5. Table 2.7 shows that there are $27+42+18+5=92$ applicants whose height is less than 71.5 inches.

The cumulative frequency at this point is 92, denoting that these applicants are shorter than 71.5 inches.

A cumulative frequency table of the heights of the applicants is shown in Table 2.7.

When **cumulative frequency is plotted against the upper class boundary** the resulting curve appears as shown in Figure 2.6.

Smoothed cumulative frequency graphs are known as ogives because of their shape.

Table 2.6 Heights of applicants for auto training course

Height (inches)	No. applicants	Relative frequency (%)
60–62	5	5
63–65	18	18
66–68	42	42
69–71	27	27
72–74	8	8
Total	100	

Table 2.7 Cumulative frequency – auto course applicants

Height (less than) (inches)	No. applicants	Cumulative frequency	
59.5	0	0	
62.5	5	5	(5+0)
65.5	18	23	(18+5+0)
68.5	42	65	(42+18+5+0)
71.5	27	92	(27+42+18+5+0)
74.5	8	100	(8+27+42+18+5+0)

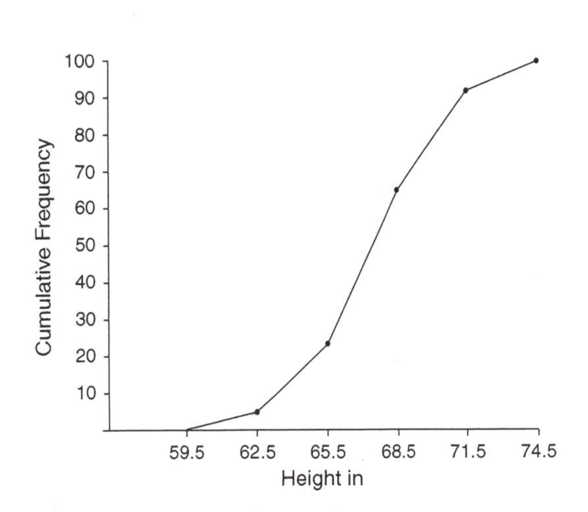

Fig. 2.6 Cumulative frequency graph

Smoothing the graph to make the ogive

Example 2.6

Table 2.8 shows the test scores of applicants for places on a training scheme for vehicle technicians.

Construct a cumulative frequency polygon and from this polygon draw the ogive.

The marks are placed along the horizontal axis and the cumulative frequency on the vertical one.

Table 2.8 Test scores for training scheme places (Example 2.6)

Mark	No. trainees (frequency)	Cumulative frequency
21–30	7	7
31–40	11	18
41– 50	21	39
51–60	34	73
61–70	25	98
71–80	13	111
81–90	4	115
91–100	1	116

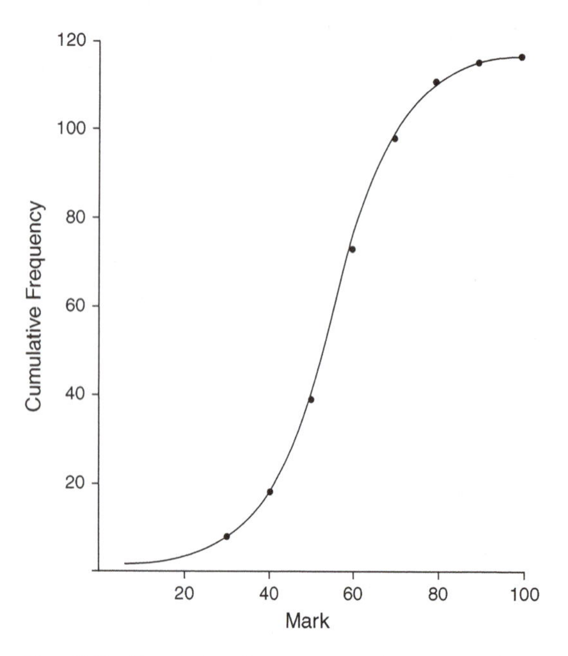

Fig. 2.7 Cumulative frequency graph in ogive form (Example 2.6)

The cumulative frequency is plotted against the upper of each class. In this case, the cumulative frequency of 18 corresponds with the class 31–40, so 18 is plotted against 40, which is the upper class boundary.

This process is repeated for the other values.

When all points have been plotted, the points can be joined together by a smooth curve to produce the ogive shown in Figure 2.7.

2.7 Interpreting statistics

Sampling

The statistics dealt with up to this point have been largely concerned with methods of using charts and diagrams to highlight particular features of data such as frequency and frequency distribution. The data that has been used has, in most cases, been derived from samples that were taken for some purpose. For example, it would not be possible to measure the height of every person in the United Kingdom. However, by taking a suitably sized sample of the population and measuring their height it is possible to obtain a reasonable height measurement that would represent the heights of all people in the UK. The theory of sampling is that a suitably selected sample will provide data that is representative of the population that it is drawn from. For example, a population means the total of all the types of bolts produced in a factory, or the total number of four-wheel drive vehicles in use on UK roads and so on.

2.8 Features of the population that are looked for in a sample

Average

An average is a value that is typical, or representative, of a set of data. Because an average value tends to be placed centrally in a set of data arranged according to magnitude, the average is called a **measure of central tendency**. The types of average that are commonly used in statistics are the **arithmetic mean**, the **median**, and the **mode**. Each of these may be used in analysing data – the choice of which is dependent on the data and the intended purpose of the calculation.

Arithmetic mean. The symbol $\bar{y}$ (called y bar) is often used to denote a mean value.

Example 2.7

The arithmetic mean of the numbers 7, 6, 5, 4 , 9, 1 and 10 is

$$\bar{y} = \frac{7+6+5+4+9+1+10}{7} = \frac{42}{7} = 6$$

The arithmetic mean is calculated by adding together all of the data values and then dividing by the number of values. The mean takes account of all of the data but it can be affected by a few very large values which may distort the average.

Mode

The mode, or modal value, of a set of numbers is the value that occurs with the greatest frequency; i.e. it is the most common value.

Example 2.8

The set of numbers 5, 6, 8, 10, 14, 9, 9, 9, 2, 3, has mode 9.

Example 2.9

The set of numbers 5, 6, 8, 9, 10, 11, has no mode.

Example 2.10

The set of numbers 4, 4, 4, 8, 7, 6, 5, 9, 9, 9, 10, 11, has two modes, 4 and 9, and is said to be **bi-modal**.

Frequency distributions that have only one mode are **uni-modal**.

Median

The median of a set of values arranged in order of magnitude is either the middle value or the arithmetic mean (average) of the two middle values.

Example 2.11

The set of numbers 3, 4, 4, 5, 6, 8, 8, 8, 11, has a median of 6. This is the middle value and there are 4 numbers below it and 4 above.

Example 2.12

Take the set of numbers 5, 5, 7, 9, 11, 12, 15, 20. Here there are two middle values so the median is $\frac{9+11}{2} = 10$

Range

In statistics **range** is a measure of spread in a batch of data. Range is the distance between the lower and upper extremes of the data.

Example 2.13

A group of technicians take a test at the end of a training session. The highest score was 95 marks and the lowest was 35.

In this case, the range $= 95-35 = 60$.

2.9 The normal distribution

When large amounts of data such as the weight of each student in a sample of 500 students is tabulated and presented as a frequency graph, the result is likely to be a bell-shaped curve of the type shown in Figure 2.8.

This type of frequency curve is produced when chance events are plotted on a frequency graph; for example, if a large sample of ball-bearings

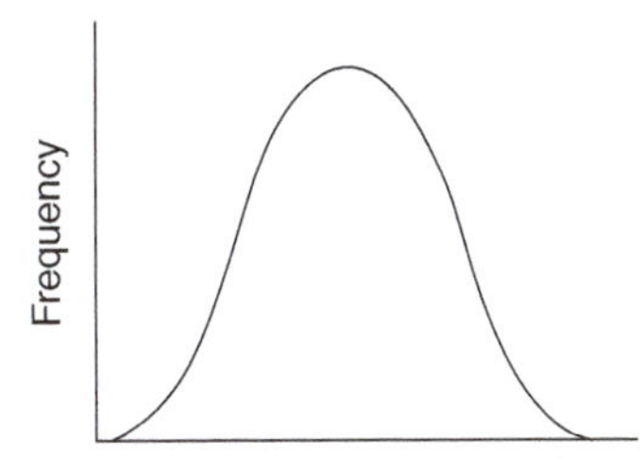

Fig. 2.8 The normal distribution curve

with a nominal size of 10 mm is checked individually for diameter the sizes are likely to be slightly different. If the diameters are plotted on a frequency graph the same type of bell-shaped curve will be produced. Because this type of bell-shaped curve normally occurs when **chance** events are plotted it is known as the **normal distribution curve**. Aspects such as the size of ball-bearings and the height of people are said to be normally distributed.

Importance of the normal distribution

Standard deviation

The standard deviation is a figure that can be calculated from a set of data. It is a figure that is used to test certain features of a sample. For example, in a normal distribution 68% of values will lie within two standard deviations of the mean.

Example 2.14

The mean (average) test score for 20 vehicle technicians who had just completed a 5 day course on automatic transmission systems was 72. The test scores were found to be normally distributed and the standard deviation worked out at 5 marks. Between what number of marks might one expect to find 68% of the scores?

Solution

The upper mark is $72+5=77$. The lower mark is $72-5=67$.

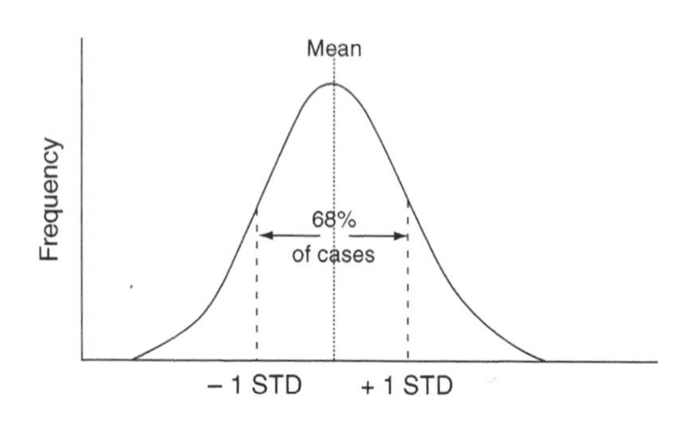

Fig. 2.9 The normal curve and two standard deviations

Skew

Distributions that produce the normal curve are symmetrical about the central value. Other distributions may be asymmetrical like those shown in Figure 2.10 (a) and (b) and they are said to be **skew**. When the tail of the distribution extends into the larger values that are plotted on the horizontal axis, the distribution is said to be positively skewed, or right skewed. When the tail of the distribution extends out to the left, the distribution is said to be negatively skewed, or left skewed.

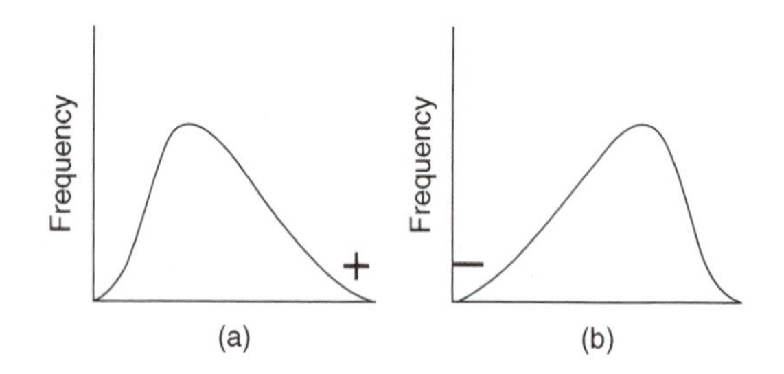

Fig. 2.10 (a) and (b) Skewness

Measures of skewness and the effect on samples

When the frequency distribution is perfectly symmetrical, the mean, median and mode coincide. That is, mean = median = mode for symmetrical distributions.

For uni-modal (just one hump) distributions that are moderately skewed, an approximate relationship between the mean, the median and the mode gives:

(mean – mode) is approximately = 3(mean – median)

Example 2.15

In one week the wages of four men were £288, £308, £317, and £770.

(a) Find the arithmetic mean of the men's weekly wage.
(b) Would you say that this mean (average) wage is typical of wages in this small group of men?

Solution

(a) The arithmetic mean $= \dfrac{288+308+317+770}{4} = £420.75$

(b) It is not typical. The mean figure is much higher than the lower three figures and much lower than the fourth figure. This highlights a disadvantage of the mean because it is strongly affected by extreme values. In this case the £770 is an extreme value.

There is a point of view that suggests the median is a better way of looking at averages in wages and salaries because the median is the figure below which 50% of the figures will be.

Other ways of viewing frequency distributions – quartiles, deciles, percentiles

If a set of data is arranged in order of magnitude – smallest to largest – the middle value (or the arithmetic mean of the two middle values) divides the set into two equal parts. In order to examine the data in greater detail it can be divided into four equal parts called quartiles.

Quartiles

Example 2.16

Table 2.9 shows the number of hours worked in 1 week by a group of 100 workers. Construct a cumulative frequency diagram and use it to determine:

(a) the lower quartile
(b) the upper quartile
(c) the interquartile range.
(d) Comment on the data in the interquartile range.

Solution

(see Figure 2.11)

(a) If the sample is large the first or lower quartile is at n/4, where n = total frequency.

In this example n = 100, so the lower quartile is at 100/4 = 25.

Table 2.9 Number of hours worked in 1 week by a group of 100 workers (Example 2.16)

Number of hours worked this week	Number of workers with these hours	Cumulative frequency
up to 15 hours	1	1
up to 20 hours	2	3
up to 25 hours	8	13
up to 30 hours	10	21
up to 35 hours	23	44
up to 40 hours	27	71
up to 45 hours	14	85
up to 50 hours	10	95
up to 55 hours	5	100

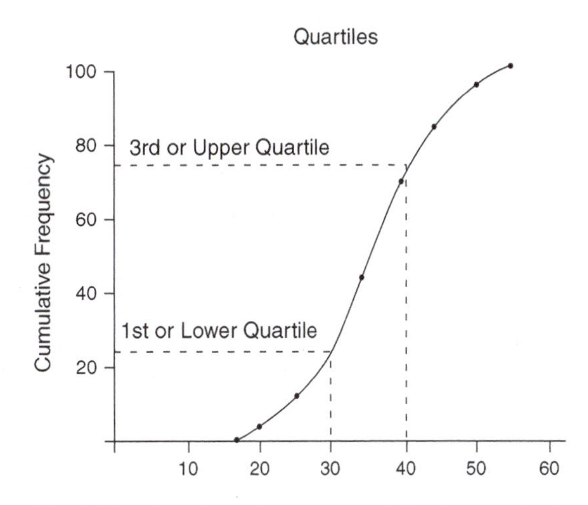

Fig. 2.11 Cumulative frequency graph (for Example 2.16)

A horizontal line is drawn from f = 25 until it touches the curve.

A vertical line is then drawn from this point until it touches the horizontal axis and this point is the lower quartile.

In this case the lower quartile = 30 hours.

(b) To determine the upper quartile a horizontal line is drawn from the point on the frequency axis where cumulative frequency = $^3/_4$, n = 75 to touch the cumulative frequency curve.

From this point on the curve a vertical line is drawn to touch the horizontal axis and this gives the upper quartile.

In this case the upper quartile = 42 hours.

(c) The interquartile range is = (upper quartile − lower quartile) = 42 − 30 = 12 hours.

(d) As the interquartile range contains 50% of the workers it is possible to say that half the workforce worked between 30 and 42 hours in the week in question.

Advantage of using the interquartile range

In any frequency distribution half the data is in the interquartile range. It is a useful guide because it only uses the half of the data that is closest to the median. It thus avoids distortions that are caused by exceptionally large or small values.

Deciles

The decile is another way of looking at data. As the name suggests, the data is divided up into ten equal parts. In Example 2.16 the deciles would correspond to cumulative frequencies of 10, 20, 30, and so on up to 90. Horizontal lines would be drawn from these cumulative frequencies to touch the curve, the points then being projected down on to the horizontal axis. The fifth decile corresponds to the median.

Percentiles

When the data is divided into 100 equal parts, the parts are called percentiles.

2.10 Summary of main points

2.11 Exercises

2.1 (a) Find the arithmetic mean and the median of this set of figures:
3, 4, 4, 5, 6, 8, 8, 10.

(b) Find the mode of this set of numbers:
2, 2, 5, 7, 9, 9, 9, 10, 10, 11, 12, 18.

(c) This set of numbers is bi-modal. What are the modes?
3, 3, 4, 4, 4, 5, 5, 7, 7, 7, 9.

(d) Find the median of this set of numbers:
5, 5, 7, 9, 11, 12, 15, 18.

(e) In a set of test marks the highest score was 98 and the lowest was 43. What is the range?

2.2 Table 2.10 shows the average petrol consumption of a range of cars that were used in a trial.

Table 2.10 Average petrol consumption of a range of cars (Exercise 2.2)

Size of car	Petrol consumption (mpg)
Small	38
Medium	32
Large	27

Construct a bar chart to show this information pictorially.

2.3 During a test on a diesel-engine car, the following emissions were collected.

CO 1 g/km
$HC + NO_x$ 1.1 g/km
SO_2 0.22 g/km

Construct a pie chart of this data.

2.4 Over the course of 1 year, a roadside recovery operator recorded the following causes of breakdowns.

Run out of fuel (320)
Electrical problems (460)
Tyre problems (280)
Fuel system problems (105)

Construct a pie chart and a bar chart to display this data.

2.5 The charge-out rate for service work on a certain vehicle is £45 per hour.

The technicians' pay rate is £9.50 per hour. Heating of the premises and business rates total the equivalent of £7.00 per hour.

Administration and other overheads amount to £23.00 per hour. Calculate the amount that represents profit and then display the breakdown of the hourly charge for service work, on a pie chart.

2.6 A company that specialises in engine reconditioning conducts dynamometer tests on all engines after rebuilding them. Use Table 2.11 to construct a histogram of the test results.

Table 2.11 Engine power on test (Exercise 2.6)

Engine power (kW)	No. engines (f)
59.4–59.6	4
59.7–59.9	18
60.0–60.2	41
60.3–60.5	23
60.6–60.8	7

2.7 A large van rental company keeps a record of repairs to each type of van that it hires out. Table 2.12 shows some details of clutch replacements and the mileage at which they were carried out on one type of delivery van.

Table 2.12 Frequency of clutch replacements (Exercise 2.7)

Mileage at which clutch was replaced (000s)	Frequency (f)
0–9	12
10–20	40
21–31	50
32–42	60
43–53	80
54–64	90

Construct a histogram and a frequency polygon of this data.

2.8 The weights of 60 aluminium alloy pistons for a certain type of four-cylinder engine are given in Table 2.13.

Table 2.13 Weights of alloy pistons for four-cylinder engine (Exercise 2.8)

Piston weight (g)	No. pistons
247	4
248	9
249	10
250	18
251	15
252	4

Draw up a cumulative frequency table and construct a frequency polygon.

2.9 The free lengths of a set of valve springs for an eight-cylinder engine with four valves per cylinder are shown in the upper part of Table 2.14.

Table 2.14 Valve spring lengths in millimetres for eight-cylinder engine (Exercise 2.9)

47.1	47.5	47.3	47.4	47.3	47.1	47.4	47.3
47.2	47.2	47.3	47.1	47.6	47.4	47.5	47.2
47.4	47.1	47.4	47.5	47.2	47.3	47.4	47.2
47.2	47.3	47.2	47.4	47.3	47.4	47.2	47.4

Free length of spring (mm)	No. springs of this length – Tally Chart	Frequency
47.1		
47.2		
47.3		
47.4		
47.5		
47.6		

Complete the accompanying tally chart.

2.10 Table 2.15 shows the number of hours worked in 1 week by a garage business that employs part-time staff as well as full-time staff.

(a) Complete this table.

(b) Draw a cumulative frequency graph using the table.

(c) How many people work less than 30 hours per week?

Table 2.15 Hours worked per week in a garage business (Exercise 2.10)

Hours worked per week	No. people	Cumulative frequency
less than 15	3	3
less than 20	2	5
less than 25	9	14
less than 30	12	
less than 40	28	
less than 50	6	

3
Algebra and graphs

3.1 Introduction

Algebra for the purposes of this book deals with equations and formulae of the type found in dealing with the science and technology encountered in automotive technicians' studies.

3.2 Formulae

Much information about vehicle technology is given in formulae and equations. A formula is normally presented in the form of an equation. For example, pressure is defined as the amount of force acting on each unit of area.

In the example of an engine cylinder that is shown in Figure 3.1 the gas in the cylinder is being compressed by the piston force of 600 N.

Pressure is calculated by dividing the force by the area over which the force is acting. If P is chosen to represent pressure, A the area , and F the force, the statement about pressure can be written as:

$$\text{Pressure} = \frac{\text{Force}}{\text{Area}}$$

Or, using the symbols, $P = \frac{F}{A}$

$P = \frac{F}{A}$ is a *formula*

In the case shown in Figure 3.1, $F = 600\,\text{N}$, $A = 0.1\,\text{m}^2$

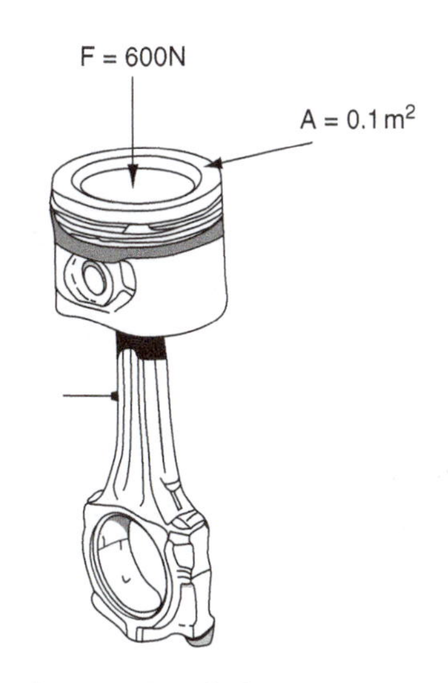

Fig. 3.1 Pressure in a cylinder

Substituting these values in the formula gives $P = \frac{600\,\text{N}}{0.1\,\text{m}^2}$ and $P = 60\,000\,\text{N/m}^2$

Thus, the formula for pressure is $P = \frac{F}{A}$

NB: Multiplication signs are not used. Division is indicated by / or ÷.

3.3 Evaluating formulae

The process of placing given values into a formula and working out the result is known as evaluating formulae. A formula that contains a range of operations such as brackets, multiplication, division, powers and roots requires that these operations are carried out in the following order: **brackets**, **powers**, **multiplication** and **division**, **addition** and **subtraction**.

Example 3.1

A formula for calculating the velocity of a vehicle that is being accelerated is:

$$v = u + at$$

where V = final velocity, a = acceleration, u = initial velocity, and t = time taken; u = 30 m/s, a = 2 m/s^2, and t = 8 seconds.

v can be calculated as follows:

$$v = 30 + 2 \times 6$$

$$v = 30 + 12 \text{ (because multiplying is done before adding)}$$

$$v = 42\,\text{m/s}$$

Example 3.2

$$s = \frac{(u+v)}{2} t$$

In this case, s = distance covered, u = initial velocity, v = final velocity, t = time taken; u = 10 m/s, v = 60 m/s, and t = 20 seconds. s can be calculated as follows.

$$s = \frac{(10+60)}{2} \times 20$$

The bracket is dealt with first. This means adding the 10 and 60 inside the bracket; the bracket is then removed to give

$$s = \frac{70 \times 20}{2}$$

$$s = \frac{1400}{2}$$

$$s = 700 \text{ metres}$$

3.4 Processes in algebra

Brackets

Brackets are used to group like terms together, for example $(x+2)+(x+2) = 2(x+2)$.

Brackets can be removed from an algebraic expression by multiplying out as follows. The number or symbol immediately outside the bracket multiplies each term inside the bracket. Two like signs multiplied together produce a plus, and two unlike signs produce a minus.

$$\text{Examples:}\quad 4(x+3) = 4 \times x + 4 \times 3$$

$$= 4x + 12$$

$$-3(4x-2) = -12x + 6.$$

Multiplying two binomial expressions in brackets

An expression such as (a + b)(a + 2) means (a + b) multiplied by (a + 2).

Example 3.3

The diagram example shows how this type of situation is dealt with.

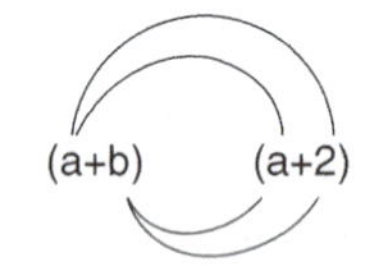

The procedure is: (a × a) = a^2, (a × 2) = 2a, (b × a) = ab, (b × 2) = 2b

The expression when multiplied out becomes $a^2 + 2a + ab + 2b$

Each term in the second bracket is multiplied by each term in the first bracket.

3.5 Algebraic expressions and simplification

Expression

A group of algebraic symbols such as $3x + 2y + 5x + 6y$ is known as an **expression**.

Each separate part such as $3x$, $2y$ is known as a **term**. In order to simplify this group of symbols, the like terms are added together as follows: $3x + 5x + 2y + 6y = 8x + 8y$.

Examples

Example 3.4

Simplify the expression $3(2x + 4y) + 3y - x$.

First, clear the brackets: $6x + 12y + 3y - x$, then group the like terms, $6x - x = 5x$ and $12y + 3y = 15y$.

The simplified expression is $5x + 15y$. The rule for the brackets is that each term inside the bracket is multiplied by what is outside the brackets; in this case it is the 3.

Example 3.5

Simplify the expression $3x(2y + 4z) + 3xy + 5xz$.

First clear the brackets, $6xy + 12xz$ then add the like terms together:

$$6xy + 3xy + 12xz + 5xz = 9xy + 17xz$$

Example 3.6

Simplify the expression $4a(2b + 4c) + 2a(b - 2c)$

Again, deal with the brackets first: $8ab + 16ac + 2ab - 4ac = 10ab + 12ac$.

3.6 Factorising

An expression such as $5x - 5y$ has two terms and 5 is common to both of them. 5 is therefore a common factor. A factor is defined as a common part of two or more terms that make up an algebraic expression. The expression $5x - 5y$ can be written as $5(x - y)$, and the factors of the expression are 5 and $(x - y)$.

Example 3.7

Factorise $ab + bc$

Each term has a common factor of b; the b can be taken outside a bracket to give:

$$ab + bc = b\ (a + c)$$

The factors of $ab + bc$ are b and $(a + c)$

Example 3.8

Factorise $3a + 4bx - 3b - 4ax$

First step: collect the similar terms together

$$3a - 3b + 4bx - 4ax$$

Second step: take out the common factors

$$3\ (a - b) + 4x\ (b - a)$$

Example 3.9

Factorise $2a^2 + 4b$

By inspection, 2 is a common factor of both terms

The factors are 2 and $(a^2 + 2b)$

$$2(a^2 + 2b) = 2a^2 + 4b$$

Example 3.10

Factorise $nx + mx - yn - ym$

Rewrite the expression as $(nx + mx) - (yn + ym)$ – the minus outside the bracket changes the sign of everything inside the bracket.

This can now be reduced to $x(n + m) - y(n + m) = (n + m)(x - y)$.

3.7 Equations

Solving equations

Solving equations, in other words 'finding the value of the unknown quantity', can be achieved by the application of some fairly simple rules. The following examples demonstrate the procedure.

The most important rule is: ***Whatever we do to one side of the equation we must do exactly the same thing to the other side.***

In the following working, LHS is used for left hand side and RHS for right hand side.

Example 3.11

Find the value of x in the equation: $x + 5 = 8$

The aim is to get x on its own on the left hand side of the equation.

By subtracting 5 from the LHS, the x will be left on its own. But because we subtracted 5 from the LHS we must also subtract 5 from the RHS. This gives:

$$x+5-5=8-5$$
$$x=3$$

Example 3.12

Find the value of x in the equation: $\frac{x}{3}=2$

Multiplying both sides of the equation by 3 gives:

$$\frac{x}{3}\times 3=2\times 3$$

The 3s cancel out on the LHS leaving $3\times 2=6$ on the RHS

$$x=6$$

Example 3.13

Find the value of x in the equation: $x-3=9$

Here the problem is to get rid of the -3; this can be done by adding 3 to the LHS and, because of doing so, 3 must be added to the RHS.

$$x-3+3=9+3$$
$$x=12$$

Example 3.14

Solve for x in the equation: $5(2x+6)=60$

First, remove the brackets: $10x+30=60$

Next, take 30 from each side of the equation: $10x+30-30=60-30$

This gives $10x=30$

Now divide both sides of the equation by 10:

$$\frac{10x}{10}=\frac{30}{10}$$

which gives : $x=3$

Example 3.15

Solve for x in the equation: $5x+3=x+15$

First, subtract x from each side of the equation

$$5x-x+3=x-x+15$$
$$4x+3=15$$

Next, subtract 3 from both sides of the equation:

$$4x+3-3=15-3$$
$$4x=12$$

Now, divide both sides by 4:

$$\frac{4x}{4}=\frac{12}{4}$$

From which $x=3$

Example 3.16

Solve for x in the equation: $\frac{x}{3}+\frac{1}{2}=2$

Here it is necessary to multiply each term of the equation by lowest common multiple (LCM) of the denominators. In this case the denominators are 3 and 2 and the LCM is 6.

Multiplying each term by 6 gives: $6\frac{(x)}{3}+6\times\frac{1}{2}=6\times 2$

This reduces to $2x+3=12$

Next, take 3 from each side: $2x+3-3=12-3$

$$2x=9$$

Now, divide both sides by 2: $\frac{2x}{2}=\frac{9}{2}$

$$x=4.5$$

Example 3.17

Solve for x: $\frac{6}{(x+3)}=\frac{5}{(x+2)}$

Step 1

The LCM of the denominators is $(x+3)\ (x+2)$

Step 2

Multiply each term of the equation by the LCM

$$\frac{(x+3)(x+2)\ 6}{(x+3)}=\frac{(x+3)(x+2)\ 5}{(x+2)}$$
$$6(x+2)=5(x+3)$$

Step 3
Remove the brackets

$$6x + 12 = 5x + 15$$

Step 4

$$6x - 5x = 15 - 12$$
$$x = 3$$

3.8 Transposition of formulae

In a formula such as $E = a + b$, E is referred to as the subject of the formula.

It often happens that the formula has to be changed around to make one of the other variables the subject. For example, it may be necessary to find the value of b when the values of E, a are known.

In this case it is necessary to make b the subject of the formula.

The process for doing this is known as transformation, or transposition, of formulae and it requires careful use of the rules for solving equations. Namely: ***Whatever we do to one side of the equation we must do exactly the same thing to the other side.***

Example 3.18
Take the formula $E = a + b$ as an example. Make b the subject.

Step 1
It is often useful to rewrite the formula so that the required subject appears on the left hand side.

In this case, $a + b = E$

Step 2
Subtract a from both sides: $a - a + b = E - a$
This leaves $b = E - a$

Example 3.19
Transpose $F = ma$ to make a the subject.

Step 1
Rewrite the formula to give $ma = F$

Step 2
Divide both sides of the formula by m:

$$\frac{ma}{m} = \frac{F}{m}$$
$$a = \frac{F}{m}$$

Example 3.20
Transpose $A = \frac{L}{B}$ to make L the subject.

Step 1
Rewrite the formula to give $\frac{L}{B} = A$

Step 2
Multiply both sides by B:

$$\frac{BL}{B} = AB$$
$$L = AB$$

Example 3.21
Transposition when the formula contains brackets:

Transpose $T = \frac{l(x - a)}{a}$ to make x the subject.

Step 1

$$\frac{l(x - a)}{a} = T$$

Step 2
Multiply both sides by a:

$$\frac{a\,l\,(x - a)}{a} = aT$$
$$l(x - a) = aT$$

Step 3
Divide both sides by l:

$$\frac{l\,(x - a)}{l} = \frac{aT}{l}$$
$$(x - a) = \frac{aT}{l}$$

Step 4
Remove the bracket:

$$x - a = \frac{aT}{l}$$

Step 5

Add a to both sides:

$$x - \mathrm{a} + \mathrm{a} = \frac{\mathrm{aT}}{l} + \mathrm{a}$$

$$x = \frac{\mathrm{aT}}{l} + \mathrm{a}$$

3.9 Graphs

Graphs are used to display numerical data in such a way that connections and relationships between variables can be seen. The following examples show a selection of graphs and the descriptions that accompany them explain the method of constructing graphs and the meanings of terms used.

Example 3.22

Table 3.1 shows the amount of carbon dioxide that is passed out through the exhaust systems of vehicles of various engine sizes. The CO_2 figures are derived from standard tests. The engine size is given in litres and the CO_2 emissions are in grams per kilometre.

To construct a graph from this data it is necessary to know which variable to plot on the vertical axis and which on the horizontal axis.

Table 3.1 Carbon dioxide emissions from vehicles – petrol engines (Example 3.22)

Engine size (l)	CO_2 emissions (g/km)
4.0	310
1.4	176
1.8	182
2.2	206
4.2	314
1.3	168
3.0	251

Variables

In this case the variables are engine size and the CO_2 emission figures.

Dependent variable

The amount of CO_2 emitted through the exhaust is dependent on engine size. In this case the CO_2 figure is said to be the **dependent variable**.

Independent variable

In this case the **independent variable** is the engine size.

When plotting a graph, the independent variable is placed on the horizontal axis and the dependent variable on the vertical axis. In this example the CO_2 emission figures are placed on the vertical axis and the engine size is placed on the horizontal axis.

Scales

Graphs are normally plotted on squared paper. Each square is chosen to represent a set amount of each of the variables. The scales used must be clearly shown on the graph.

The graph of the data in Table 3.1 is shown in Figure 3.2.

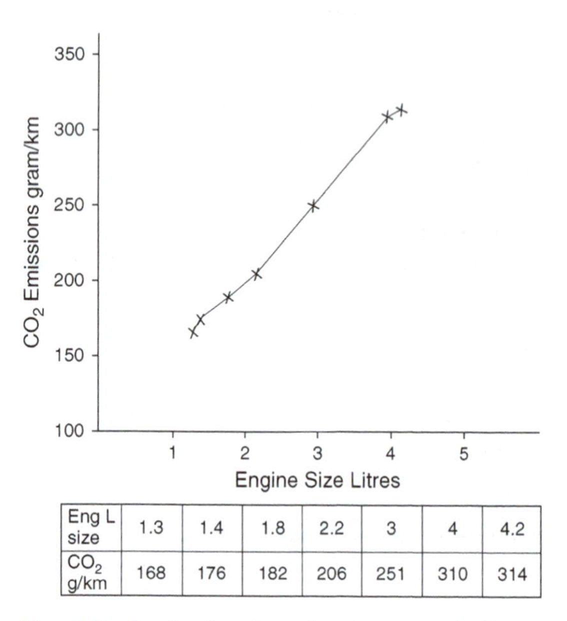

Eng L size	1.3	1.4	1.8	2.2	3	4	4.2
CO_2 g/km	168	176	182	206	251	310	314

Fig. 3.2 Graph of carbon dioxide emissions (Example 3.22)

From the graph of carbon dioxide emissions versus engine size it can be seen that there is a connection between engine size and the amount of CO_2 emitted through the exhaust system. In broad terms, the larger the engine size, the more CO_2 is emitted.

Coordinates

When points are plotted on a graph the values of each variable are read on the scales. In the case of carbon dioxide emissions the 4-litre engine has emissions of 310 grams of carbon dioxide per kilometre. Therefore, 4 litres is read from the horizontal axis and 310 g/km from the vertical axis; imaginary lines are drawn from these points and where these lines cross a point is marked. The values of 4 litres and 310 g/km are called rectangular coordinates. In graphical work it is customary to place these rectangular coordinates in brackets which, in this example, would appear as (310, 4). The point formed by the intersection of the imaginary lines from 4 and 310 is referred to as (310, 4). In general, graphs have y values plotted on the vertical axis and x values on the horizontal axis. When rectangular coordinates are given, the x value appears first inside the brackets, like this (x, y).

Example 3.23

Table 3.2 shows the load raised and the effort required to raise it for a simple lifting machine. Plot a graph with effort E on the vertical axis and load L on the horizontal axis.

Table 3.2 Effort and load for a simple lifting machine (Example 3.23)

Effort (newtons)	Load (newtons)
5	90
10	180
15	270
20	360
25	450
30	545

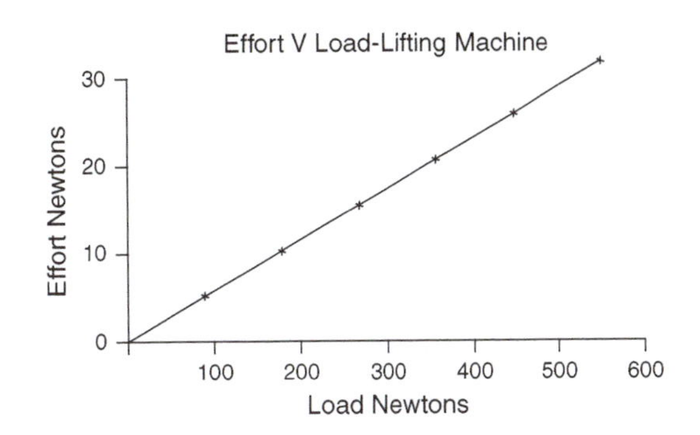

Fig. 3.3 Graph of load plotted against effort (Example 3.23)

In the case of the lifting machine the effort is placed on the vertical scale because it is dependent on the load applied to the machine. The scales chosen are:

one large square = 10 N for the effort
one large square = 100 N for the load.

When the points are joined together they form a neat straight line, which shows that there is a definite mathematical relationship between effort and load.

Example 3.24

Table 3.3 shows the amount of valve lift produced by rotation of an engine cam. Plot a graph of these values with cam angle on the horizontal axis and lift on the vertical axis.

Table 3.3 Valve lift and angle of cam rotation (Example 3.24)

Cam angle (degrees)	0	5	10	13	20	30	40	50	55
Valve lift (mm)	0	0.43	1.73	2.9	5.81	9.13	11.4	12.56	12.7

In the case of valve lift, the amount of lift is placed on the vertical axis and the degrees of cam rotation on the horizontal axis. When the points are joined together a smooth curve results. This also arises from a mathematical relationship between cam rotation and valve lift. The data for

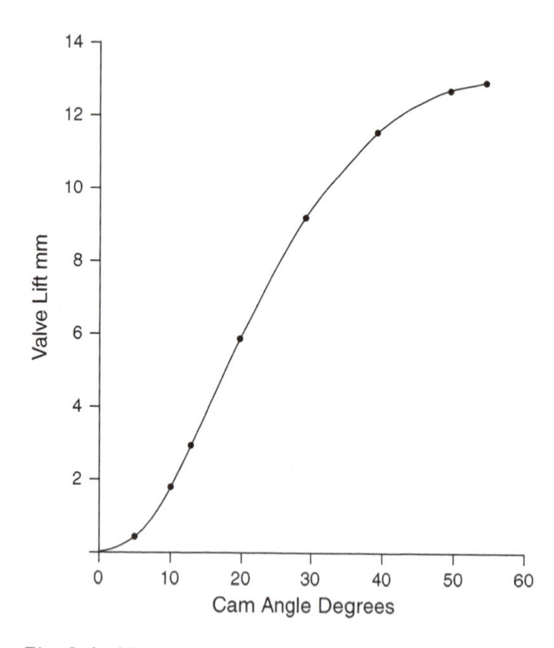

Fig. 3.4 Valve lift and cam angle (Example 3.24)

a graph of cam angle and valve lift may be readily obtained with the aid of a 360° protractor placed on the end of the camshaft to measure cam angle of rotation and a dial gauge indicator placed at the valve to measure valve lift.

3.10 Graphs and equations

When a graph such as that for effort and load shown in Figure 3.3 produces a perfect straight line, it demonstrates that there is a mathematical relationship existing between the values that are plotted. In such cases the relationship between the values can be expressed in the form of an equation of the form $y = mx + c$.

The straight-line graph

Example 3.25
Table 3.4 shows a set of values of x and y that are connected by an equation of the form $y = mx + c$. Plot the graph and determine the value of the constants m and c.

The constant m is the gradient, or slope, of the straight line that is formed when the points are joined together. The angle that the line makes with the horizontal is denoted by the symbol θ and m, the gradient, is the tangent of the angle θ. The value of m is found as shown in Figure 3.5. A triangle is constructed at convenient points and x and y values are read from the scales. Using these values the constant m, which is the tangent of angle θ, is determined as follows:

$$m = \tan\theta = \frac{y_2 - y_1}{x_2 - x_1}$$

Table 3.4 Values of x and y connected by $y = mx + c$ (Example 3.25)

x	8	14	18	22	30
y	22	34	42	50	66

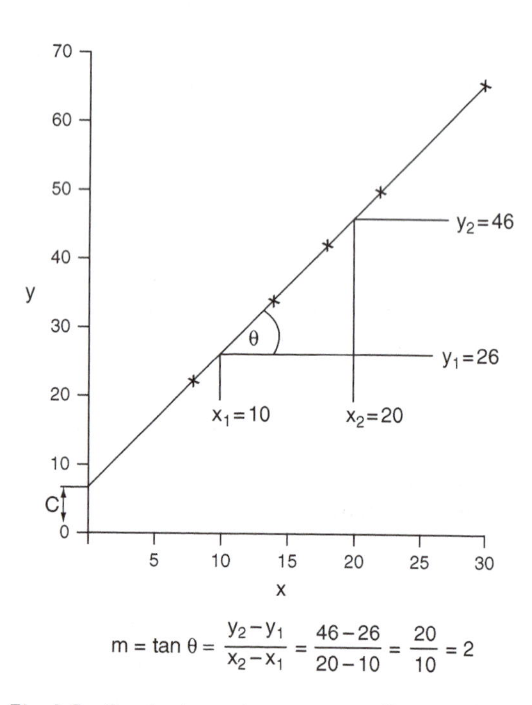

Fig. 3.5 Graph of equation $y = mx + c$ (Example 3.25)

Reading the values from the graph shown in Figure 3.5 the values are:

$y_2 = 46$

$y_1 = 26$

$x_2 = 20$

$x_1 = 10$

$$m = \frac{y_2 - y_1}{x_2 - x_1} = \frac{46 - 26}{20 - 10} = \frac{20}{10} = 2$$

The constant c is known as the intercept and it is the y value where the straight line meets the y axis; provided that the x and y values start at 0, the value of c may be read directly from the graph. **In this case c = 6.**

The equation of this graph is $y = 2x + 6$.

Example 3.26

The results in Table 3.5 were obtained in an experiment on a simple worm and wheel lifting machine.

Plot a graph of the results given in Table 3.5 and determine the law of the machine in the form E = mL + C (where E = effort; L = load). Plot E on the vertical axis.

In this case the variables are E and L instead of x and y. The procedure for obtaining the slope m and the intercept C is the same as that used in Example 3.4. The graph is plotted, with both values starting at zero. As it is a straight line, the values used to determine m and C may be read from the graph.

$$\text{The slope m} = \frac{E_2 - E_1}{L_2 - L_1} = \frac{29.5 - 16}{400 - 200}$$

$$= \frac{13.5}{200} = 0.0675$$

$$\mathbf{m = 0.0675}$$

Table 3.5 Load and effort for a simple worm and wheel machine (Example 3.26)

Load (L) (newtons)	100	200	300	400	500	600
Effort (E) (newtons)	8.5	16	23	29.5	36.5	43.5

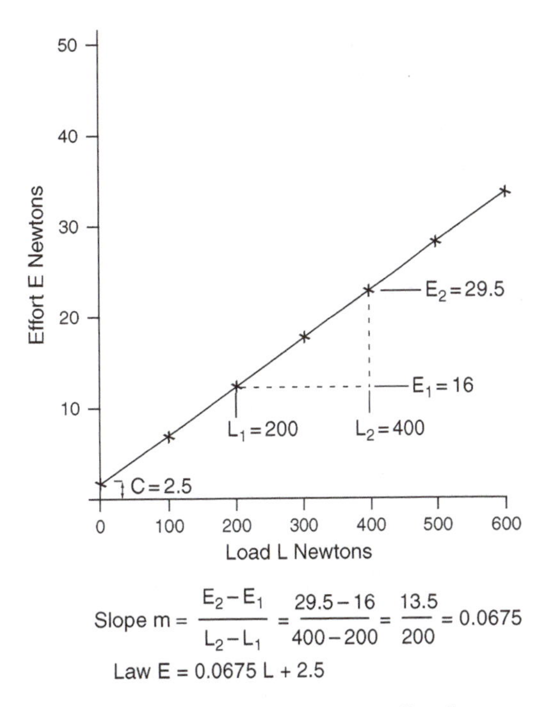

Fig. 3.6 Graph of load plotted against effort for worm and wheel (Example 3.26)

Using the same procedure as in Example 3.4, the intercept C is read from the E scale and this gives a value of **C = 2.5 newtons**.

The law of the machine is E = 0.0675 L + 2.5.

3.11 Summary of main points

Multiplication sign is not used. LM means L × M

Division sign not used. $\frac{L}{M}$ or L/M means L ÷ M

Like signs multiplied together give a +

Unlike signs multiplied together give a −

Like signs divided give a +

Unlike signs divided give a −

When simplifying an expression remove the brackets first and then group the like terms together.

A minus sign outside brackets changes the sign of everything inside the brackets.

When solving an equation, the most important rule is: ***Whatever we do to one side of the equation we must do exactly the same thing to the other side***.

Dealing with mixed operations: **Use BODMAS, as is done in arithmetic**.

3.12 Exercises

Exercises – Section 3.3

1. $S = ut + \frac{1}{2}at^2$. Calculate S when u = 20, t = 8, a = 5.
2. $R = \dfrac{R_1\,R_2}{R_1 + R_2}$. Calculate R when $R_1 = 4$ and $R_2 = 4$.
3. $V = 4/3\pi R^3$. Calculate V when R = 5 and $\pi = 22/7$.
4. $T = \mu P\dfrac{(R_1 + R_2)}{2}$. Calculate T when $\mu = 0.4$, P = 500, $R_1 = 10$, $R_2 = 18$.
5. $Rc = \dfrac{Vs + Vc}{Vc}$. Determine value of Rc when Vs = 500 and Vc = 50.
6. $V = 4/3\,\pi r^3$. Calculate V when r = 7 and $\pi = 22/7$.
7. $S = (\mu gR)^{1/2}$. Calculate S when $\mu = 0.8$, g = 9.8 and R = 50.
8. $T = \pi fs/16\,\dfrac{(D^4 - d^4)}{D}$. Calculate T when D = 50, d = 45 and fs = 5000.
9. $P_b = \dfrac{2\pi\,T\,N}{60}$. Calculate P_b when T = 150 and N = 3000.
10. $Pi = \dfrac{P\,l\,a\,N}{60 \times 100}$ Calculate Pi when P = 150000, l = 0.2, a = 0.04, N = 2000.

Exercises – Section 3.4

1. In the following expression, multiply out the brackets and simplify as far as possible. $3a(b + 4c) + 3a(2b - 3c) - 5bc$
2. Simplify each of the following expressions. (a) $4a(3b - 2c) + 6ab$; (b) $2(4x + 3y) + 3(2x + 5y)$; (c) $5(6x - 3) - 3(8x - 12)$
3. $5x - 3(x - 10) + 2(4 - x)$

Exercises – Section 3.5

Simplify the following (multiply to remove the brackets).

(a) $(3 + x)(2 + y)$
(b) $(x + 2)(x + 3)$
(c) $(x - 2)(x + 3)$
(d) $(x - 4)(x - 2)$

Exercises – Section 3.6

Factorise the following expressions.

(a) $2x^2 - 4x$
(b) $ax + 2a + 2bx + 4b$
(c) $ax^2 + ax$

Exercises – Section 3.7

Solve the following equations.

1. $x + 4 = 8$
2. $6y + 11 = 29$
3. $0.6d = 18$
4. $2(x + 5) = 20$
5. $\dfrac{8}{x} = 2$
6. $4(x - 5) = 15 - (3 - 2x)$
7. $\dfrac{1}{x} + \dfrac{2}{3x} = 5$
8. $\dfrac{3}{(x - 3)} = \dfrac{4}{(x + 1)}$

Exercises – Section 3.8

1. Transpose $y = mx + c$ to make x the subject.
2. Transpose $T = 4Mn^2L^2$ to make L the subject.
3. Transpose $C = \pi d$ to make d the subject.
4. Transpose $V = I\,R$ to make I the subject.
5. Transpose $V = \dfrac{\pi d^2 L}{4}$ to make L the subject.

Exercises – Section 3.10

3.10.1
Table 3.6 shows the results of a brake power test on a small engine. Plot a graph with power

Table 3.6 Engine speed and power output (Exercise 3.10.1)

Engine speed (rev/min) (N)	1000	1500	2000	2500	3000	3500	4000	4500
Power (kW) (P)	5.2	8.1	11	13.75	15.75	16.8	17.3	17.4

in kW on the vertical axis and engine speed in rev/min on the horizontal axis. Compare the shape of the graph with the performance data that vehicle manufacturers supply.

3.10.2

Table 3.7 shows the wind and other resistance that a vehicle encounters at various speeds. Plot a graph of resistance to motion against vehicle speed. Place the resistance to motion on the vertical axis. How do you think that the resistance changes as the vehicular speed increases?

Table 3.7 Wind and other resistance encountered by vehicles (Exercise 3.10.2)

Vehicle speed (km/h)	20	40	60	70	80
Total resistance to motion (kN)	0.45	0.9	1.6	2.1	2.7

3.10.3

The figures in Table 3.8 were obtained in an experiment on a thin steel wire. The cable is secured at one end and a series of weights (loads) is placed at the free end. The extension produced by each load is recorded.

Plot a graph of these results and find the extension produced by a load of 230 N.

Table 3.8 Load and extension of a steel cable (Exercise 3.10.3)

Load (N)	50	100	150	200	250	300
Extension (mm)	0.81	1.61	2.43	3.22	4.02	4.82

3.10.4

Table 3.9 shows the piston speed of an engine at various positions of the crank as the piston moves from top dead centre to bottom dead centre on one stroke.

Plot a graph of piston speed in m/s on the vertical axis and crank angle on the horizontal axis. Use the graph to determine piston speed when the crank has turned through 90° from top dead centre.

Table 3.9 Piston speed and crank angle at 2500 rev/min (Exercise 3.10.4)

Crank angle (degrees)	0 TDC	20	40	60	80	100	120	140	160	180 BDC
piston speed (m/s)	0	3.4	6.22	8	8.6	7.99	6.55	4.57	2.34	0

3.10.5

The performance of a four-post vehicle lift is tested by measuring the power required to lift vehicles of various weights. The results are shown in Table 3.10.

Table 3.10 Power and vehicle weight for a four-post lift (Exercise 3.10.5)

Lift power (P) (kW)	2.7	3.08	3.45	3.83	4.2	4.58	4.95
Vehicle weight (W) (tonnes)	1	1.25	1.5	1.75	2	2.25	2.5

It is believed that the vehicle weight and the power required to lift it on the lift are connected by a law of the form $P = mW + C$.

Plot a graph of power P on the vertical axis and vehicle weight W on the horizontal axis. Using this graph:

(a) decide if the power and weight are related by the type of equation suggested;
(b) if the values are connected by an equation of the form $P = mW + C$, determine the values of m and C and hence write down the equation.

4
Geometry and trigonometry

4.1 Angles

An angle is formed when two straight lines meet at a point, as shown in Figure 4.1.

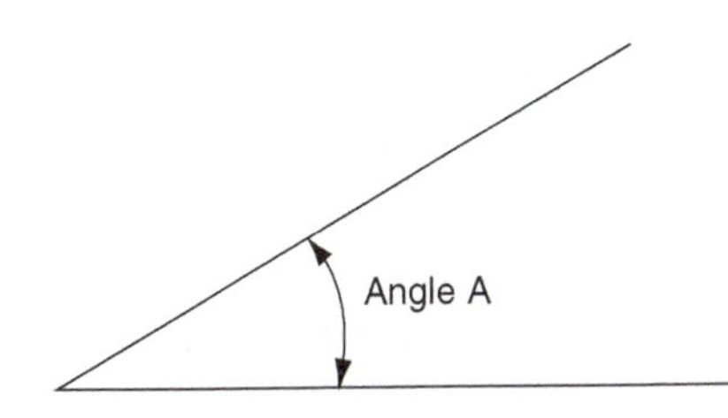

Fig. 4.1 The angle between two lines meeting at a point

Angular measurement

The basis unit of angular measurement is the degree. A degree may be subdivided into smaller parts known as minutes and seconds. One degree is equal to 60 minutes and one minute is equal to 60 seconds. The symbols used to denote angles are shown in Example 4.1. The small circle° indicates degrees, the single dash ′ indicates minutes, and the double dash ″ indicates seconds.

Example 4.1
A certain angle is stated to be $22^\circ\ 15'\ 12''$. In words this is 22 degrees, 15 minutes and 12 seconds.

The radian

A radian is a unit of angular measurement that is used in some aspects of engineering science. A radian is the angle that is covered at the centre of a circle by an arc equal in length to one radius on the circumference of the circle. As there are 2π radiuses (radii) in the circumference it follows that there are **2π radians** in 360°. This means that one radian $= 57^\circ\ 18'$, as shown in Figure 4.2.

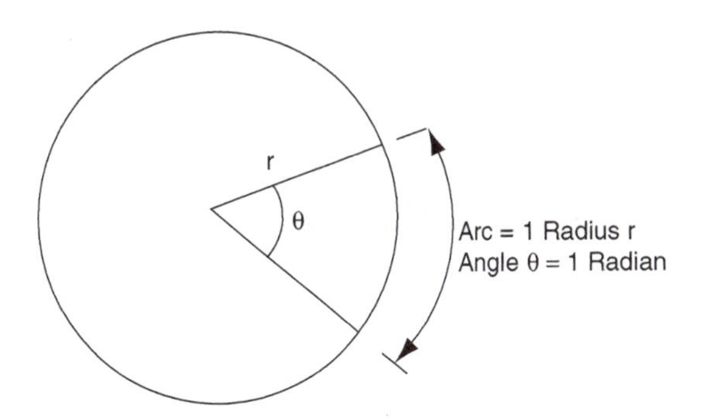

Fig. 4.2 The radian

Angles and rotation

When a straight line is rotated about a fixed point O, as shown in Figure 4.3, the amount of movement may be expressed as an angle. If the

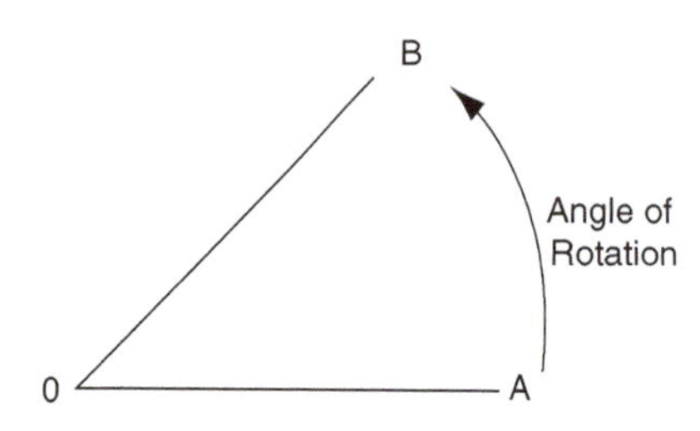

Fig. 4.3 Angular movement

line OA is rotated so that it makes one complete revolution, the angle turned through is 360°, or 2π radians.

Example 4.2

Convert an engine speed of 3600 rev/min to angular velocity of the crankshaft in radians per second.

Solution

3600 rev/min = 3600/60 = 60 rev/sec.

There are 2π radians in one revolution, which means that 60 rev/s = $60 \times 2\pi$ rad/s = 377 rad/s.

The angular velocity of the crankshaft = 377 rad/s.

4.2 Examples of angles in automotive work

Figure 4.4 shows a number of examples of the use of angles in motor vehicle systems. These angles are determined at the design stage and are set during the manufacturing process. In all cases the accuracy of the angular settings is a critical element in the performance of the vehicle. It is the function of the service technician to ensure that the angular settings are properly maintained throughout the working life of the vehicle. Wear and accidental damage are the most likely causes of these angles being altered and special equipment is used to check the angles so that remedial

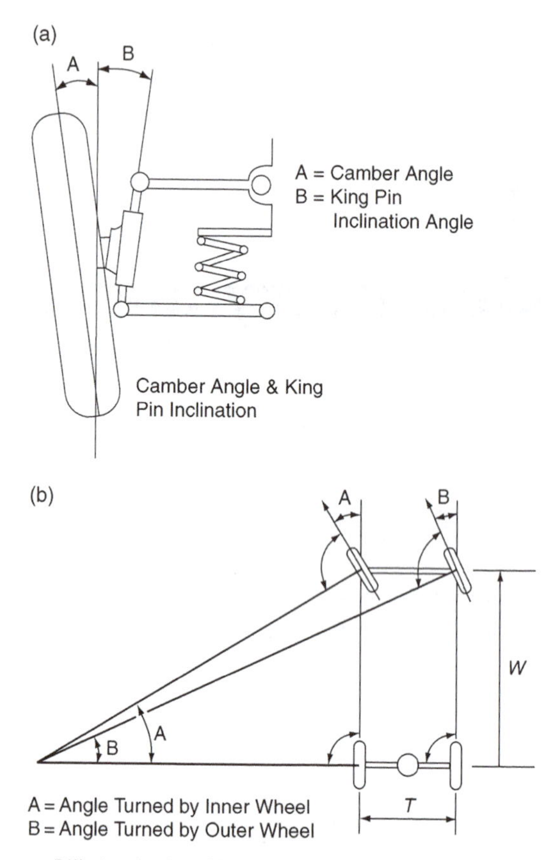

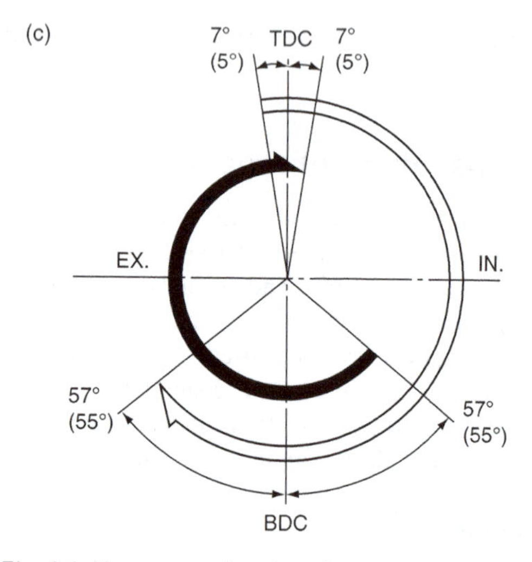

Fig. 4.4 Some examples of angular settings in automotive systems

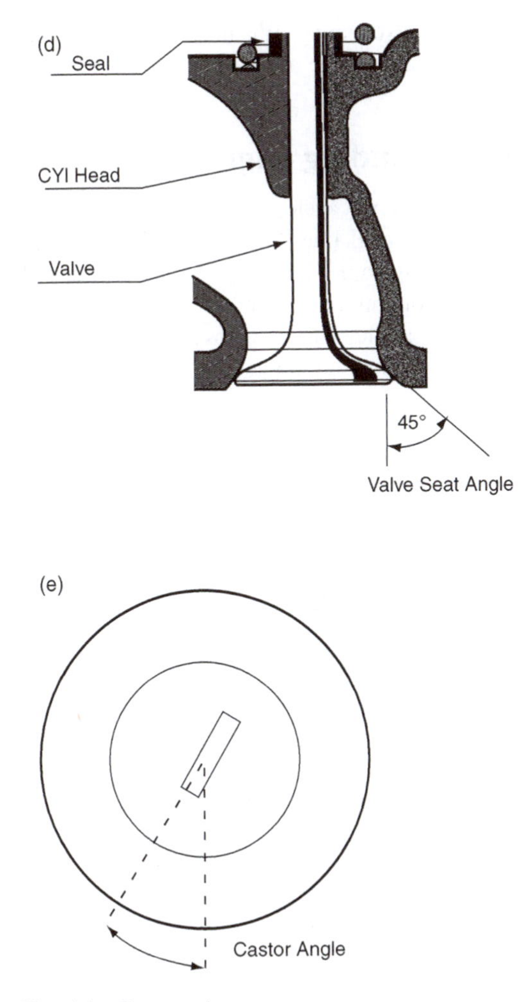

Fig. 4.4 Continued

work can be performed to restore settings to the recommended figures.

Angles and lines

Types of angle

An acute angle is 0° to 90°
An obtuse angle is 90° to 180°
A reflex angle is greater than 180°
A right angle is 90°

These angles are shown in Figure 4.5.

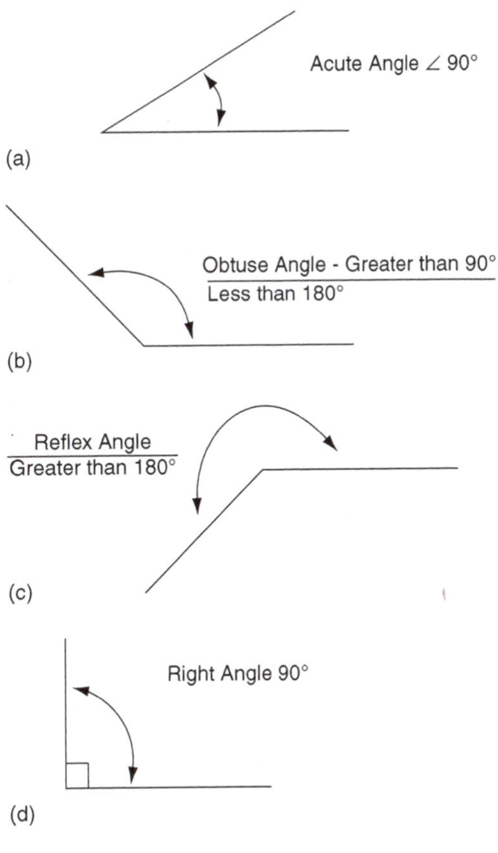

Fig. 4.5 Types of angle

Adding and subtracting angles

Example 4.3

Adding and subtracting angles.

(a) Add together 49° 54′ and 31° 23′.

Set the corresponding parts of each angle in line, as shown, for example:

49° 54′
31° 23′
81° 17′

Adding the minutes together gives 77′. This is 1° 17′. Adding the 1° to the 31° makes it 32° and $32° + 49° = 81°$, so the total angle = **81°17′**.

(b) Subtract 21° 15′ from 46° 12′.

$$\begin{array}{l} 46^\circ\ 12' \\ \underline{21^\circ\ 15'} \\ \mathbf{24^\circ\ 57'} \end{array}$$

15′ cannot be subtracted from 12′ so it is necessary to borrow 1° from the 46° making it 45°. The 12′ now becomes $12' + 60' = 72'$. The 15′ is now subtracted from 72′ to give **57′**. The 21° is now subtracted from 45° (because 1° was borrowed to make 72′), which gives 24°. The final result is 24° 57′, as shown.

4.3 Types of angle

Adjacent angles

The angles A and B shown in Figure 4.6 are known as adjacent angles. Angles A and B added together total 180°.

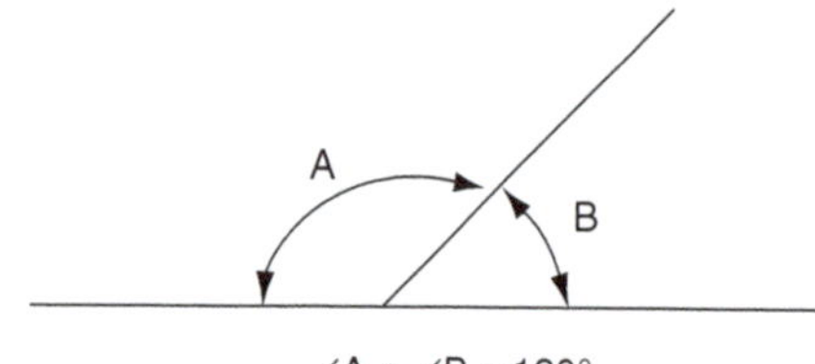

Fig. 4.6 Adjacent angles

Opposite angles

Figure 4.7 shows some opposite angles. In this figure angle C is opposite angle A. These angles are equal. Angle D is opposite angle B, and these angles are also equal. When two straight lines intersect, the opposite angles are equal.

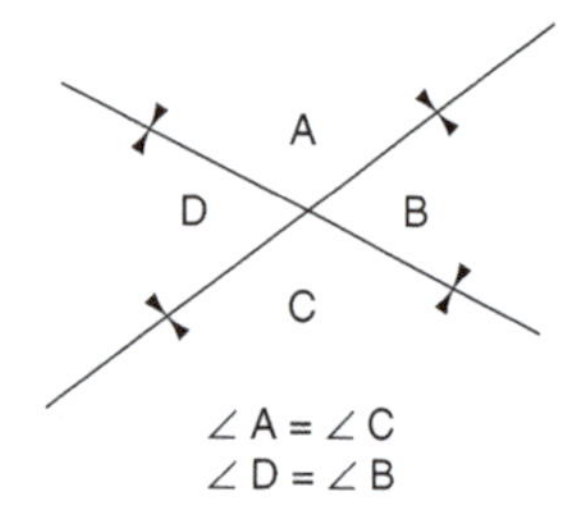

Fig. 4.7 Opposite angles

Corresponding angles

Figure 4.8 shows a sloping line that is crossed by two horizontal lines. The angles B and E are corresponding angles and they are equal. A and H are corresponding angles and they are also equal. When two parallel lines are crossed by a transversal, the corresponding angles are equal.

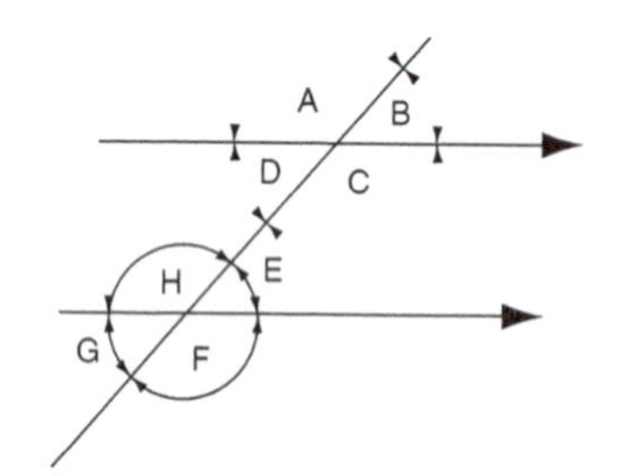

Corresponding, Angles are Equal A = H; B = E; C = F, D = G.
Alternate Angles are Equal D = E; C = H.

Fig. 4.8 Corresponding and alternate angles

Alternate angles

The angles D and E are alternate angles. These angles are equal. Angles and C and H are also alternate angles and they are equal. When two parallel lines are crossed by a transversal, the alternate angles are equal.

Supplementary angles

Supplementary angles are angles whose sum is 180°.

Complementary angles

Complementary angles are angles whose sum is 90°.

4.4 Types of triangle

The following types of triangles and their properties are illustrated in Figure 4.9.

Acute angled triangle

All angles of an acute angled triangle are less than 90°.

Obtuse angled triangle

An obtuse angled triangle has one angle that is greater than 90°.

Equilateral triangle

All sides and angles of an equilateral triangle are equal.

Isosceles triangle

An isosceles triangle has two sides that are of equal length and the angles opposite these sides are also equal.

Scalene triangle

All three sides of a scalene triangle are of different length.

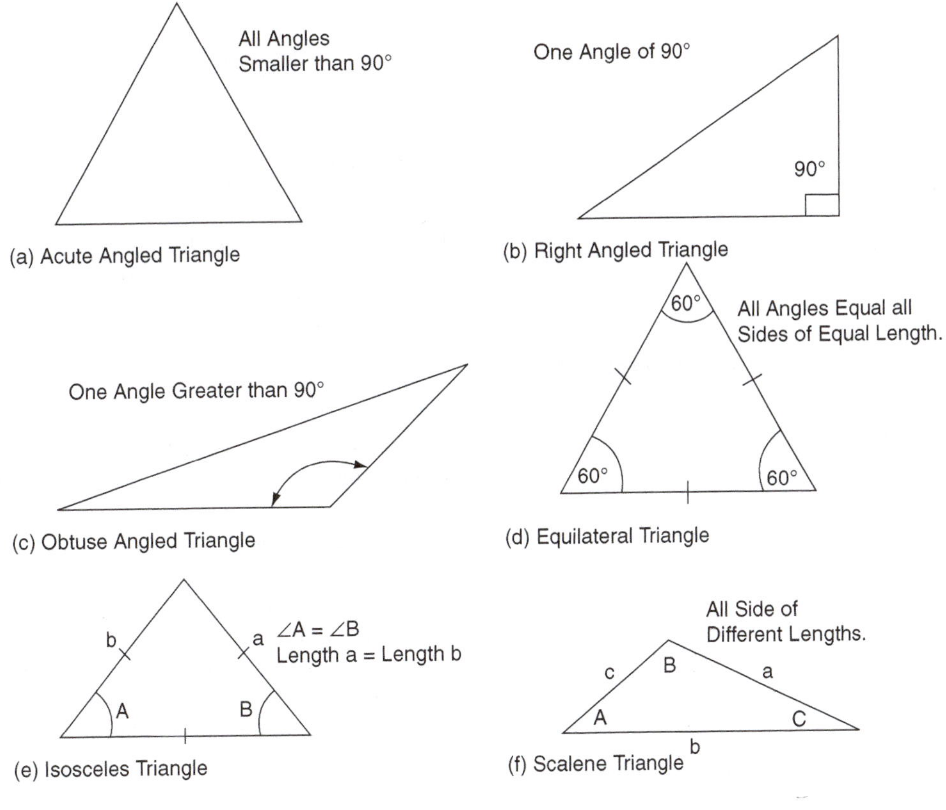

Fig. 4.9 Types of triangle

Right angled triangle

A right angled triangle has one angle that is equal to 90°. The side opposite the right angle is the longest side of the triangle and it is known as the hypotenuse.

Labelling sides and angles of a triangle

The angles of a triangle are denoted by capital letters and the side opposite each angle is denoted by the corresponding small letter, as shown in Figure 4.10.

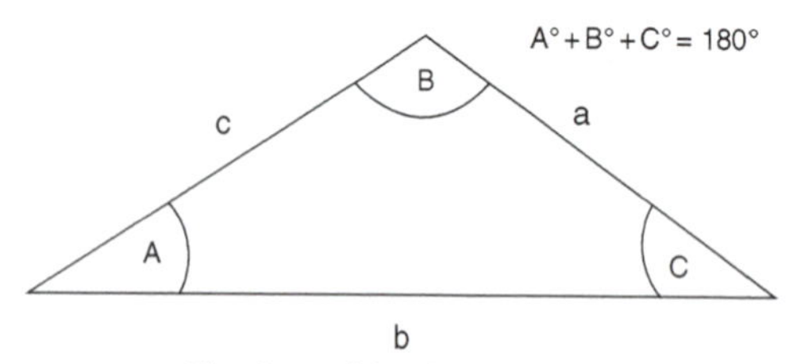

Fig. 4.10 Lettering the sides of a triangle

Sum of the three angles of a triangle

As shown in Figure 4.10, the sum of the three angles = 180°.

4.5 Pythagoras' theorem

In a right angled triangle, the square on the hypotenuse is equal to the sum of the squares on the other two sides. Figure 4.11 shows a right angled triangle with sides a, b and c. Pythagoras' theorem states that $c^2 = a^2 + b^2$.

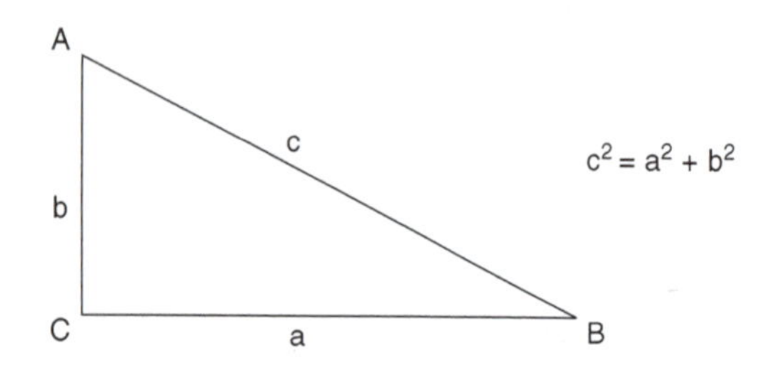

Fig. 4.11 A right angled triangle

Example 4.4

Figure 4.12 shows a right angled triangle. Use Pythagoras' theorem to calculate the length of the hypotenuse x.

By Pythagoras' theorem

$$x^2 = 15^2 + 8^2$$
$$x = \sqrt{(15^2 + 8^2)}$$
$$x = \sqrt{(225 + 64)}$$
$$x = \sqrt{289}$$
$$x = 17$$

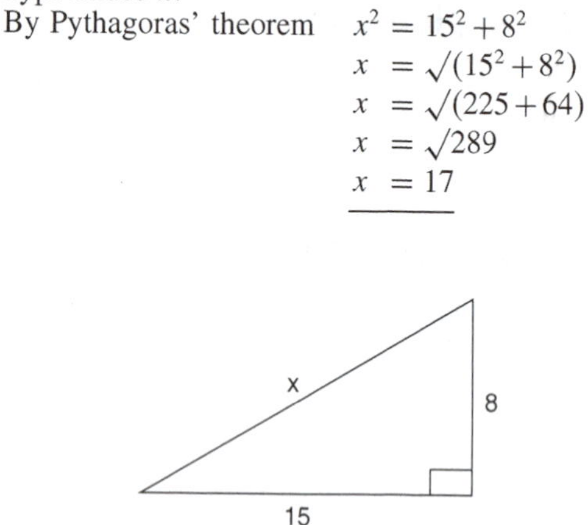

Fig. 4.12 A right angled triangle (Example 4.4)

4.6 Circles

Figure 4.13(a–c) shows features of circles that are used here and elsewhere in this book.

Diameter: the length of a straight line drawn from one side of a circle to the other and passing through the centre.

Radius: a straight line running from the centre of a circle to any point on the circumference.

Circumference: the length of the boundary of a circle.

Arc: the length of a continuous section of the circumference.

Chord: a straight line joining two points on the circumference of a circle.

Sector: a part of a circle bounded by two radii and an arc.

Segment: the region of a circle formed by an arc and a chord.

Tangent: A straight line drawn at right angles to the radius of a circle.

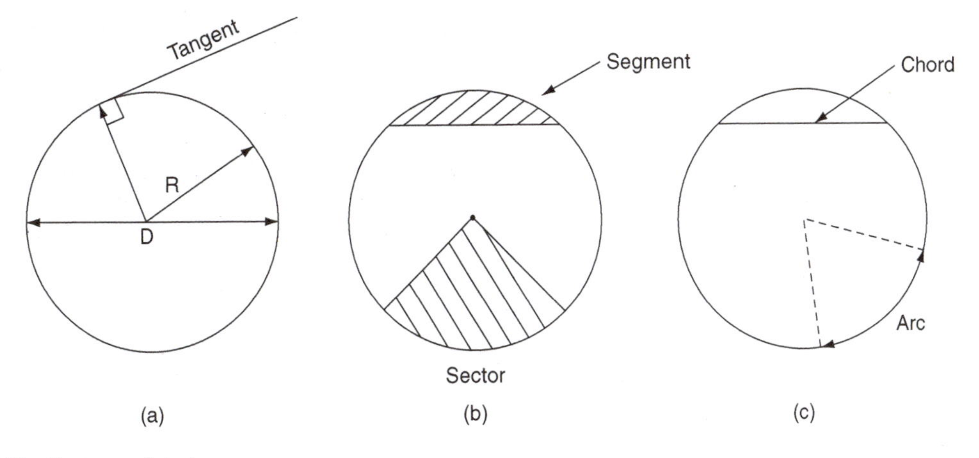

Fig. 4.13 Features of circles

Ratio of diameter and circumference π

When the circumference of a circle is divided by the radius, the result is always approximately 3.14159 (which can be refined to additional decimal places according to the degree of accuracy required); the figure thus obtained is called pi and is denoted by the Greek letter π. For simple calculations π is often taken as 22/7 and on other occasions the value $\pi = 3.142$ is used.

The length of the circumference of a circle $= \pi \times \text{diameter} = 2\pi \times \text{radius}$ because the diameter $= 2 \times \text{radius}$.

Fig. 4.14 Timing marks

Length of arc

The length of an arc of a circle $= \dfrac{\theta \times \pi \times \text{diameter}}{360°}$, where θ is the angle of the arc.

4.7 Timing marks

Engine timing marks are often stamped on the outer edge of crankshaft pulleys and flywheels, as shown in Figure 4.14. These marks are used to ensure that the piston is in the correct position for setting ignition timing, fuel pump timing, and valve timing.

Example 4.5

As a result of wear, the timing marks on an engine flywheel have been lost and it has become necessary to reinstate them. The top dead centre position has been determined and the rim of the flywheel marked accordingly. The task is to place a mark on the flywheel rim that indicates 15° before TDC.

The diameter of the flywheel at the place where the timing mark is to be made is 350 mm. An arc of length that represents 15° of crank rotation is calculated as follows:

$$\text{length of arc} = \frac{\text{angle}}{\text{circumference } 360°}$$

From this, length of arc $= \text{circumference} \times \frac{\text{angle}}{360°}$

$$\text{length of arc} = \frac{22 \times 350}{7} \times \frac{15°}{360°} = 45.8\,\text{mm}$$

4.8 Wheel revolutions and distance travelled

When a wheel and tyre on a moving vehicle makes one complete revolution the distance moved by the vehicle in a straight line is equal to the circumference of the tyre, as shown in Figure 4.15.

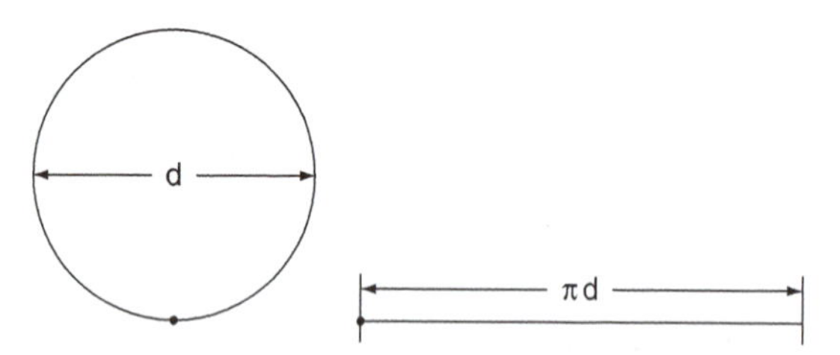

Fig. 4.15 Distance moved by a point on the rim of a wheel and tyre

Example 4.6

The rolling diameter of a large tyre for a commercial vehicle is 943.5 mm. How many metres will the vehicle move when the wheel and tyre rotate 10 times? Take $\pi = 3.142$.

Solution

Circumference of tyre $= \pi d = 3.142 \times 0.9435 = 2.964\,\text{m}$

In 10 revolutions the distance moved $= 10 \times 2.964 = 29.64\,\text{m}$.

4.9 Valve opening area

Figure 4.16 shows part of a poppet valve and its seat. The effective area through which air or mixture can pass when the valve is fully open is formed by a hollow cylinder that arises from the circumference of the valve head and the amount of valve lift.

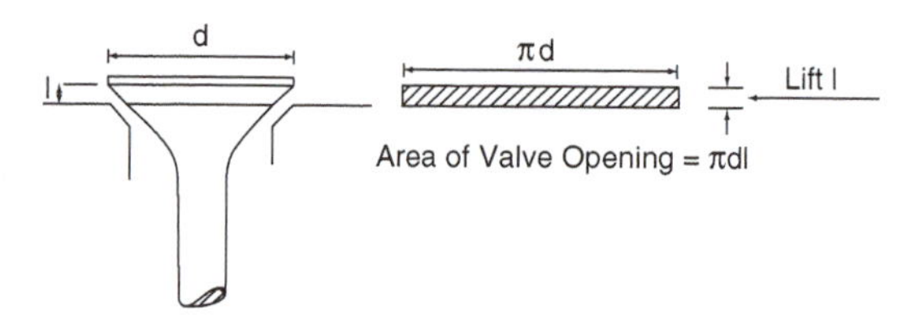

Fig. 4.16 Valve opening area

Example 4.7

A poppet valve with a diameter of 40 mm has a lift of 15 mm. Calculate the area through which air or mixture can pass on its way to the engine cylinder.

Solution

Effective area $= \pi \times \text{diameter of valve} \times \text{valve lift}$
$= 3.142 \times 40 \times 15 = 1885\,\text{mm}^2$

4.10 Trigonometry

In a right angled triangle the following names are given to the sides of the triangle and these names are used to define the trigonometrical ratios: **sine**, **cosine** and **tangent**. It is standard practice to use the following abbreviations for these ratios.

- sine – abbreviated to sin;
- cosine – abbreviated to cos;
- tangent – abbreviated to tan;
- opposite – the side opposite the given angle;
- hypotenuse – the side opposite the right angle and it is the longest side of the triangle;
- adjacent – the side next to the given angle.

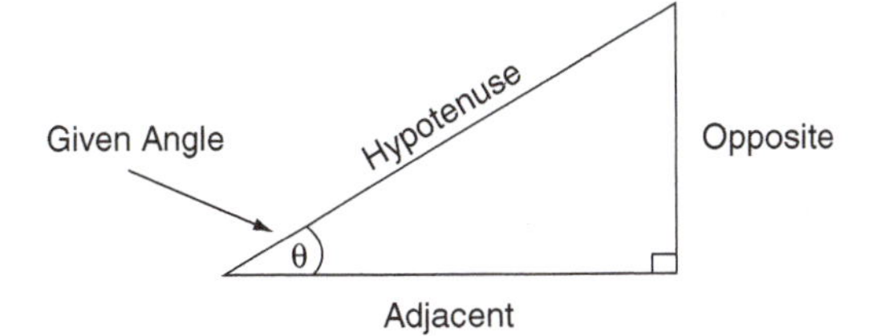

Fig. 4.17 The sides of a right angled triangle

In a right angled triangle the sides and angles are related by three trigonometrical ratios, which are shown in Table 4.1.

Table 4.1 Right angled triangle: trigonometrical relationship between sides and angles

Trig ratio	Abbreviation	Sides of triangle
sine	sin	opposite ÷ hypotenuse
cosine	cos	adjacent ÷ hypotenuse
tangent	tan	opposite ÷ adjacent

4.11 Using sines, cosines and tangents

The values of the trigonometrical ratios are a standard facility on scientific calculators and they are also given in mathematical tables. A scientific calculator is used in the examples and exercises given below. The procedure varies according to the type of calculator being used and it is important to have the instruction book to hand so that the correct procedure is followed.

Example 4.8
Use your scientific calculator to find sin 24°.

On the calculator that I am using the sequence of operations is:

Switch on
Press AC
Enter 24
Press sin

The reading is 0.4067366.

The value of sin 24° = 0.4067 correct to 4 decimal places.

Example 4.9
Find cos 30°.

The steps are:

Switch on
Press AC
Enter 30
Press cos

The readout is 0.8660254

The value of cos 30° = 0.8660 correct to 4 decimal places.

Example 4.10
Find tan 40°.

The steps are:

Switch on
Press AC
Enter 40
Press tan

The readout is 0.8390996

The value of tan 40° = 0.8391 correct to 4 decimal places.

Sines

Example 4.11
Use a calculator to find the angle θ in the triangle shown in Figure 4.18(a).

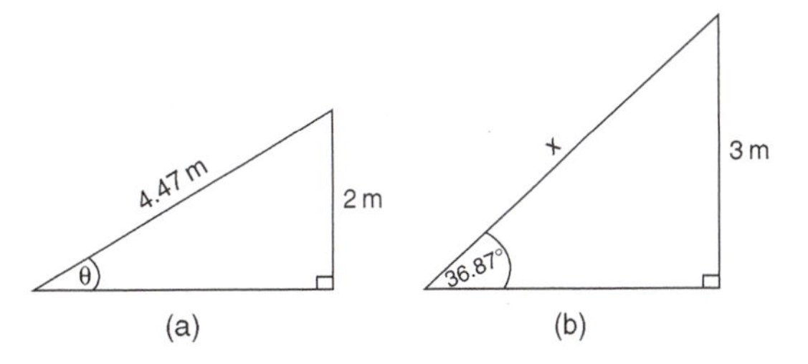

Fig. 4.18 Sines (Examples 4.11 and 4.12)

Solution

The first step is to find the sin of angle θ.

$$\text{Sin}\,\theta = \frac{\text{opposite}}{\text{hypotenuse}} = \frac{2}{4.47} = 0.4474$$

The next step is to use the $\sin^{-1}$ function on the calculator to determine the angle θ.

The inverse notation is a shorthand way of saying 'the angle whose sine is ...?'

In this case the angle is θ and its sin is 0.4474. To find the size of the angle θ in degrees use the following procedure.

With the value 0.4474 displayed on the calculator, press the ***shift*** key followed by the $\boldsymbol{sin}^{-1}$ key.

$\text{Sin}^{-1}\ 0.4474 = 26.58°$. Angle $\theta = 26.58°$.

Example 4.12

Determine the length of side x in Figure 4.18(b).

Use the equation for $\sin\theta = \dfrac{\text{opposite}}{\text{hypotenuse}}$

In this case, the opposite = 3 m; the hypotenuse $= x$. The angle $\theta = 36.87°$

$\sin 36.87° = \dfrac{3}{x}$

Transposing to make x the subject, gives $x = 3 \div \sin 36.87°$

Using the calculator the procedure is:

Enter 36.87
Press sin

Result 0.6000

Now divide 3 by 0.6

Result: $x = 5$.

Example 4.13

Find the value of sin 24° 15′.

Solution

The calculator works in degrees and decimal parts of a degree so it is necessary to convert 15′ into a decimal. As there are 60′ in one degree, $15' = 15/60 = 0.25°$.

$24°\ 15' = 24.25°$.

The procedure is:

Make sure that the calculator is on the correct setting

Enter 24.25
Press sin

The result is 0.4107.

The value of sin 24° 15′ is 0.4107.

Example 4.14

A formula for the approximate velocity of a piston in a reciprocating engine gives

$$v = \omega r\{\sin\theta + (\sin 2\theta)/2n\}$$

where v = piston velocity in m/s, ω = angular velocity of the crankshaft in radians/s, r = radius of the crank throw in metres and n = ratio of connecting rod length ÷ radius of crank throw. Calculate piston velocity when $\omega = 300\,\text{rad/s}$, $r = 0.05\,\text{m}$, $n = 4.5$ and $\theta = 30°$.

Solution

$$v = 300 \times 0.05\ \{\sin 30° + (\sin 60°)/9\}$$
$$= 15\ \{0.5 + 0.8660/9\}$$
$$= 15 \times 0.15962\,\text{m/s}$$
$$\therefore v = 2.39\,\text{m/s}$$

Cosines

Example 4.15

Use a calculator to find the angle θ in the triangle shown in Figure 4.19.

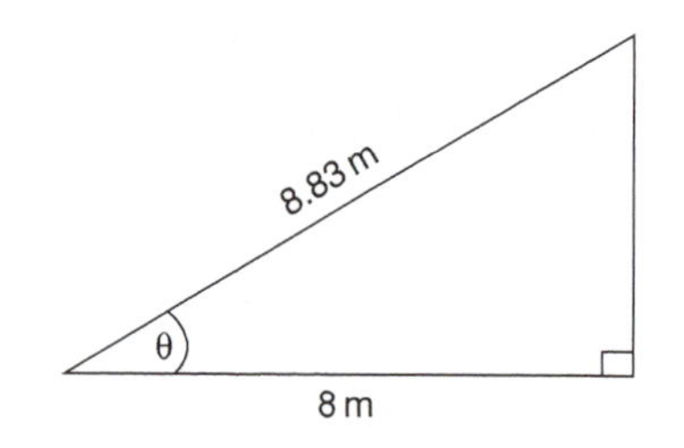

Fig. 4.19 Calculation of angle θ (Example 4.15)

Solution

The first step is to find the cos of angle θ.

$$\cos\theta = \frac{\text{adjacent}}{\text{hypotenuse}}$$

The next step is to use the $\cos^{-1}$ function on the calculator to determine the angle θ.

The inverse notation is a shorthand way of saying 'the angle whose cosine is ...?'

In this case the angle is θ and its cos is $= 8/8.83$. To find the size of the angle θ in degrees use the following procedure.

Write down $\cos\theta = \frac{\text{adjacent}}{\text{hypotenuse}}$

Write in the values that are given, i.e. $\cos\theta = \frac{8}{8.83}$

Enter $8 \div 8.83$
Press =; this gives 0.9060
Press *shift*
Press $\cos^{-1}$

Result: 25.04; the answer is $\theta = 25.04°$.

Example 4.16

The acceleration a of an engine piston is given by the following formula.

$$a = \omega^2 r\{\cos\theta + (\cos 2\theta)n\}$$

Where a = linear acceleration of piston in m/s^2, ω = angular velocity of crankshaft in radians per second, θ is the angle in degrees that the crank has turned from the top dead centre position, and n = ratio of connecting rod length ÷ radius of crank throw.

Calculate linear acceleration a, when $\omega = 50\,\text{rad/s}$, $\theta = 30°$, $r = 0.05\,\text{m}$, and $n = 4.5$.

Solution

$a = \omega^2 r\{\cos\theta + (\cos 2\theta)n\}$

$\omega = 50\,\text{rad/s}$, $\theta = 30°$, $r = 0.05\,\text{m}$, and $n = 4.5$.

Placing these values in the equation gives:

$$a = 50 \times 50 \times 0.05\left\{\cos 30° + \frac{\cos 60°}{9}\right\}$$

$$= 125\left(0.8660 + \frac{0.5}{9}\right)$$

$$= 125 \times 0.9216$$

$$\therefore a = 115.19\,\text{m/s}^2.$$

Tangents

Example 4.17

A road vehicle has a track width of 1.52 m and its centre of gravity is at a height of 1.17 m above the road surface. To what angle may the vehicle be tilted before it will turn over?

Solution

A vehicle will turn over when a vertical line through the centre of gravity passes through the point where the wheel makes contact with the road, as shown in Figure 4.20.

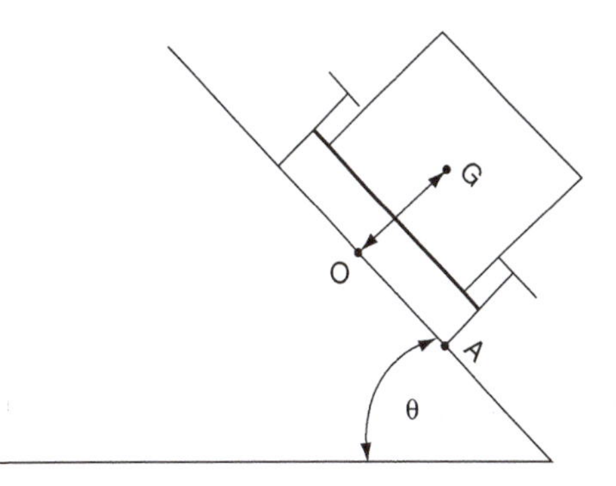

Vehicle will Tip over When G Passes Vertically over A. This Occurs When $\tan\theta = \frac{OA}{OG}$

Fig. 4.20 Vehicle turning over (Example 4.17)

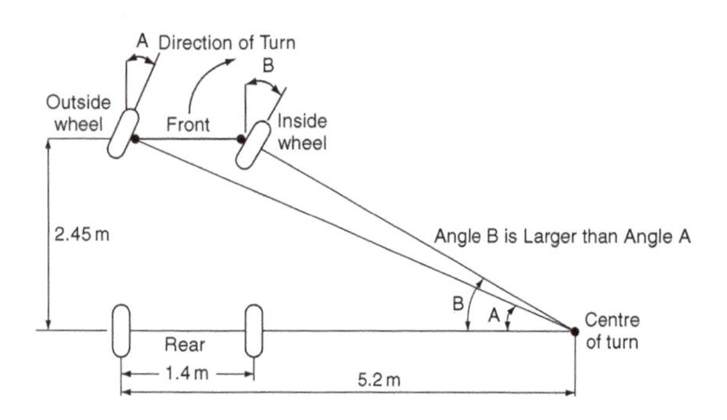

Fig. 4.21 Steered angles (Example 4.18)

From Figure 4.20 it will be seen that $\tan\theta = \frac{OA}{OG} = \frac{0.76}{1.17} = 0.65$

Using the calculator, enter 0.65; then press *shift* followed by $\tan^{-1}$. The result is 33°.

The vehicle will turn over if the angle of tilt reaches 33°.

Example 4.18

Determine angles A and B in Figure 4.21.

Solution

$$\tan A = \frac{2.45}{5.2} = 0.471$$

Angle A = 25.22°

$$\tan B = \frac{2.45}{3.8} = 0.645$$

Angle B = 32.82°

4.12 Summary of formulae

Pythagoras' theorem: 'In a right angled triangle the square of the hypotenuse is equal to the sum of the squares on the other two sides'.

For right angled triangles:

$$\sin = \frac{\text{opposite}}{\text{hypotenuse}}$$

$$\cos = \frac{\text{adjacent}}{\text{hypotenuse}}$$

$$\tan = \frac{\text{opposite}}{\text{adjacent}}$$

Circles:

Circumference = $\pi \times$ diameter

Area of circle $= \pi r^2 = \frac{\pi d^2}{4}$

Radian: There are 2π radians in 360°; 1 radian = 57.3°.

4.13 Exercises

4.1 It takes two complete revolutions of the crankshaft to complete the 4-stroke cycle of operations. How many degrees is this?

4.2 A shaft rotates at 3000 rev/min. Convert this to radians per second.

4.3 Convert 36.40° to degrees and minutes.

4.4 On a certain vehicle the specified steered angle of the outer wheel when measured on a turntable is 31° 10′ plus or minus 55′. What are the largest and smallest angles that are acceptable?

4.5 The steered angles of a vehicle as measured on turntables are 31° 10′ on the outer wheel and 33° 55′ on the inner wheel. How much larger is the inner wheel angle than the outer wheel angle?

4.6 On a certain engine the inlet valve opens 10° before top dead centre and closes 45° degrees after bottom dead centre. Determine the number of degrees of crank rotation for which the inlet valve is open.

4.7 On a certain high-performance petrol engine the inlet valve opens 15° before top dead centre and the exhaust valve closes 20° after top dead centre. How much valve overlap is there?

Exercises – Section 4.2

4.8 Figure 4.22(a) shows the point angle of a twist drill. Determine the size of the angle α.

4.9 The angle A° shown in Figure 4.22(b) is: (a) acute, (b) obtuse, or (c) reflex?

4.10 Determine the size of angle θ in Figure 4.22(c).

4.11 Determine the size of the angle α in Figure 4.22(d).

4.12 Determine the angle of king pin inclination KPI in Figure 4.22(e).

4.13 Determine the valve seat angle θ in Figure 4.22(f).

Exercises – Section 4.3

4.14 Figure 4.23(a) shows the outline of a crank and connecting rod. The distance x is the distance that the piston has moved from TDC when the connecting rod is at an angle of 90° to the crank throw. Use Pythagoras' theorem to calculate the distance x.

4.15 Figure 4.23(b) shows a section of a shaft that has had a flat surface machined on it. Use Pythagoras' theorem to calculate the distance d across the flat surface.

4.16 Figure 4.23(c) shows a driver's foot pressing on a brake pedal. Calculate the distance x.

4.17 The distance across the corners of the hexagonal bolt head shown in Figure 4.23(d) is 55 mm. Calculate the distance d across the flats of the bolt head.

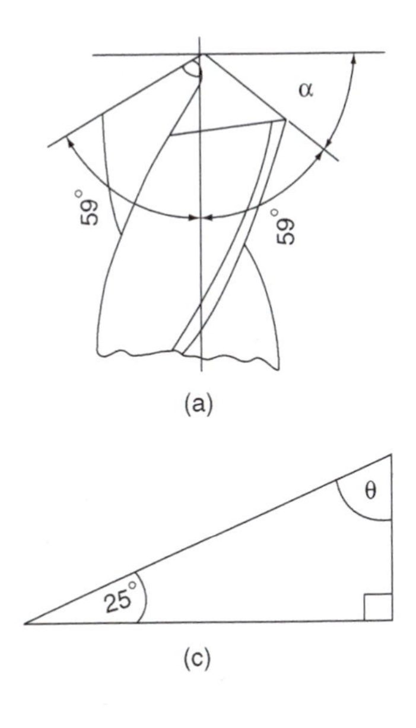

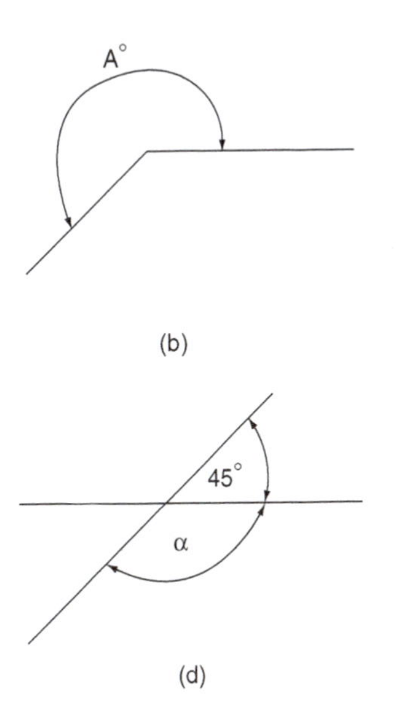

Fig. 4.22 Calculation of angles (Exercises 4.8–4.13)

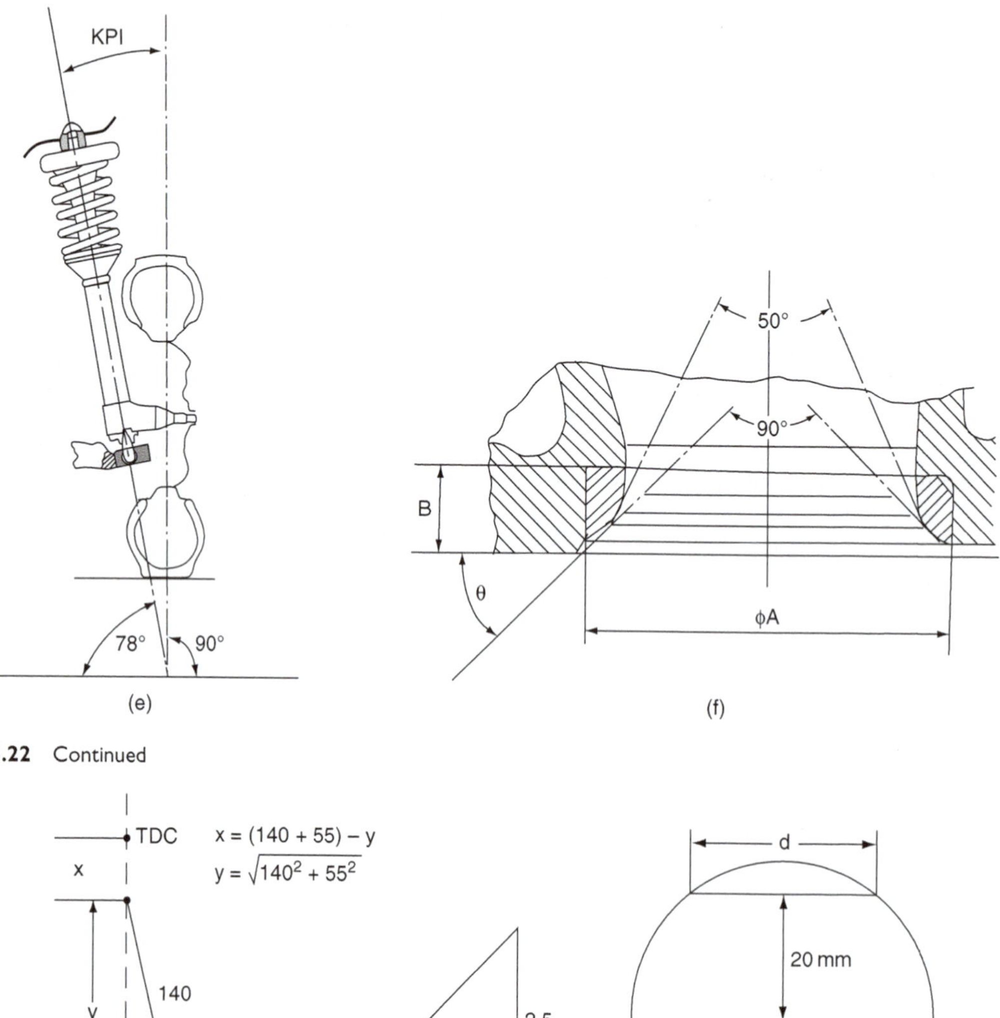

Fig. 4.22 Continued

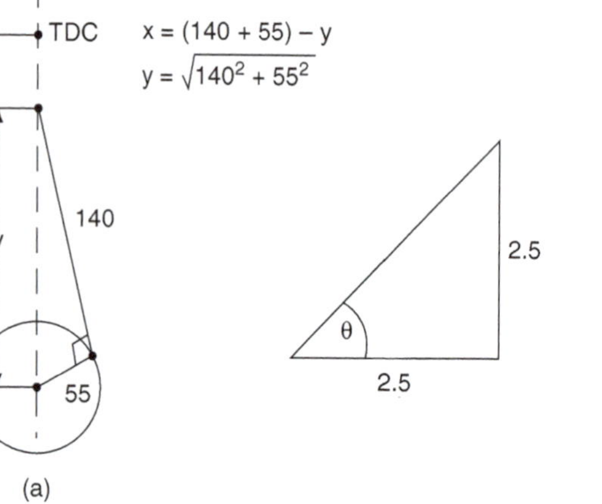

Fig. 4.23 Calculation of distances (Exercises 4.14–4.17)

Exercises – Section 4.4

4.18 A vehicle is fitted with tyres that have a rolling radius of 250 mm. How far will the vehicle travel when the wheels make 1000 revolutions?

4.19 A poppet valve has a diameter of 35 mm and a lift of 12 mm. Calculate the area through

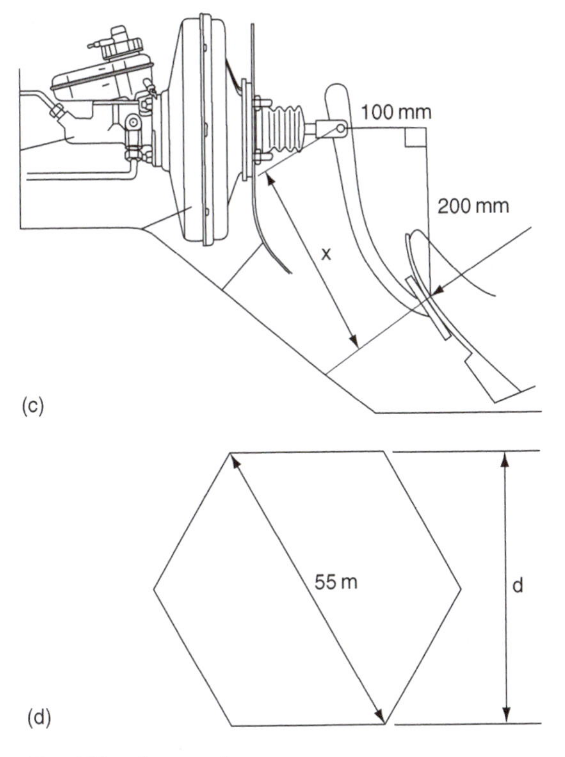

Fig. 4.23 Continued

which mixture passes on its way to the engine cylinder.

4.20 A certain vehicle is fitted with tyres that have a rolling radius of 300 mm. When the vehicle is travelling at a certain speed the wheels rotate at 600 rev/min. Determine the speed at which the vehicle is travelling, in km/h.

4.21 In a project to increase the power output of a 4-stroke engine, the inlet valve ports were polished and increased in diameter by 0.5 mm. The valve head diameter was increased from 30 mm to 30.5 mm and high lift cams were fitted that increase the maximum valve lift from 14 mm to 15 mm. Calculate the percentage increase in the area through which air or mixture can enter the cylinders through the inlet valves.

4.22 A technician is required to mark a point on the rim of a flywheel to indicate the point at which fuel injection is to start. The injection point is 25° before TDC. If the flywheel diameter at the point where the mark is to be made is 400 mm, calculate the distance around the circumference from the TDC position that the injection timing mark is to be made.

Exercises – Section 4.5

4.23 Using a calculator find the values of:
(a) $\sin 42°$; (b) $\sin 86°$; (c) $\sin 29°$; (d) $\sin 15°$; (e) $\sin 23°\ 15'$; (f) $\sin\ 30'$.

4.24 Use a calculator to determine the angle:
(a) $\sin^{-1} 0.8660$; (b) $\sin^{-1}\ 0.1500$; (c) $\sin^{-1}$ 0.5000.

4.25 Using the formula given in Example 4.14, calculate the velocity of a piston when $r = 0.06$ m, engine crank speed is 6000 rev/min, $\theta = 45°$, and $n = 4$. (Note there are 2π radians in one revolution.)

4.26 Find the length of the side x in the three triangles shown in Figure 4.24.

4.27 The velocity of an object whose motion is a simple harmonic is given by the formula $v = \omega r \sin\theta$, where v = velocity in m/s, r = radius in m, ω = angular velocity in rad/s, and θ = angular displacement. Calculate v when $\omega = 50$ rad/s, $r = 0.1$ m,k $\theta = 30°$.

Exercises – Section 4.6

4.28 Find the dimension x in each of the three right angled triangles shown in Figure 4.25.

4.29 Find the angle θ in each of the three triangles shown in Figure 4.26.

4.30 Use a calculator to find the angles whose tangents are:
(a) 0.5010; (b) 0.7500; (c) 0.5333; (d) 1.3300

4.31 Find the length of side x in each of the three right angled triangles shown in Figure 4.27.

4.32 Figure 4.28 shows the layout of a crank and connecting rod mechanism. Calculate the angle θ; also calculate the dimension x and hence determine the distance that the piston has travelled from TDC.

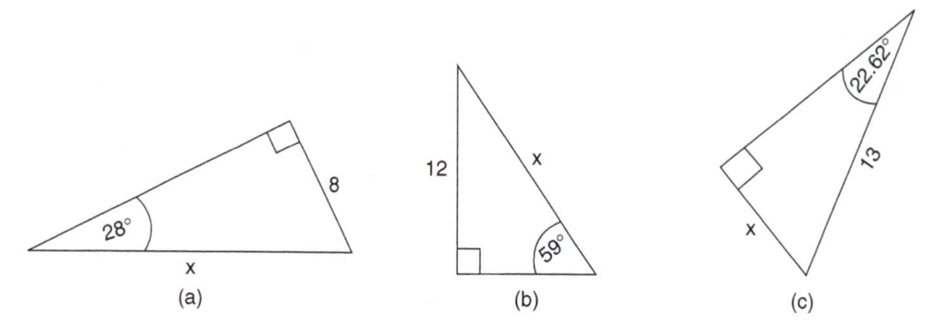

Fig. 4.24 Calculation of hypotenuse lengths (Exercise 4.26)

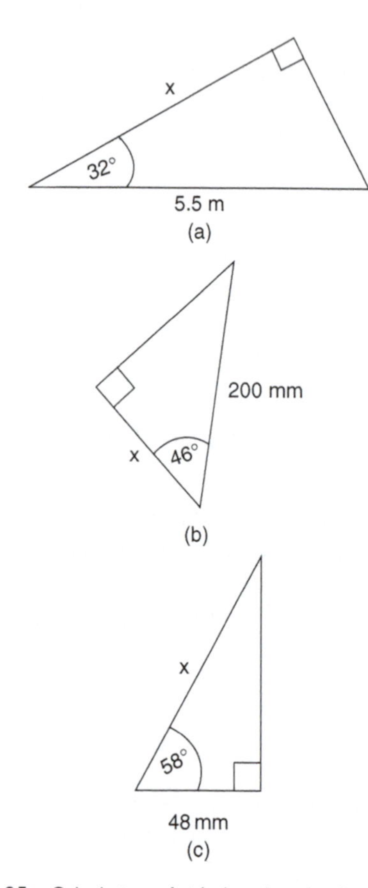

Fig. 4.25 Calculation of side lengths of right angled triangles (Exercise 4.28)

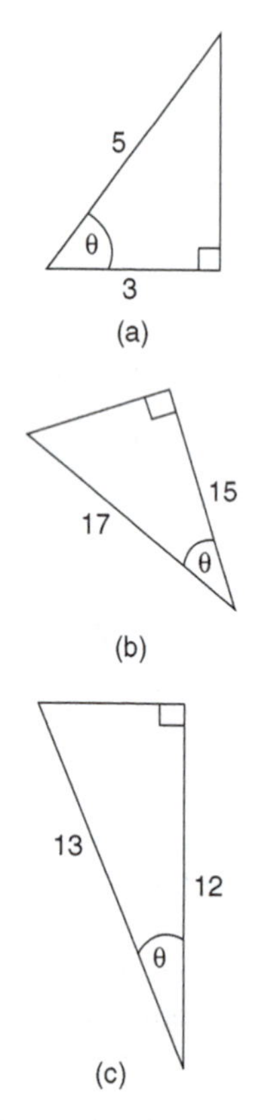

Fig. 4.26 Calculation of angles in right angled triangles (Exercise 4.29)

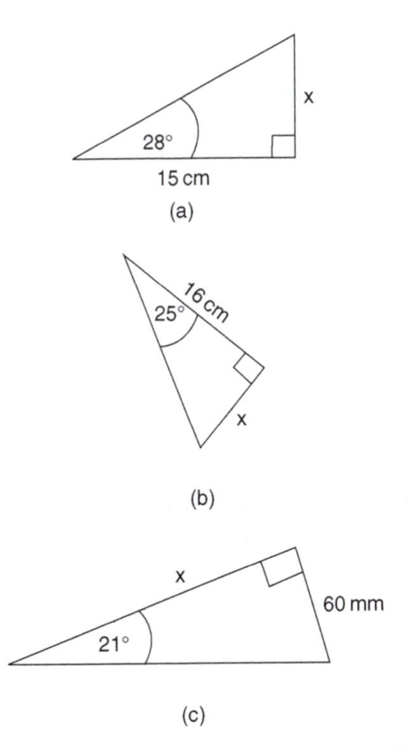

Fig. 4.27 Calculation of side lengths of right angled triangles (Exercise 4.31)

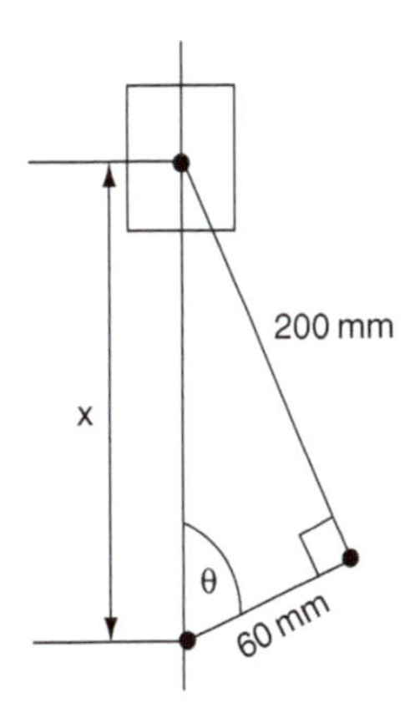

Fig. 4.28 Calculations based on crank and connecting rod (Exercise 4.32)

5

Forces

5.1 Force

Force is an effect that changes, or tends to change, the state of rest or uniform motion of an object. For example, if a steel block is resting on a flat surface it can be moved by the application of a force. If the steel block is bolted to the surface the application of the same force will tend not to move it. If an object such as a trolley is moving at steady speed across a surface the application of force in the direction of travel will cause it to move faster. The unit of force is the newton (N). 1 newton is the force that will produce an acceleration of $1\,m/s^2$ when applied to a mass of 1 kg that is free to move.

5.2 Types of force – examples

Direct forces – these may be a push or a pull. In the case of the steel block, the pull may be exerted by means of a cord attached to the block – in the case of a push, the force would be applied by means of a solid rod.

Attractive forces – the collective force exerted by a magnet or the gravitational effect of the earth. The force of gravity is exerted by the earth on all objects on, or near, its surface and tends to pull such bodies towards the earth's centre. The weight of an object is a measure of the gravitational force acting on it.

Explosive forces – the collective force exerted by the rapidly expanding gases in an engine cylinder following combustion.

5.3 Describing forces

In order to fully describe a force the following must be known.

- magnitude – number of newtons (N);
- direction – 50° to the horizontal, due west, for example;
- sense – push or pull;
- point of application.

5.4 Graphical representation of a force

All of the features of a force can be represented graphically by a line. The line is drawn so that its length represents the magnitude of the force. The direction of the line, for example 50° to the horizontal, represents the direction of the force. An arrow drawn on the line represents the sense – push or pull.

Example

Represent the following forces acting at a common point (see Figure 5.1).

(a) 30 N pulling away from the point, direction due north;
(b) 20 N pushing toward the point, direction due west;
(c) 15 N pulling away from the point, direction south west;
(d) 40 N pushing towards the point, direction 40° west of north.

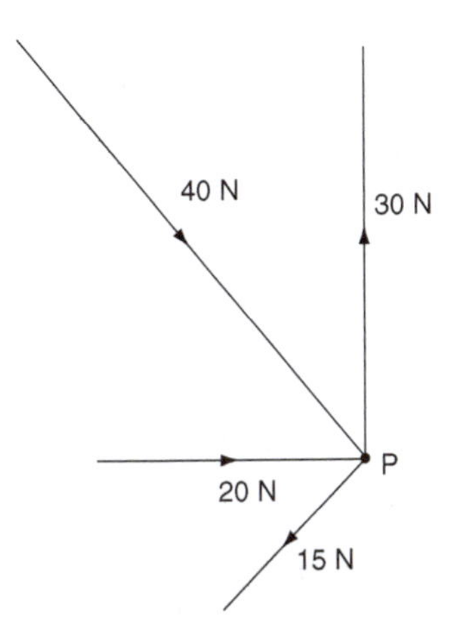

Fig. 5.1 Graphical representation of forces

5.5 Addition of forces

When two forces act at a point so that both are directly in line their combined effect can be found by adding the forces together.

In Figure 5.2(a) the combined effect (**resultant**) of the two forces of 50 N is 100 N in the same direction.

In Figure 5.2(b) the forces are in opposition 50 N horizontally to the left and 80 N horizontally to the right. The combined effect (**resultant**) is a force of 30 N to the right. The resultant of two or more forces acting at a point is the single

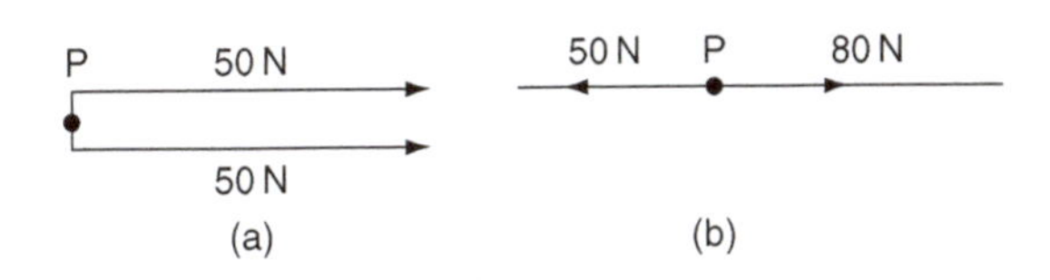

Fig. 5.2 Addition of forces – the resultant

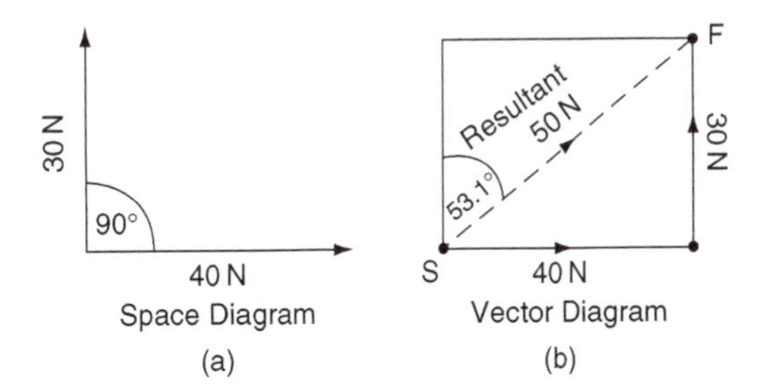

Fig. 5.3 Vector diagram – adding forces

force that would replace the others and produce the same effect.

When the forces are not in line as shown by the space diagram in Figure 5.3(a), the procedure for finding the resultant is slightly more complicated and requires the use of vectors. A vector is a line that represents a quantity such as a force. The length of the line represents the size of the force, the arrow represents the sense of the force, push or pull, and the angle of the line represents the direction of action of the force.

To add vectors in order to find the resultant the vectors are drawn parallel to the lines of action of the forces. An S is placed at the point where the drawing of the vector diagram starts. From S the first vector is drawn in the required direction, the next vector is drawn from the end of this first vector, and when the last vector has been drawn an F is placed at the end of the final vector. A line is then drawn from S to F, an arrow pointing from S to F is placed on this line and this line (vector) represents the resultant. In the example shown in Figure 5.3(b) the resultant is a 50 N pull acting at an angle of 53.1° east of north.

5.6 Parallelogram of forces

If three concurrent, coplanar forces are in equilibrium, two of the forces may be represented by two sides of a parallelogram. If the parallelogram is completed by drawing in the other two sides, the diagonal drawn from the angular point formed by the adjacent sides will represent the third force in magnitude and direction. (see Figure 5.4).

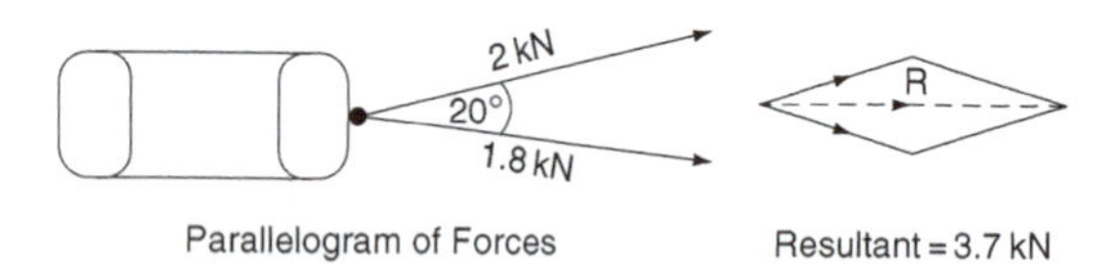

Fig. 5.4 Parallelogram of forces

Example 5.1
Find the resultant of two forces acting at a point: (1) a 45 kN pull acting in a horizontal direction and (2) a 30 kN pull acting at an angle of 60° to the horizontal (see Figure 5.5).

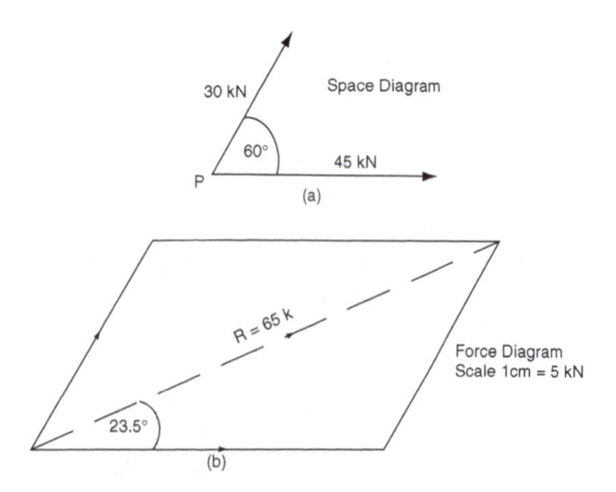

Fig. 5.5 Parallelogram of forces (Example 5.5)

Solution
First draw the space diagram as shown in Figure 5.5a.

Next select a scale for the force diagram. The author has chosen 0.5 cm = 5 kN.

Draw a horizontal line 4.5 cm long from point P to represent the 45 kN force.

At 60° above this line and from point P, draw a line 3 cm long to represent the 30 kN force, as shown in Figure 5.5b.

Then complete the parallelogram by drawing a horizontal line from the end of the 3 cm line. From the end of the 4.5 cm line draw parallel to the 3 cm line and this will complete the parallelogram.

From point P, draw in the diagonal of the parallelogram and measure it. The length of the diagonal in centimetres multiplied by the scale of 5 kN/cm will give the numerical value of the resultant of the two forces.

The angle of the resultant can then be determined with the aid of a protractor.

In this case the resultant is a pull of 65 kN at 23.5° to the horizontal.

5.7 Triangle of forces

When three concurrent (acting at a point), coplanar (acting in the same plane) forces are in equilibrium the forces may be represented by the three sides of a triangle. In this case there is no resultant because the three forces balance each other out. The triangle of forces law can be used to solve force problems when three forces acting at a point are known to be in equilibrium and the magnitude and direction of one force are known and the directions of action of the other two forces are also known.

Example 5.2
Figure 5.6a shows the details of a simple wall crane. The tie is in tension and it exerts a force in the direction shown. The jib is in compression and its force pushes away from the wall. The load of 60 kN acts vertically downward as shown. Find the forces in the jib and the tie. The 3 forces are in equilibrium so the triangle of forces law can be used.

The letters A,B,C in the spaces between the jib, the tie and the load cable are used to identify

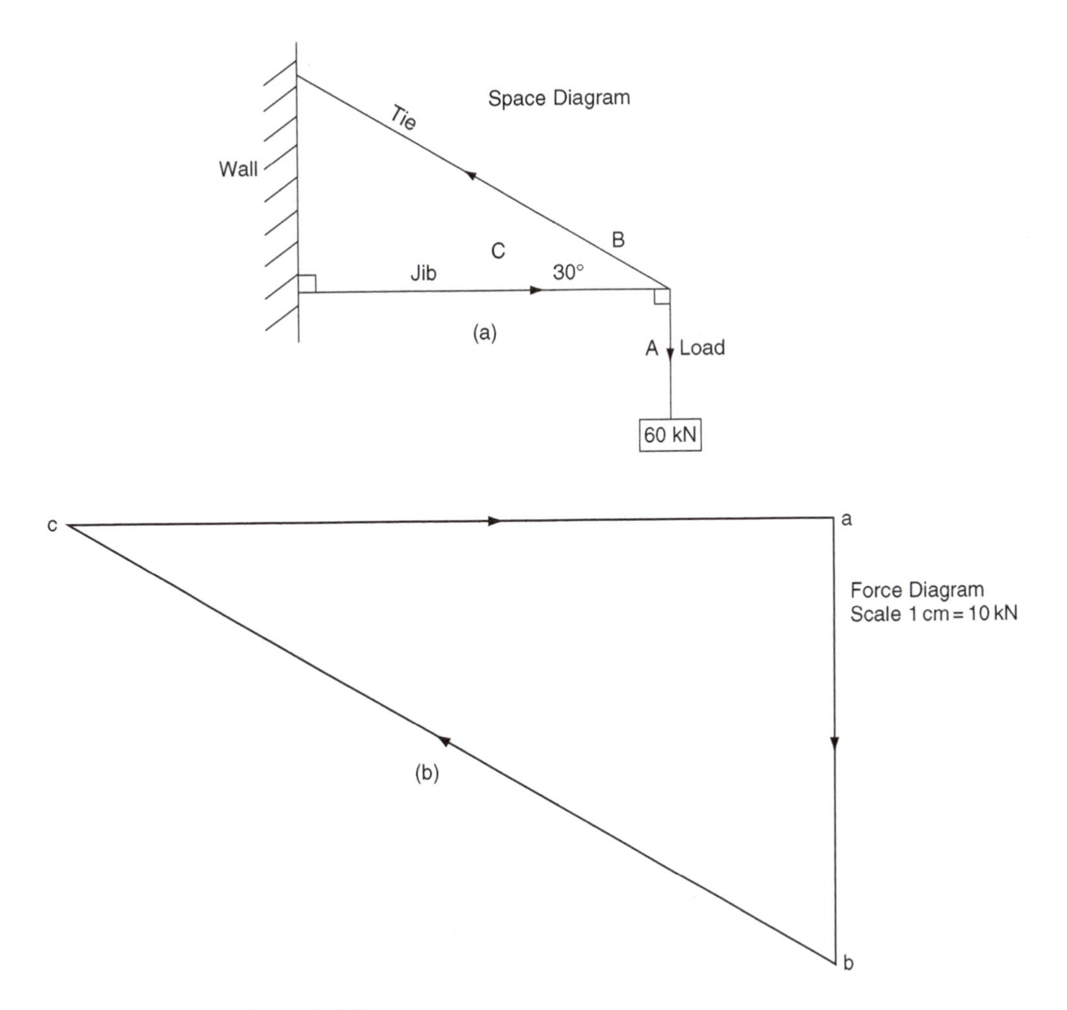

Fig. 5.6 Triangle of forces (Example 5.2)

each force. This procedure of lettering the spaces is known as Bow's notation.

Solution
Choose a suitable scale for the force diagram – in this case the author has chosen 1 cm = 10 kN.

Draw a line 6 cm long parallel to the load cable. Start this line at point a and end it at point b. Write these letters on the force diagram.

From b draw a line parallel to the tie – in the same direction as the arrow in the space diagram.

From point a on the force diagram draw a line parallel to the tie – in the same direction as the arrow inn the space diagram.

Measure the length of a to c on the force diagram. It is 10.4 cm, which means that the force in the jib is = 104 kN.

Measure the length of b to c on the force diagram. It is 12 cm, which means that the force in the tie is 120 kN.

5.8 Resolution of forces

In dealing with some problems it is often helpful to break down a single force into two components.

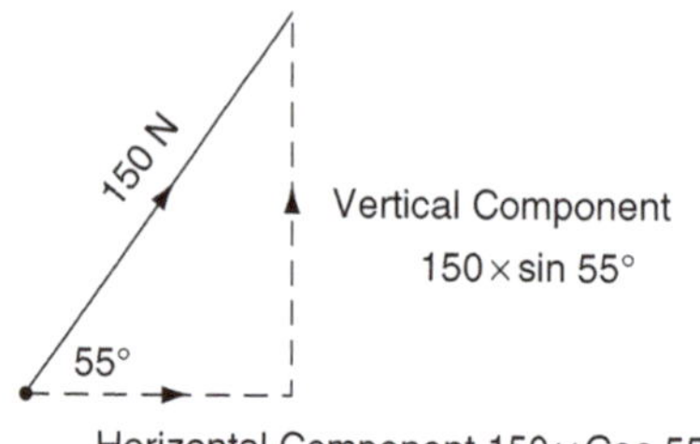

Fig. 5.7 Resolution of a force (Example 5.3)

Example 5.3 Resolution of forces

In the example shown in Figure 5.7 the force of 150 N acting pulling away from P at angle of 35° east of north has been resolved into a horizontal and a vertical component. This operation can be performed by drawing or, more often, by calculation.

To perform the calculation the 150 N vector is taken as the hypotenuse of a right angled triangle.

The vertical component is $150 \sin 55° = 150 \times 0.8191 = 122.87$ N

The horizontal component is $150 \cos 55° = 150 \times 0.574 = 86.04$ N.

5.9 Mass

The mass of an object is the quantity of matter contained in the object. The basic unit of mass is the kilogram.

Weight is the force that gravity exerts on an object. The force that gravity exerts on a mass of 1 kg = 1 × 9.81 N, because the acceleration due to gravity is taken to be $9.81\,\text{m/s}^2$.

5.10 Equilibrium

A body is said to be in equilibrium when it is completely at rest.

Stable equilibrium. A body is said to be in stable equilibrium when the turning moment of its weight acts as a righting moment that tends to restore the body to its original position after slight displacement from this position.

Unstable equilibrium. If the moment is an overturning moment that tends to capsize the body then the body is in unstable equilibrium.

Overturning. A body will overturn when a vertical line drawn through its centre of gravity falls outside the boundary lines formed by joining the extreme points of its contact with the surface on which the body is placed.

The various states of equilibrium are shown in Figure 5.8.

5.11 Pressure

When substances such as gases and fluids are in a cylinder or similar container, and they have a force applied to them, they are said to be under pressure, or pressurised. In SI units, the unit of pressure is the newton per square metre. $1\,\text{N/m}^2$ is known as a pascal. A pascal is a very small unit in vehicle engineering terms and it is common practice to use a larger unit of pressure which is known as a bar: $1\text{ bar} = 100\,000\,\text{N/m}^2$.

In vehicle engineering it has been common practice to set the operating pressure of diesel engine fuel injectors in units of pressure known

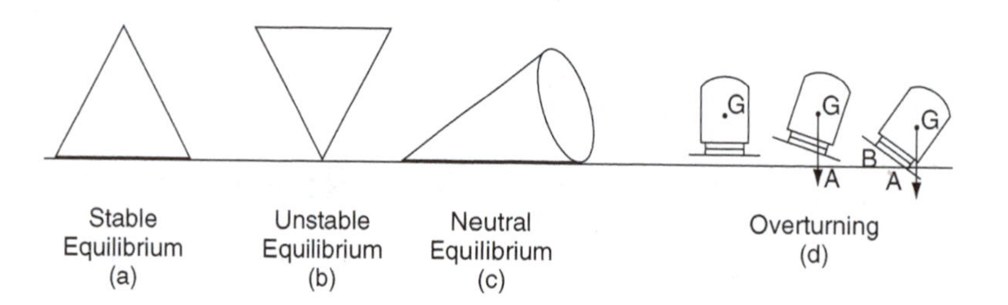

Fig. 5.8 Equilibrium – states of

as atmospheres. One atmosphere is approximately 14.7 pounds per square inch and is approximately equal to the pressure exerted by the atmosphere, at sea level. One bar is approximately equal to one atmosphere.

Pressure calculation. Pressure = Force in newtons (N) divided by area in square metres (m^2).

Example 5.4

A force of 600 N is applied to a piston which has a cross-sectional area at the crown of $0.01\ m^2$, as shown in Figure 5.9. Calculate the pressure that this force creates in the gas in the cylinder.

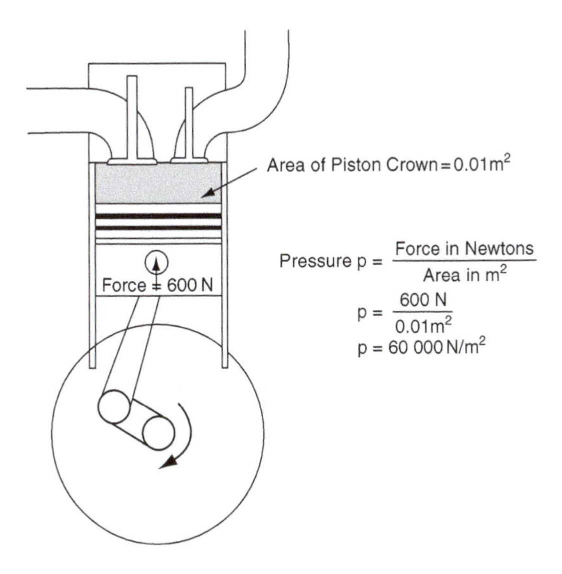

Fig. 5.9 Definition of pressure

Table 5.1 Examples of approximate pressures found in vehicle engineering

Compression pressure in a diesel engine	25–30 bar
Engine oil pressure	2–3 bar
Brake fluid pressure	20 bar
Air pressure in compressed air braking system	7 bar
Fuel gallery pressure in a petrol engine	3 bar
Tyre pressure – car	2–3 bar
Tyre pressure – large truck	6–7 bar
Diesel injection	120–200 bar
Diesel common rail injection	1000–1500 bar

Pressures encountered in vehicle engineering vary from very high pressures of hundreds of bar in diesel fuel injection systems to low pressures, below atmospheric, in the induction system of a petrol engine.

Examples of pressures are shown in Table 5.1. These examples are approximate only; they are given as a guide. *In all cases it is essential that the data relating to a specific application is used when performing tests and adjustments.*

5.12 Pressure in hydraulic systems

Hydraulic systems rely on the fact that the pressure inside a closed vessel is constant throughout the vessel. Figure 5.10(a) shows a closed vessel that

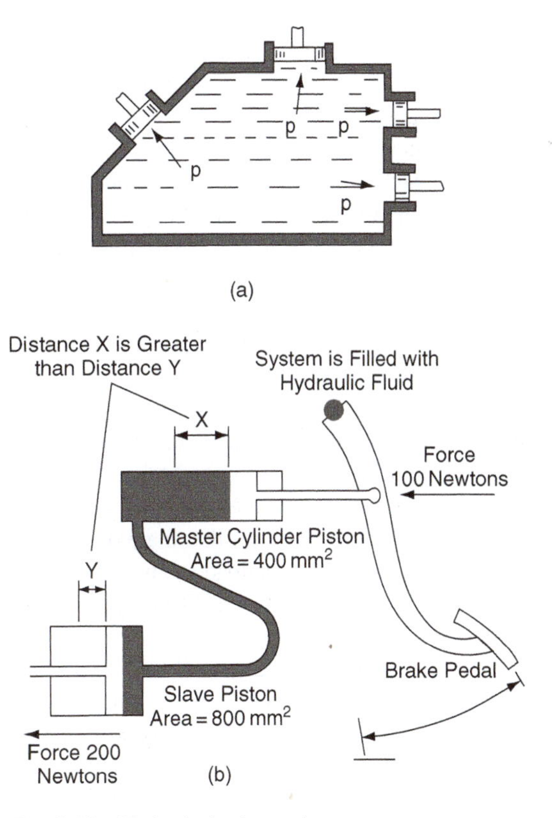

Fig. 5.10 Hydraulic brakes and pressure

has a number of pistons located in it. The pressure on each piston (not the force) is equal.

Example 5.5

Figure 5.10(b) shows the basic structure of a hydraulic braking system.

The pressure at the master cylinder = $\dfrac{100\,\text{N}}{400\,\text{mm}^2} = 0.25\,\text{N/mm}^2 = 2.5\,\text{bar}$

This is also the pressure at the slave cylinder.

The force at the slave cylinder = pressure × area = $2.5\,\text{N/mm}^2 \times 800\,\text{mm}^2 = 200\,\text{N}$.

5.13 Hooke's law

Figure 5.11(a) shows a simple apparatus that is used to demonstrate the effect of force applied to an elastic body. In this case the elastic body is a metal coil spring. The spring is suspended from a secure support and at its lower end there is a weight container to which various weights may be added and removed. A pointer is attached to the lower end of the spring and this permits readings of spring extension to be recorded as weights are added. When a series of weights and corresponding spring extension have been recorded, the weights are removed and the spring will return to its original unloaded length.

A typical set of results for this type of demonstration are shown in the table in Figure 5.9(b) and, when these are plotted on squared paper, they produce a straight-line graph as shown in the figure. This result shows that the extension is directly proportional to the load producing it. This is known as Hooke's law and it is named after Robert Hooke (1635–1703).

The relationship between load and extension is known as the stiffness of the spring, or the spring rate. Expressed mathematically, stiffness = load/extension.

When the spring is in compression, stiffness = load/compression.

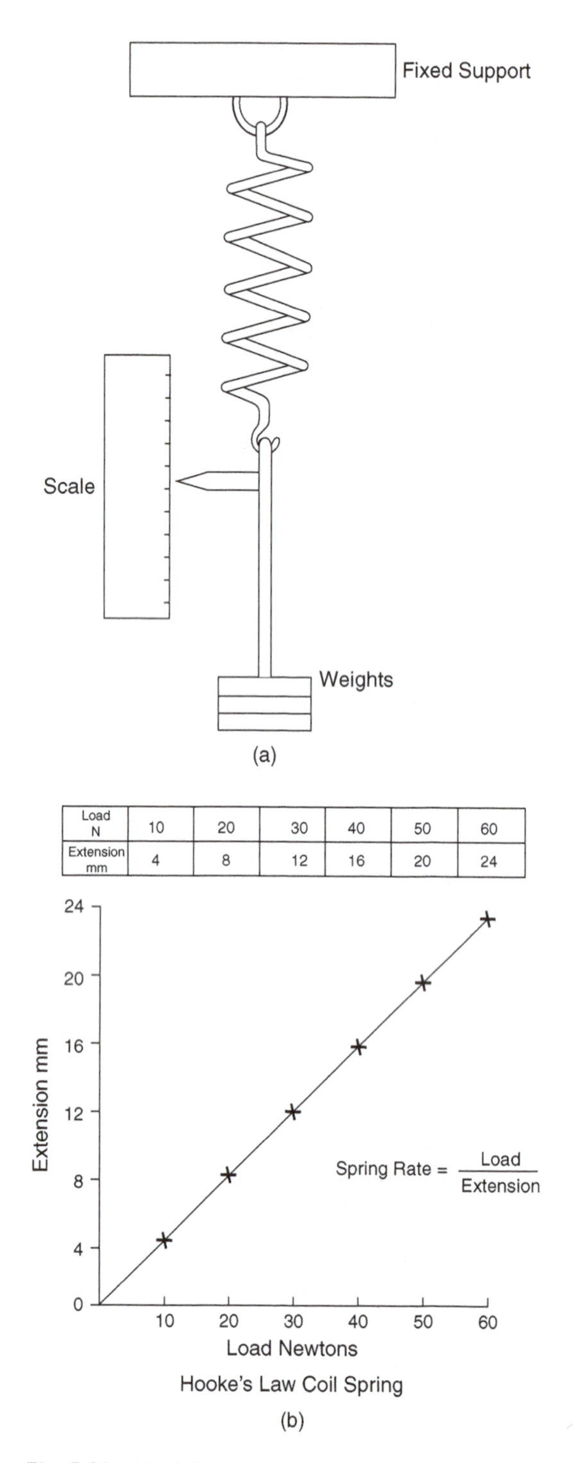

Load N	10	20	30	40	50	60
Extension mm	4	8	12	16	20	24

Fig. 5.11 Hooke's law apparatus and results

Example 5.6

A front suspension spring of a certain vehicle is compressed by 30 mm when it carries a load of 240 N. Calculate the spring stiffness (rate).

Solution

Calculate the spring stiffness (rate).

$$\text{Spring stiffness} = \frac{\text{load}}{\text{compression}}$$

$$240/30 = 8\,\text{N/mm}$$

5.14 Practical applications

Figure 5.12 shows one of the many automotive applications where factors relating to spring rate are likely to be encountered.

Valve spring rate is critical in ensuring the valves operate correctly. Valve springs can deteriorate in service and the dimensions shown here can be used to check that valve springs are suitable for further use.

The type of equipment used for testing valve springs is shown in Figure 5.12.

The force applied to the spring is shown on the dial gauge and a calibrated scale to the right, under the operating handle, indicates the amount of spring compression.

5.15 Summary

Hooke's law: Within the elastic limit for a material the extension produced is directly proportional to the load producing it.

$$\text{Stiffness of a spring} = \frac{\text{force}}{\text{extension}}$$

$$\text{Pressure} = \frac{\text{force}}{\text{area}}$$

Pressure units: $1\,\text{N/m}^2 = 1\,\text{Pa}$ (pascal).

$1\text{ bar} = 10^5\,\text{N/m}^2$

The hydraulic pressure inside a closed container is constant throughout the container.

5.16 Exercises

5.1 In order to carry out work on a front suspension system, a spring compressor is used to compress the spring axially by 250 mm. The spring rate is 9 N/mm.

Is the spring force that is held back by the spring compressor:

(a) 2250 N
(b) 27.8 N
(c) 225 kN
(d) 278 N

5.2 The coil spring in a ball and spring type of oil pressure relief valve exerts a force of 24 N on the ball valve when it is compressed by 8 mm. Is the spring rate:

(a) 192 N/mm
(b) 30 N/mm

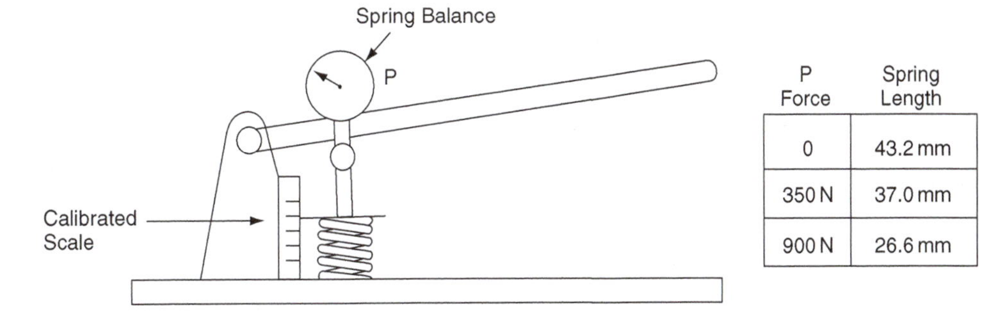

P Force	Spring Length
0	43.2 mm
350 N	37.0 mm
900 N	26.6 mm

Fig. 5.12 Testing a valve spring

(c) 3 N/mm
(d) 0.33 N/mm

5.3 Might a weak spring in an oil pressure relief valve cause
(a) the oil pressure at normal engine running speed to be too low?
(b) the oil pressure at normal engine running speed to be too high?
(c) not cause any difference in the oil pressure?
(d) the effect can be overcome by fitting a new oil pump?

5.4 Find the magnitude and direction of the resultant of the two forces A and B shown in Figure 5.13. What alteration is required to be made to this resultant in order to make it become the equilibrant of the two forces A and B?

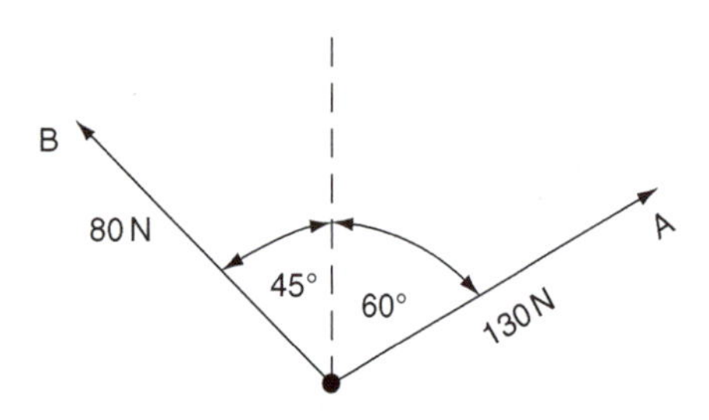

Fig. 5.13 Calculation of forces (Exercise 5.4)

5.5 Two forces OA and OB of 50 N and 70 N respectively, pull on a body at point O. The angle AOB is 70°. Find the magnitude and direction of the third force acting at point O that will balance the other two forces.

5.6 State Hooke's law. A compression spring has a stiffness of 40 kN/m. Calculate (a) the amount of compression when a force of 280 N is applied to the spring, (b) the force required to compress the spring 5 mm.

5.7 When a vehicle is being lifted by a crane as shown in Figure 5.14 the force F, in the lifting cable, is producing two effects. One effect is to attempt to lift the vehicle in a vertical direction and the other is to attempt to move the vehicle in a horizontal direction. Use a force diagram to determine the magnitude of the horizontal component of force F.

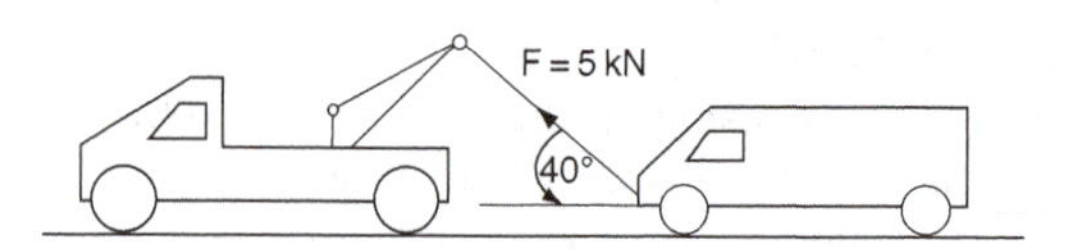

Fig. 5.14 Vertical and horizontal components of a force

5.8 Find the forces in the jib and the tie of the wall mounted crane shown in Figure 5.15(a).

5.9 Find the side thrust F acting on the cylinder wall of the engine mechanism shown in Figure 5.15(b).

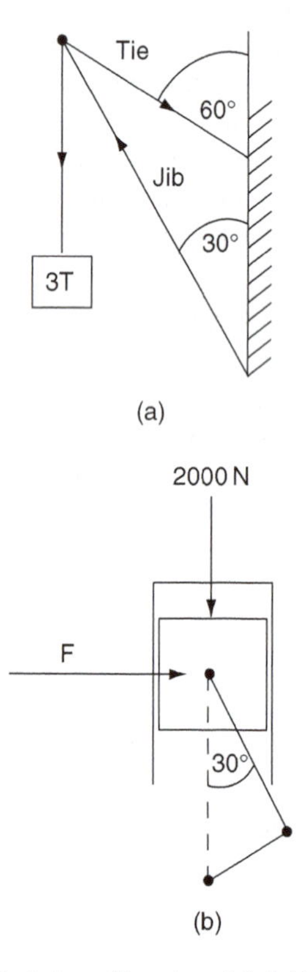

Fig. 5.15 Calculations (Exercises 5.8–5.11)

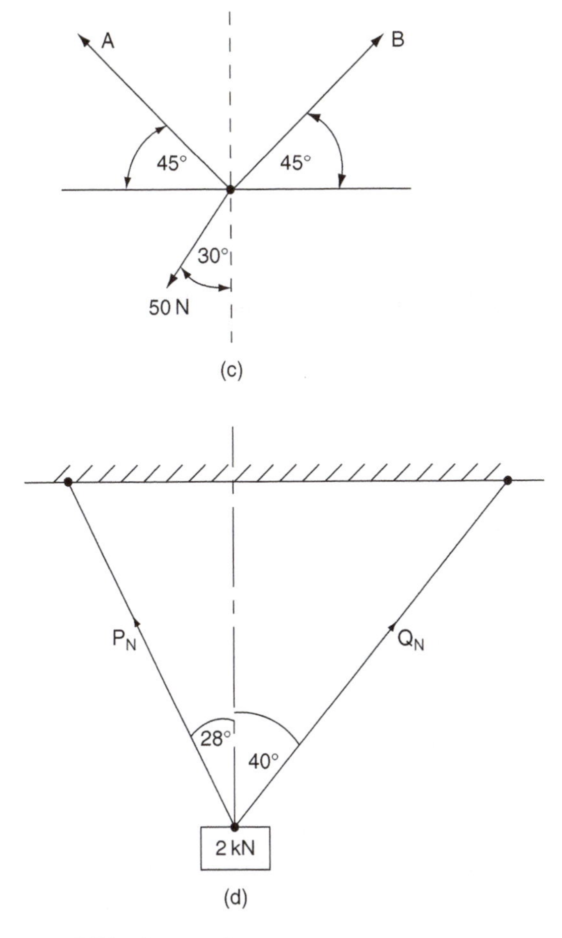

Fig. 5.15 Continued

5.10 Three coplanar forces are acting at a point as shown in Figure 5.15(c). Determine the sizes of forces A and B.

5.11 A 2 kN weight is suspended from two ropes as shown in Figure 5.15(d). Determine the sizes of forces P and Q.

5.12 A certain petrol engine piston crown has a cross-sectional area of $0.012\,m^2$. If the compression force on the piston is 30 kN, calculate the pressure acting on the piston crown.

15.13 If the gravitational constant is $9.81\,m/s^2$, is the gravitational force acting on a vehicle of 1 tonne mass:

(a) 981 kg
(b) 98.1 N
(c) 9.81 kN
(d) 981 N

15.14 Is an object whose centre of gravity rises when a force is applied and falls back to its original position when the force is removed:

(a) in unstable equilibrium?
(b) in stable equilibrium?
(c) in neutral equilibrium?
(d) lacking in stability?

6

Materials – Stress, strain, elasticity

6.1 Introduction

Many of the materials that are used in motor vehicle engineering are elastic, that is to say that they change shape under load and then return to their original shape when the load is removed. Elasticity is one of the properties that allow materials and components to experience much stress and strain during their working lives. For example, coil springs used in suspension systems compress under load and then expand when the load is removed; this type of action occurs thousands of times and the springs are expected to function properly for many years. Other components such as bolts, cables, rods, etc., experience equally severe operating conditions but the elastic effect is not so obvious. A change of dimensions called strain occurs when materials are stressed; for example when a cylinder head bolt is tightened to its final torque the bolt will stretch a little. Much of the energy that is used to tighten the bolt is stored as strain energy in the bolt and it is this energy that gives rise to the force that clamps the cylinder head in position. When the cylinder head bolts are released (undone) the strain energy is released and the bolt returns to its original size because the material is elastic. However, if the bolt is over-tightened, the bolt may stretch or break. In fact it will most probably stretch first. To see why this is so it is useful to have an insight into some of the basics about the behaviour of materials.

6.2 Stress

Forces that tend to stretch, or pull something apart, are known as tensile forces and they produce two important effects:

1. In trying to pull the bolt apart, internal resisting forces are created and these internal forces are known as **stress**.
2. The length of the bolt will increase, and this change in the bolt's dimensions is known as **strain**.

Stress is calculated by dividing the applied force by the cross-sectional area of the bolt.

$$\text{Stress} = \frac{\text{Force}}{\text{Cross-sectional area}} \qquad \textbf{(6.1)}$$

Types of stress

There are three basic forms of stress:

1. tensile stress;
2. compressive stress;
3. shear stress – torsional stress is a form of shear stress.

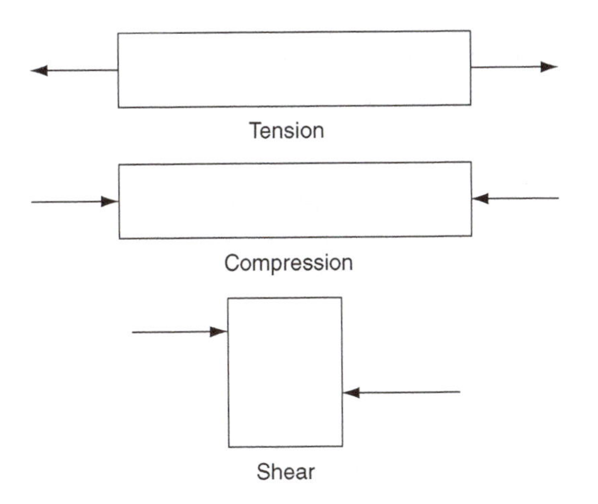

Fig. 6.1 The three basic stresses

Examples of the three basic stresses in automotive components

Figure 6.2(a) shows a cylinder head bolt. This is a component that is highly stressed in tension.

Figure 6.2(b) shows a piston and connecting rod assembly. On the compression and power strokes the connecting rod is subject to large compressive stress. The gas force on the gudgeon pin attempts to shear the pin on each side of the small end of the connecting rod. In this case the gudgeon pin is said to be in double shear because two cross-sectional areas are subject to the shearing force.

6.3 Tensile test

In materials testing, a standard size specimen of steel is stretched in a tensile testing machine until it breaks. During the test, the applied force and the corresponding extension are recorded. Figure 6.3 shows a typical graph obtained during a tensile test on a steel specimen.

The graph shows that up to point A the extension is proportional to the load; this is the elastic limit, or limit of proportionality. After point A,

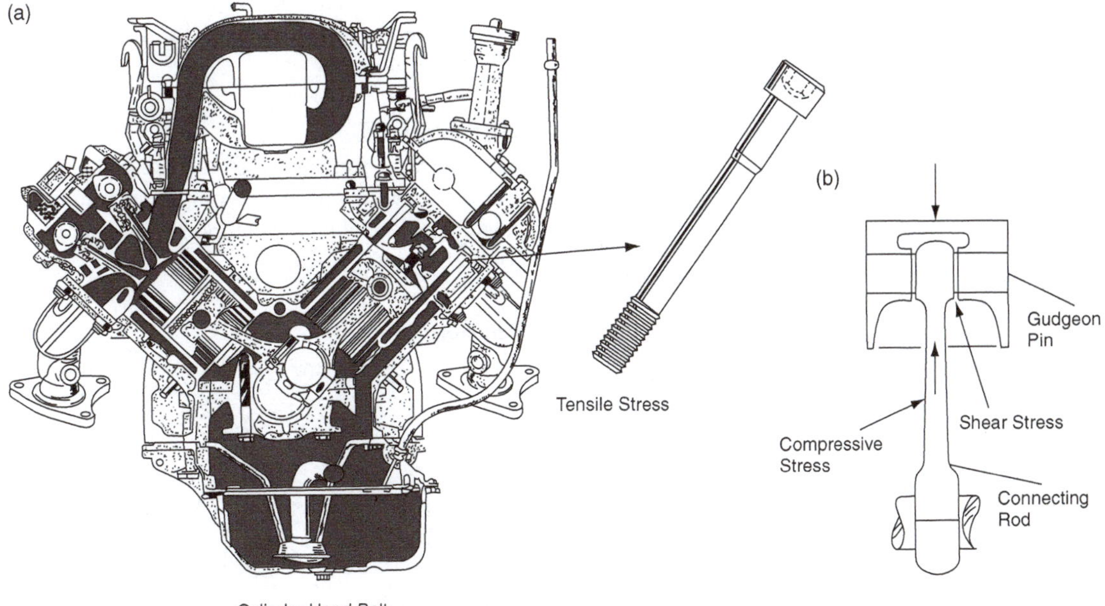

Fig. 6.2 Components under stress

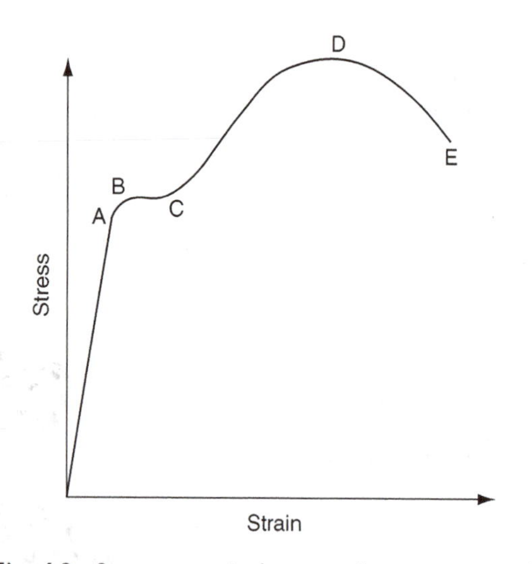

Fig. 6.3 Stress vs strain for a tensile test on a steel specimen

the amount of extension becomes greater because the material is no longer elastic; the material is in a plastic state and, when the load is removed, the material will not return to its original shape. Point B is known as the yield point, because this is the point at which the material stretches without any increase in the load on it. At point C the mechanical resistance of the specimen increases and extra force (load) is required to produce further extension. The texture of the steel changes at this point and the shape of the specimen changes, a 'waist' appearing at the central portion of the specimen, as shown in Figure 6.4. The cross-sectional area of the specimen decreases and this means that the stress at the waist increases without any further increase in the load. The point of maximum load occurs at point D and the test proceeds until the specimen fractures at point E.

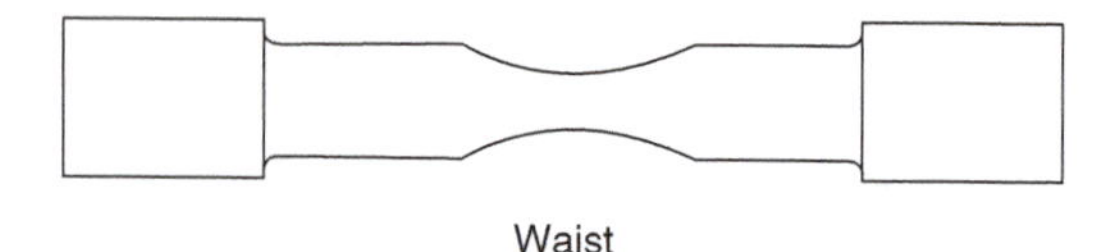

Fig. 6.4 The shape of a strained specimen

6.4 Examples of stress and strain

Example 6.1

A cylinder head bolt with an effective diameter of 15 mm carries a tensile load of 10 kN. Calculate the tensile stress in the bolt.

Solution

From Equation 6.1,

$$\text{Stress} = \frac{\text{Force}}{\text{Cross-sectional Area}}$$

$$\text{Cross-sectional area of the cylinder head bolt} = \frac{\pi d^2}{4} = \frac{3.142 \times 15 \times 15}{4} = \underline{176.7\,\text{mm}^2}$$

$$\text{Stress} = \frac{10000\,\text{N}}{176.7\,\text{mm}^2} = 56\,\text{N/mm}^2 = 56\,\text{MN/m}^2$$

Stress is normally quoted in kN/m^2, or MN/m^2. Stress may also be stated in Pa (pascals); $1\,\text{Pa} = 1\,\text{N/m}^2$.

Example 6.2

A connecting rod has a cross-sectional area of $200\,\text{mm}^2$ and it carries a compressive force of 2.4 tonnes. Calculate the compressive stress in the connecting rod.

Solution

$$\text{Stress} = \frac{\text{Force}}{\text{Cross-sectional area}} = \frac{2400}{200} = 12\,\text{N/mm}^2$$

$$\text{Compressive stress in connecting rod} = 12\,\text{MN/m}^2$$

Example 6.3

The hand brake linkage shown in Figure 6.5 carries a tensile force of 600 N. Calculate the

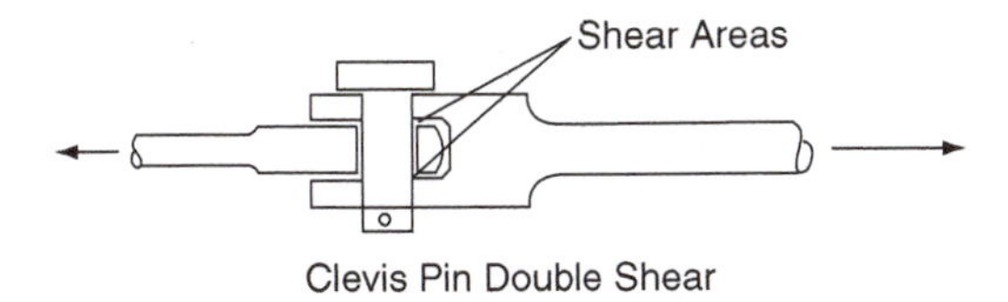

Fig. 6.5 Clevis pin in double shear (Example 6.3)

shear stress in the clevis pin, which is 12 mm in diameter.

In this case the shearing action is attempting to shear the clevis pin across two cross-sectional areas.

The cross-sectional are of the clevis pin

$$= \frac{\pi \times 12 \times 12}{4}$$

$$= \frac{3.142 \times 12 \times 12}{4}$$

$$= 113.1\,\text{mm}^2$$

In this case the shearing action is attempting to shear the clevis pin across two cross-sectional areas. The area resisting shear $= 2 \times 113.1$

$$= 226.2\,\text{mm}^2$$

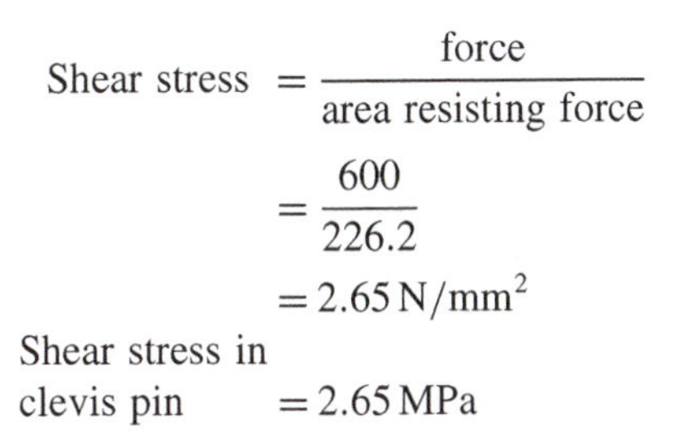

$$\text{Shear stress} = \frac{\text{force}}{\text{area resisting force}}$$

$$= \frac{600}{226.2}$$

$$= 2.65\,\text{N/mm}^2$$

Shear stress in clevis pin $= 2.65\,\text{MPa}$

Example 6.4

A propeller shaft coupling of a truck is secured by four bolts of 14 mm diameter that are equally spaced at a radius of 50 mm from the centre of the propeller shaft. Calculate the shear stress in each bolt when the shaft is transmitting a torque of 500 N m.

Solution

Figure 6.6 shows the layout of the coupling flange.

Torque transmitted by propeller shaft $T = 500\,\text{N m}$

$$\text{Force} = \frac{T}{R}$$

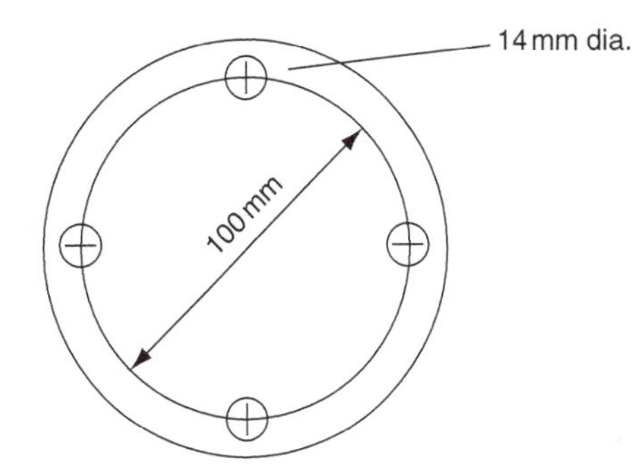

Fig. 6.6 Propeller shaft coupling flange (Example 6.4)

Radius at which force acts $= 50\,\text{mm}$

$$\therefore F = \frac{500\,\text{N m}}{0.05\,\text{m}}$$

$$= 10\,000\,\text{N}$$

This force is evenly shared by the four coupling bolts

(csa = cross-sectional area)

$$\text{Force on each bolt} = \frac{10\,000}{4}$$

$$= 2500\,\text{N}$$

$$\text{Shear stress in each bolt} = \frac{\text{Force}}{\text{csa}}$$

$$\text{csa of bolt} = \frac{\pi \times 0.014 \times 0.014}{4}$$

$$= 0.000154\,\text{m}^2$$

$$\text{Stress} = \frac{2500}{0.000154}$$

$$= 16.23\,\text{MN/m}^2$$

Stress in each coupling bolt $= 16.23\,\text{MN/m}^2$

6.5 Stress raisers

Sharp corners and notches of the type shown in Figure 6.7 cause regions of high stress that can lead to component failure. These stress raisers can

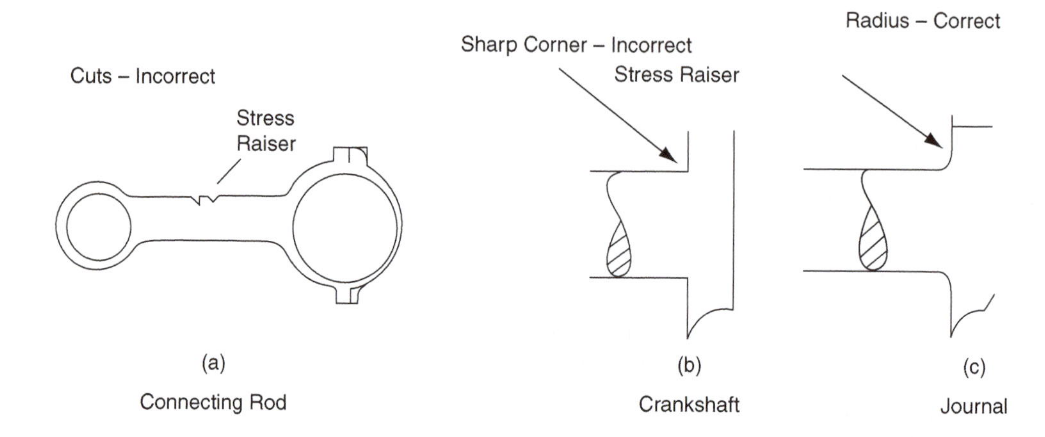

Fig. 6.7 Stress raisers

be caused by bad repair practices such as placing saw cuts to identify a component, incorrect machining when regrinding a crankshaft, drilling holes in components when fixing attachments to some part of a vehicle, etc.

6.6 Strain

When a load is applied to a metal test bar a change of shape takes place. A tensile load will stretch the bar and a compressive load will shorten it. This change of shape is called **strain**. The three basic types of strain are shown in Figure 6.8.

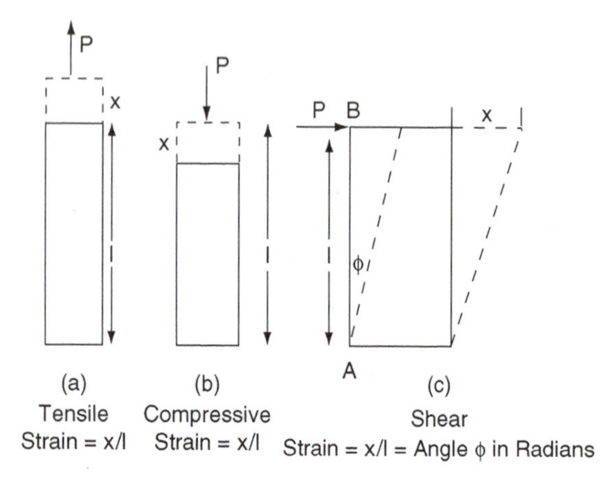

Fig. 6.8 The three types of strain

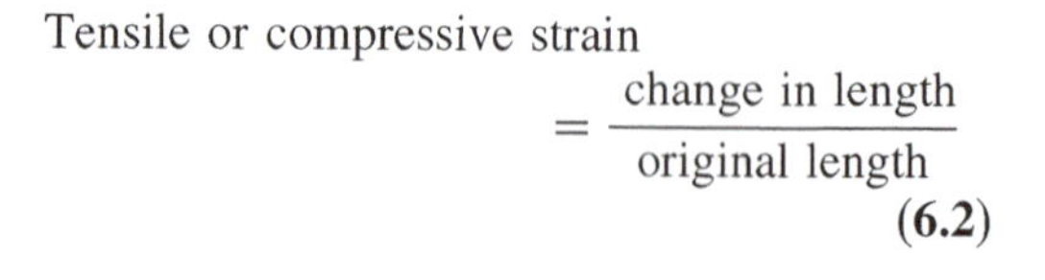

$$\text{Tensile or compressive strain} = \frac{\text{change in length}}{\text{original length}} \qquad \textbf{(6.2)}$$

Example 6.4

A steel rod 200 mm in length stretches by 0.12 mm when it is subjected to a tensile load of 2 tonnes. Determine the strain.

Solution

$$\text{Strain} = \frac{\text{change in length}}{\text{original length}}$$

$$= \frac{0.12\,\text{mm}}{200\,\text{mm}}$$

Tensile strain in the steel rod $= 0.0006$

Note: strain does not have any units.

Shear strain

Figure 6.8c shows a rectangular block of material to which a shearing force P is applied. The bottom edge of the rectangular block is secured to an immovable surface. The action of force P causes the top edge of the block to move a distance x in the direction of force P, which makes the edge AB take up an angle of ϕ radians. This angle is used as the measure of shear strain.

6.7 Elasticity

Stress, strain, elasticity

In terms of stress and strain, Hooke's law states that the strain produced in an elastic material is directly proportional to the stress that produces it.

Expressed mathematically this means that

$$\frac{\text{stress}}{\text{strain}} = \text{constant}$$

This constant is known as the modulus of elasticity, or Young's modulus. The symbol E is normally used to denote modulus of elasticity.

$$\text{For tensile or compressive stress } E = \frac{\text{stress}}{\text{strain}} \quad \textbf{(6.3)}$$

For **shear** stress the term modulus of rigidity is used and this is represented by the symbol G.

$$\text{For shear stress } G = \frac{\text{stress}}{\text{strain}} \quad \textbf{(6.4)}$$

Example 6.5

Determine the extension of a metal rod of 14 mm diameter and 200 mm length when it is placed under a tensile load of 2 tonnes. Take $g = 10\,m/s^2$. The modulus of elasticity E for the material $= 200\,GN/m^2$.

Solution

$$\text{Cross-sectional area (csa) of the rod} = \frac{\pi d^2}{4}$$

$$= \frac{3.142 \times 14 \times 14}{4}$$

$$\text{csa} = 154\,\text{mm}^2$$

$$\text{stress} = \frac{\text{load}}{\text{csa}}$$

$$= \frac{2000 \times 10}{154}$$

$$= 129.8\,\text{N/mm}^2$$

$$\text{Stress} = 129.8\,\text{MN/m}^2$$

$$E = \frac{\text{stress}}{\text{strain}}$$

$$\text{Strain} = \frac{\text{stress}}{E}$$

$$= \frac{129.8 \times 10^6}{200 \times 10^9}$$

$$= 0.649 \times 10^{-3}$$

$$\text{strain} = \frac{\text{extension}}{\text{original length}}$$

$$\text{Extension} = \text{strain} \times \text{original length}$$

$$= 0.649 \times 10^{-3} \times 200\,\text{mm}$$

$$\text{Extension} = 0.13\,\text{mm}$$

Some typical approximate values for E are shown in Table 6.1.

Table 6.1 Typical approximate values for Young's modulus

Material	**Young's modulus (E) GN/m² (GPa)**
Cast iron	110
Copper	117
Aluminium	71
Mild steel	210
Magnesium	44
Phosphor bronze	120
Nickel alloy	110

6.8 Tensile strength

The tensile strength of a material is the maximum tensile stress that the material can bear before breaking.

$$\text{Tensile strength} = \frac{\text{maximum load}}{\text{original csa of the material}} \quad \textbf{(6.5)}$$

(csa = cross-sectional area)

Example 6.6

Determine the tensile strength of a 15 mm diameter steel rod that is able to withstand a maximum tensile load of 74 kN.

Solution

Crosss-sectional area of rod $= \frac{\pi d^2}{4} = 0.7854 \times 0.015 \times 0.015 = 0.0001767\,m^2$

Tensile load = 74 kN

$$\text{Tensile strength} = \frac{\text{maximum load}}{\text{original csa of the material}}$$

$$= \frac{74 \times 1000}{0.0001767}$$

$$= 419\,MN/m^2$$

6.9 Factor of safety

In order to preserve the strength and working properties of a component, the maximum load to which a material is subjected is restricted to a figure which is below the tensile strength of the material. This maximum working load gives rise to the allowable working stress and it is a fraction of the tensile strength of the material.

$$\text{The allowable working stress} = \frac{\text{tensile strength}}{\text{factor of safety}} \quad \mathbf{(6.6)}$$

Example 6.7

A steel tie rod in a suspension system is designed with a factor of safety of 6. The tensile strength of the steel is $420\,MN/m^2$. Calculate the allowable working stress.

Solution

$$\text{Allowable working stress} = \frac{\text{tensile strength}}{\text{factor of safety}}$$

$$= \frac{420\,MN/m^2}{6}$$

$$= 70\,MN/m^2$$

6.10 Torsional stress

The twisting action that arises when torque is applied to an object such as a shaft gives rise to shear stress. The following formula relates the various factors involved when torque is applied to a shaft.

$$\frac{T}{J} = \frac{G\theta}{lR} = f_s \quad \mathbf{(6.7)}$$

Where T (in N m) is the torque applied to the shaft, J is the polar moment of inertia which is derived from the dimensions of the shaft (this is in m^4), G is the modulus of rigidity in N/m^2, l is the length of the shaft in metres, f_s is the shear stress in N/m^2, θ is the angle of twist in radians, and R is the radius where the stress is greatest, normally the outer radius of the shaft.

Example 6.8

A solid steel shaft of 20 mm diameter and 200 mm length is subjected to a torque of 220 N m. If J for the shaft $= 1.57 \times 10^{-8} m^4$, calculate the shear stress.

Solution

$$\frac{f_s}{R} = \frac{T}{J}$$

$$fs = \frac{T \times R}{J}$$

$$= \frac{220\,N\,m \times 0.010\,m}{1.57 \times 10^{-8}\,m^4}$$

$$= 140\,MN/m^2$$

Shear stress produced by 220 N m torque = $140\,MN/m^2$

Table 6.2 Modulus of rigidity G (shear modulus)

Material	Approx. values of modulus of rigidity (G) in GPa
Steel	80
Copper	25
Aluminium alloy	20–30

6.11 Strain energy

When a force is applied to an elastic component, such as steel spring or a bolt, a change in shape occurs. In the case of a tensile force, the change of shape will be in the form of an increase in the length of the component. The force has moved through a distance, which means that work has been done. As the force (weight) increases, the extension increases proportionately and the force versus distance graph has the shape shown in Figure 6.9.

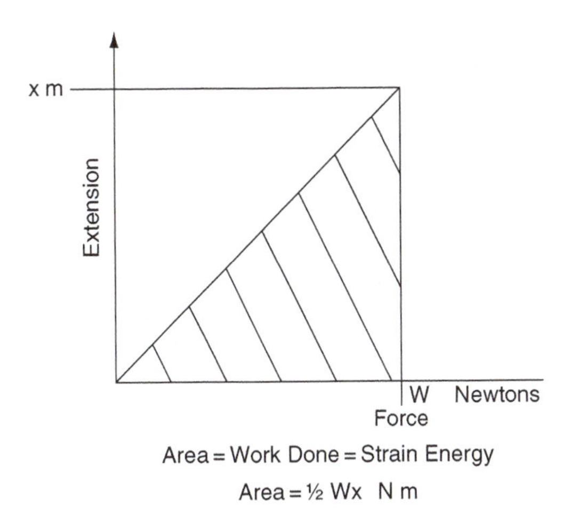

Fig. 6.9 Strain energy

The work done in stretching the spring = area under the graph

$$\text{Strain energy} = \tfrac{1}{2}\text{W}x\,\text{N m} \qquad \textbf{(6.8)}$$

Where W = the maximum force in newtons and x = extension of spring in metres.

Example 6.9
Determine the strain energy stored in a valve spring that is compressed by 15 mm under a load of 750 N.

Solution

$x = 0.015\,\text{m}$

$\text{W} = 750\,\text{N}$

$$\text{Strain energy} = \frac{\text{W}x}{2} = \frac{750 \times 0.015}{2} = 5.625\,\text{N m} = 5.625\,\text{J}$$

6.12 Strength of materials

For practical purposes, components are designed to withstand forces and loads that a device is designed for and, so long as the instructions for use and maintenance, such as safe loads and tightening torques, are observed, problems should not be experienced.

6.13 Other terms used in describing materials

Hardness

A hard material is one that resists indentation or abrasion by another material. The Brinell test is one method that is used to measure hardness and it operates by pressing a hardened metal carbide ball under a standard load into the surface of the specimen. The dimensions of the impression made are used to calculate the Brinell Hardness Number (BHN).

Hardening

Steels that contain 3% carbon or above may be hardened by heating them to a temperature of approximately 750°C (cherry red) and then quenching them in cold water. The temperature for this process varies according to carbon content. When quenching in cold water is considered too severe the steel is quenched in special quenching oil.

Tempering

Tempering is the process of reheating hardened steel to a temperature below the critical range, in order to reduce the hardness.

Annealing

This involves heating the material to a suitable temperature and holding it at that temperature for a suitable period before allowing it to cool. The purpose is to soften the material, improve the working properties, and reduce internal stresses after working.

Case hardening

Case hardening is a process that is used to impart a hard surface to steel. It is used in components that require a surface that is resistant to wear whilst retaining a tough inner structure. A crankshaft is an example of a component that is normally case hardened on the crankpins and journals. A process that is used in the manufacture of some crankshafts is known as nitriding. In nitriding, special types of steel are heated to approximately 520°C in an atmosphere of gaseous ammonia to produce a hard surface of approximately 0.5 mm depth.

Toughness

A material is said to be tough when a large amount of energy is required to fracture it.

Brittleness

Materials that break without undergoing local distortion and are unable to withstand sharp blows are said to be brittle. Most types of cast iron are brittle.

Ductility

A material that can be drawn out by tensile force is said to be ductile. The steel sheet that is used in the construction of motor car panels is of a type known as deep drawing steel and this is a ductile material.

Malleability

Metals that can hammered and bent without cracking are said to be malleable. Lead is an example of a malleable material.

6.14 Non-ferrous metals

These are mainly alloys that contain no iron. Commonly used non-ferrous alloys are those made from copper, lead, tin, aluminium or magnesium. Non-ferrous alloys are used extensively in automotive engineering.

6.15 Non-metallic materials

Synthetic materials, commonly known as plastics, are readily formed into quite intricate shapes and this makes them suitable for many applications in motor vehicle construction. In general, plastics are resistant to corrosion and a range of plastics have been developed to suit many applications on vehicles where metals were once used. Plastics are sometimes described as thermosets, which means that the plastic does not soften when re-heated, or thermoplastic, which means that the plastic does soften when re-heated.

Kevlar

Kevlar is the DuPont trade name for a synthetic material that is formed from high-strength aramid fibres. It has tensile strength equivalent to that of steel, it is lighter than steel and it is resistant to corrosion (see Table 6.3).

Table 6.3 Some properties of Kevlar

Material	Density (kg/m^3)	Tensile strength (GN/m^2)
Kevlar 49	1440	3.1
Steel piano wire	7860	3.0

6.16 Recycling of materials

Disposal of vehicles that have reached the end of their useful lives is a factor that has received much attention in recent years. Steel, copper, lead and other metallic materials are often recycled many times and in most areas facilities exist for dealing with them. Other materials are dealt with by special processing methods and facilities for dealing with all vehicle materials are being developed both locally and nationally in the UK. Vehicles that have reached the end-of-life stage are known as end-of-life vehicles (ELVs).

6.17 Summary of main formulae

$$\text{Stress} = \frac{\text{Force}}{\text{Cross-sectional area}}$$

$$\text{Tensile or compressive Strain} = \frac{\text{Change in length}}{\text{Original length}}$$

$$\text{For tensile or compressive stress, } E = \frac{\text{stress}}{\text{strain}}$$

$$\text{For shear stress } G = \frac{\text{stress}}{\text{strain}}$$

$$\text{Tensile strength} = \frac{\text{maximum load}}{\text{original csa of the material}}$$

$$\text{The allowable working stress} = \frac{\text{tensile strength}}{\text{factor of safety}}$$

$$\text{For torsion } \frac{T}{J} = \frac{G\theta}{l} = \frac{f_s}{R}$$

Strain energy = ½ W x N m

The polar moment of inertia $J = \frac{\pi d^4}{32}$ for a solid shaft

$$J = \frac{\pi (D^4 - d^4)}{32} \text{ for a hollow shaft}$$

6.18 Exercises

6.1 A steel bolt with an effective diameter of 12 mm and length of 50 mm carries a tensile load of 10 kN. Calculate the extension in length of the bolt under this load, given that E = 200 GN/m^2.

6.2 (a) Define 'strain', 'stress', 'modulus of elasticity'.
(b) A steel tie rod used in a suspension system is 400 mm long with a diameter of 15 mm. Determine the stress in the tie rod when a tensile force of 600 N is applied to it under braking.
(c) If E for the rod = 200 GN/m^2 calculate the extension of the rod caused by this force.

6.3 The shackle pin in a certain leaf spring has a diameter of 18 mm. When the vehicle is fully loaded the shackle pin carries a load of 2.5 tonnes. Calculate the shear stress in the shackle pin. Take G = 10 m/s^2.

6.4 An axle shaft 0.5 m long is subjected to a torque of 1200 N m. Given that the polar moment of inertia J for the shaft = 0.000041 m^4, calculate the angle of twist in the shaft. Take the modulus of rigidity G = 85 GN/m^2.

6.5 A piston and connecting rod assembly carries a direct gas force of 8 kN at the top dead centre position. The gudgeon pin is hollow with an outside diameter of 18 mm and an internal diameter of 12 mm if the minimum cross-sectional area of the connecting rod is 300 mm^2.
Calculate:
(a) the stress in the gudgeon pin;
(b) the stress in the connecting rod.

6.6 A certain engine has a bore diameter of 100 mm. The gudgeon pin has an outside diameter of 20 mm and an inside diameter of 14 mm. If the cylinder gas pressure is 10 bar at the top dead centre position, calculate the stress in the gudgeon pin.

6.7 A truck engine develops a maximum torque of 400 N m. The lowest gearbox ratio is 8.5:1

and the drive axle ratio is 5:1. When in the low gear the transmission efficiency is 90%. Given that the final drive axle shafts are 40 mm in diameter, calculate the shear stress in each axle shaft when transmitting maximum torque; the drive is equally balanced on each side of the differential gear.

For the axle shafts $J = \frac{\pi d^4}{32}$.

6.8 Determine the maximum torque that can be transmitted by a propeller shaft if the maximum permissible shear stress is limited to 15 MPa. The outside diameter of the shaft = 50 mm and the polar moment of inertia J for the shaft = $2.62 \times 10^{-7} m^4$.

6.9 Give three examples in each case of vehicle components that are subjected to the following: (a) shear stress; (b) tensile stress; (c) compressive stress.

Calculate the tensile force on a bolt 15 mm in diameter when the tightening torque produces a tensile stress of 60 MPa.

6.10 The figures in Table 6.4 were recorded in an experiment to determine the value of Young's modulus of elasticity for a steel wire. Plot a graph with extension on the horizontal axis and select a suitable point on the graph that will enable you to select values of load and extension from which to calculate Young's modulus. Calculate Young's modulus of elasticity for the steel wire.
Initial length of wire = 1.8 m
Initial diameter of wire = 1 mm

Table 6.4 Calculation of Young's modulus (Exercise 6.10)

Tensile force applied to the wire (N)	Extension (mm)
20	0.3
40	0.6
60	0.8
80	1.1
100	1.4
120	1.7
140	2.0
160	2.2
180	2.5
200	2.8
220	3.3

6.11 An air brake actuator has a diaphragm of 120 mm diameter. The clevis pin that transmits the braking force from the diaphragm rod to the brake lever has a diameter of 15 mm. Calculate the stress in the clevis pin when the air pressure on the diaphragm is 6 bar.

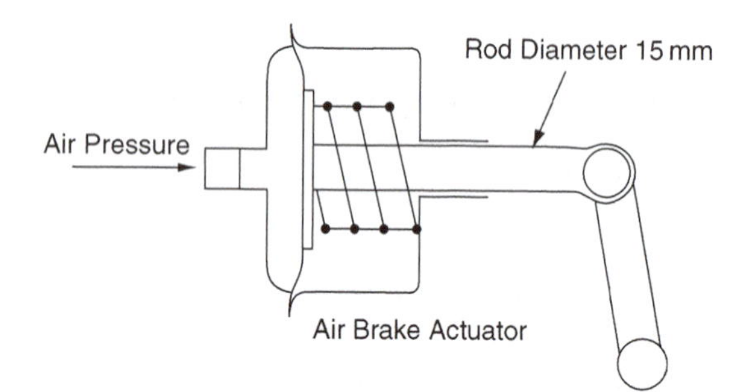

Fig. 6.10 Strain energy

6.12 The diaphragm in a vacuum servo has an effective area of $0.03 m^2$. Under certain conditions the pressure difference a between the vacuum side of the servo and the air pressure side is 50 kPa.

Calculate (a) the force that the servo diaphragm exerts on the master cylinder push rod, and (b) the stress in the master push rod that has a diameter of 10 mm, under these conditions.

6.13 Calculate the maximum power that can be transmitted by a tubular propeller shaft at 4200 rev/min if the shear stress is not to exceed $20 MN/m^2$. The outer diameter of the shaft is 50 mm and the inner diameter is 44 mm.

7
Levers and moments, torque and gears

7.1 Levers

Leverage and the use of levers occurs in the use of tools such as spanners, pry bars, pliers etc., and in many vehicle mechanisms such as clutch and brake pedals, throttle linkages and suspension units.

7.2 Principles of leverage

The basic principles of leverage are covered by a rule that is known as the principle of moments.

7.3 The principle of moments

The moment of a force is defined as the product of the force and the perpendicular distance from the fulcrum to the line in which the force is acting.

Example 7.1
In this example (see Figure 7.1) a beam that is 80 cm long rests on a pivot (fulcrum) that is 20 cm from one end. There are forces of 60 newtons and 20 newtons acting at right angles to the beam, as shown. This system of a beam, the forces and the fulcrum are in equilibrium, i.e. the beam and forces are balanced about the fulcrum. When a beam is in equilibrium like this, the principle of moments states that the clockwise moments about the fulcrum are equal to the anti-clockwise moments about the fulcrum.

In this case the clockwise moment is equal to

$$20\,\text{N} \times 60\,\text{cm} = 1200\,\text{N cm}$$

And the anti-clockwise moment is equal to

$$60\,\text{N} \times 20\,\text{cm} = 1200\,\text{N cm}$$

For convenience the clockwise moment is often abbreviated to CWM and the anti-clockwise moment to ACWM.

According to the principle of moments, the simple beam problem shown in Figure 7.1 may be written as:

$$\text{CWM} = \text{ACWM}$$

$$60\,\text{N} \times 20\,\text{cm} = 20\,\text{N} \times 60\,\text{cm}$$

$$1200\,\text{N cm} = 1200\,\text{N cm}$$

In simple lever problems it is usual to be given the size of one force and both distances, and then to be required to work out the size (magnitude) of the other force, as shown in the following example.

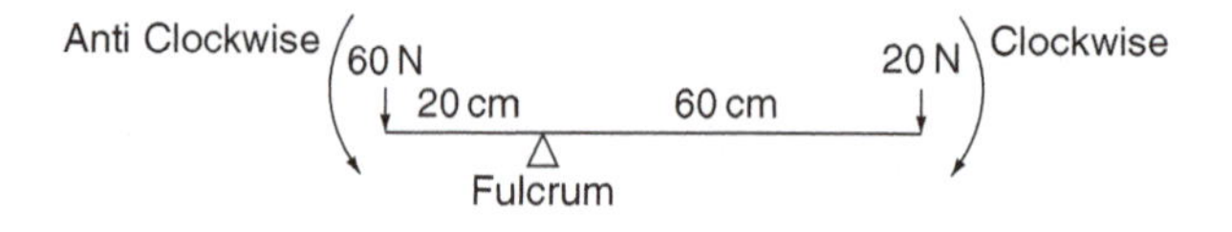

Fig. 7.1 Simple beam problem (Example 7.1)

Example 7.2

Use the principle of moments to determine the size of force F on the beam in Figure 7.2.

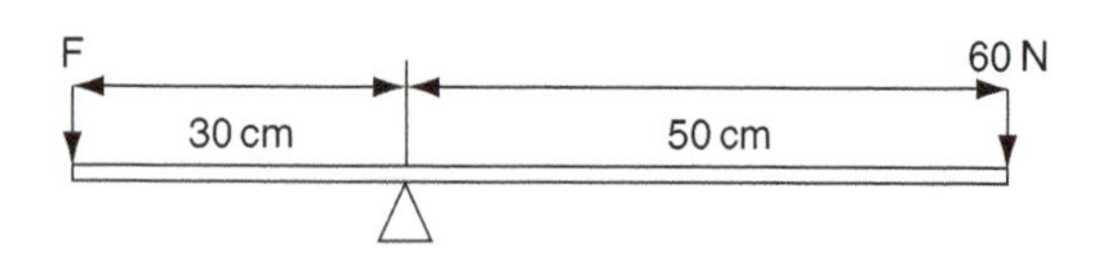

Fig. 7.2 Turning moments (Example 7.2)

Solution

Taking moments about the fulcrum:

$$\text{CWM} = \text{ACWM}$$

$$60\,\text{N} \times 50\,\text{cm} = \text{F} \times 30\,\text{cm}$$

$$3000\,\text{N cm} = \text{F} \times 30\,\text{cm}$$

$$3000\,\text{N cm} \div 30\,\text{cm} = \text{F}$$

$$100\,\text{N} = \text{F}$$

$$\text{Answer : force(F)} = 100\,\text{N}$$

Example 7.3

Vehicle controls, such as brake pedals and clutch pedals, are often hinged at one end, force from the driver's foot is applied at the other end, and the operating is hinged at some point in between as shown in Figure 7.3.

For our current purposes the brake pedal may be considered as a simple lever, as shown in Figure 7.4. Here the pedal has been substituted by a straight lever that has a length which is the same as the perpendicular distance between the driver's foot and the brake pedal fulcrum. This type of lever is called a cantilever.

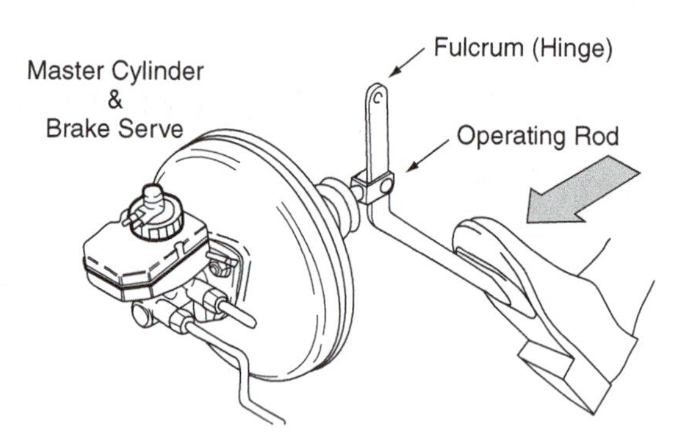

Fig. 7.3 Brake pedal – pendant type lever (Example 7.3)

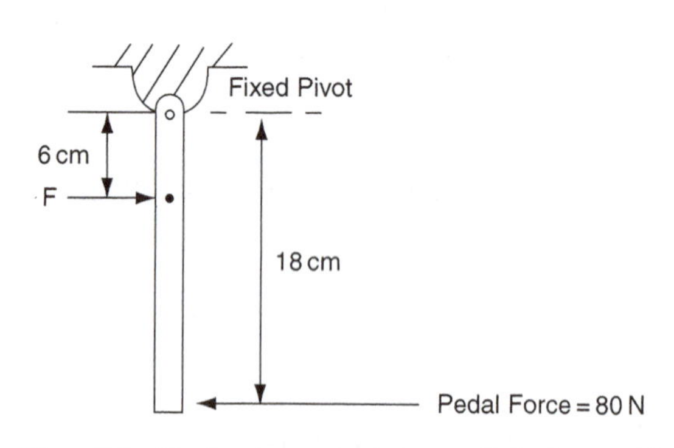

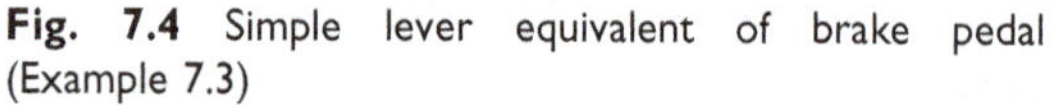

Fig. 7.4 Simple lever equivalent of brake pedal (Example 7.3)

Taking moments about the fulcrum, which is the pedal hinge:

$$\text{CWM} = \text{ACWM}$$

$$80\,\text{N} \times 18\,\text{cm} = \text{F} \times 6\,\text{cm}$$

$$\text{F} = \frac{80 \times 18}{6}$$

$$\text{F} = 240\,\text{N}$$

7.4 The bell crank lever

The bell crank lever, as shown in Figure 7.5, is used in many vehicle systems and the principle of moments may be applied to the solution of numerical problems associated with it.

When the bell crank lever is in equilibrium (balanced), the clockwise turning moment about the fulcrum (pivot) is equal to the anti-clockwise moment about the fulcrum. It is important to note that the turning moment is the product of the force and the *perpendicular distance* from the fulcrum to the *line in which the force is acting*. This is an important point to note because when a given force is applied at an angle other than at a right angle it produces a smaller turning moment.

In Figure 7.5 the distances of the forces from the fulcrum are the perpendicular distances, so the turning moments are:

$$\text{CWM} = \text{ACWM}$$

$$18\,\text{N} \times 10\,\text{cm} = 15\,\text{N} \times 12\,\text{cm}$$

$$180\,\text{N cm} = 180\,\text{N cm}$$

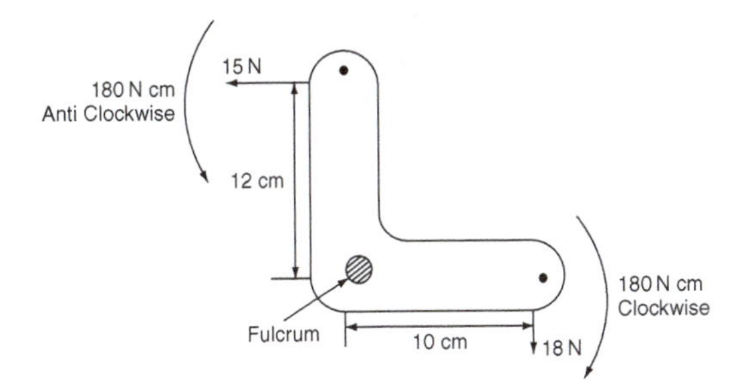

Fig. 7.5 A bell crank lever

A practical application of the bell crank lever

The handbrake lever shown in Figure 7.6 is hinged on the fulcrum pin. A force applied by hand at the operating end causes that end of the lever to move upwards and this results in forward movement of cable that applies the brakes at the rear

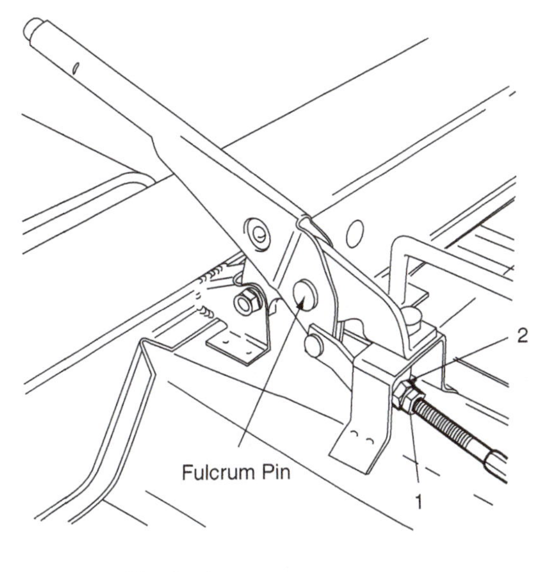

Fig. 7.6 Handbrake lever

wheels. An adjustment is provided to compensate for wear. Should the movement on the handbrake lever become excessive it becomes increasingly difficult to apply the hand operating force at right angles to the lever and this leads to a weakening of the pull that is exerted on the handbrake cable.

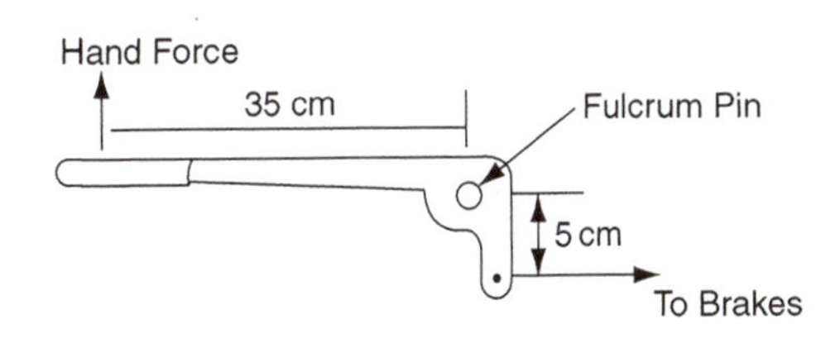

Fig. 7.7 Skeleton lever – handbrake (Example 7.4)

For purposes of calculation, the mechanics of the handbrake lever can be represented by a skeleton bell crank lever, as shown in Figure 7.7.

Example 7.4

In the handbrake lever mechanism shown in Figures 7.6 and 7.7, a brake cable force of 420 N is required to apply the brakes. Calculate the hand force, applied at right angles to the long part of the lever, that is required to produce this force in the cable.

Applying the principle of moments gives:

$$\text{CWM} = \text{ACWM}$$

$$\text{F} \times 35\,\text{cm} = 420\,\text{N} \times 5\,\text{cm}$$

$$\text{F} = \frac{420 \times 5}{35}$$

$$\text{F} = 60\,\text{N}$$

A force of 60 N is required at the operating end of the handbrake lever.

Example 7.5

Figure 7.8 shows a horizontal beam that is supported at each end. The loads on the beam are in tonnes and the distances between the loads are in millimetres. Determine the size of the reactions R_1 and R_2 at the supports.

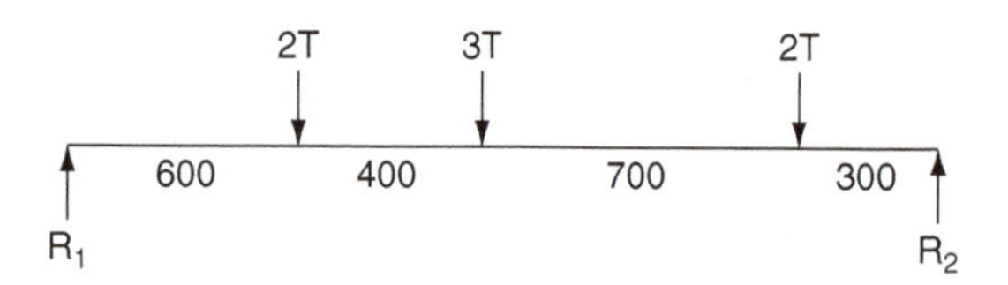

Fig. 7.8 Simply supported beam (Example 7.5)

Take moments about the right-hand support:

$$\text{CWM} = \text{ACWM}$$

$$R_1 \times 2000 = (2 \times 300) + (3 \times 1000) + (2 \times 1400)$$

$$R_1 = \frac{6400}{2000}$$

$$R_1 = 3.2 \text{ tonnes}$$

Because the beam is in equilibrium the upward forces are equal to the downward forces so R_2 is found by subtracting R_1 from the total downward force:

$$R_2 = 7 - 3.2 = 3.8 \text{ tonnes}$$

7.5 Axle loadings

Example 7.6

Figure 7.9 shows a vehicle of total weight = 1.4 tonnes. The wheelbase of the vehicle is 2.5 m and the centre of gravity is positioned 1.2 m behind the centre of the front axle.

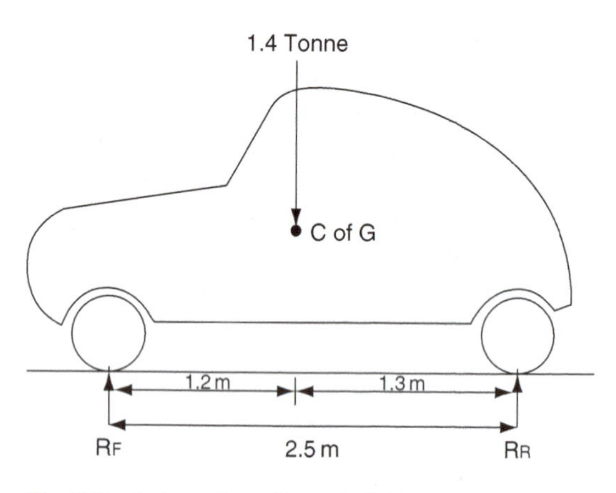

Fig. 7.9 Axle loadings (Example 7.6)

The principle of moments may be used to determine the axle loadings R_f and R_r.

Take moments about R_f:

$$\text{CWM} = \text{ACWM}$$

$$1.4 \times 1.2 = R_r \times 2.5$$

$$R_r = \frac{1.4 \times 1.2}{2.5}$$

$$R_r = 0.672 \text{ tonne}$$

The total weight pushing down = force pushing up

$$1.4 \text{ tonnes} = 0.672 \text{ tonne} + R_f$$

$$R_f = 1.4 - 0.672$$

$$R_f = 0.728 \text{ tonne}$$

The axle loadings are: front axle 0.728 tonne, rear axle 0.672 tonne.

7.6 Torque

Torque is a turning effect; a clear demonstration of the meaning of torque may be seen in the widely used torque wrench that is used for tightening nuts and bolts to a preset value, as shown in Figure 7.10.

Torque is the product of force and perpendicular distance from the centre of rotation to the line in which the force is acting.

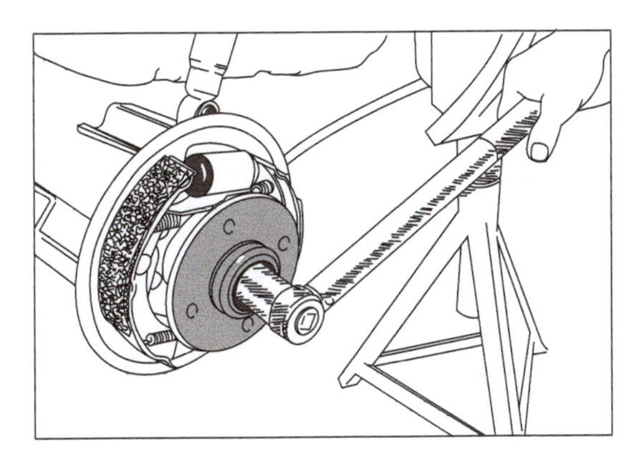

Fig. 7.10 Torque wrench

Example 7.7

In the example of the torque wrench, the torque may be calculated as follows.

Assume that the hand force is applied at right angles to the shaft of the torque wrench and that the force is applied at a distance of 40 cm from the centre of the axle. In this particular case the torque required is 25 N m. Calculate the force F required to produce this amount of torque. (See Figure 7.11.)

In this case we have: torque (T) = 25 N m; radius (R) = 40 cm = 0.4 m.

$$T = F \times R$$

$$F = T \div R$$

$$F = \frac{25\,\text{N m}}{0.4\,\text{m}}$$

$$F = 62.5\,\text{N}$$

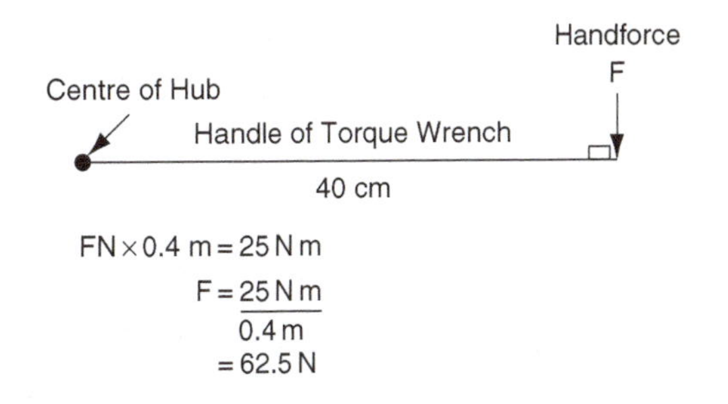

Fig. 7.11 Torque calculation (Example 7.7)

7.7 Engine torque

Engine torque is produced by the action of the gas force on the piston which is transmitted to the crankshaft through the connecting rod. The principle is shown in Figure 7.12.

Example 7.8

In this case, the effective force on the connecting rod is 2000 N and the connecting rod is making a right angle with respect to the crank throw. The throw (radius) of the crank is 60 mm. The connecting rod is at a right angle to the crank throw.

Here, the torque produced is:

Torque (T) = Force (F) × Radius R

F = 2000 N, R = 60 mm = 0.06 m

$$T = F \times R$$

$$T = 2000\,\text{N} \times 0.06\,\text{m}$$

$$T = 120\,\text{N m}$$

Crankshaft torque T = 120 N m

It is important to note that the torque is calculated from the force and the perpendicular distance from the centre of rotation to the line in which the force is acting.

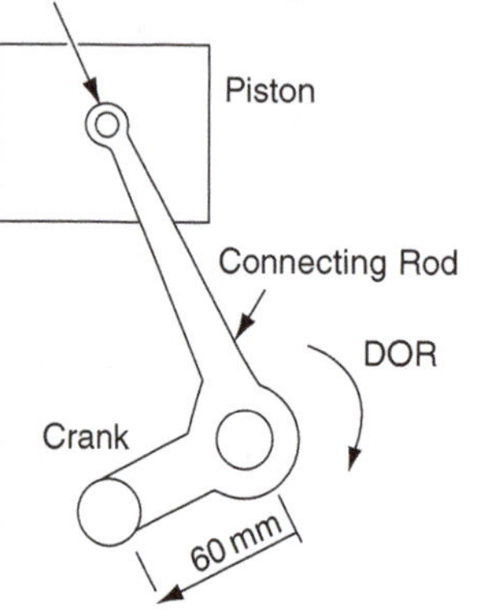

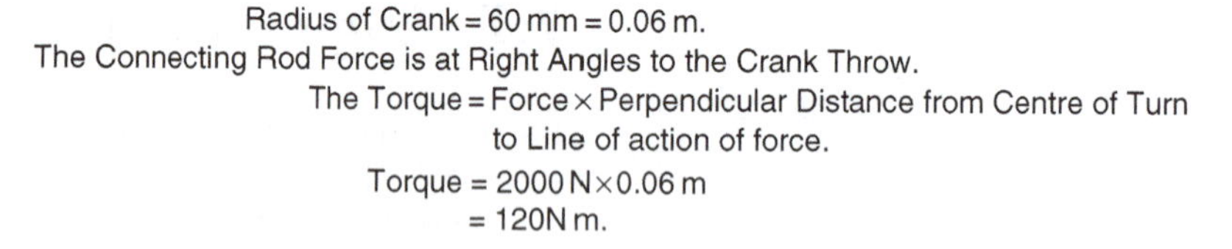

Fig. 7.12 Engine crankshaft torque

7.8 Leverage and gears

Torque multiplication

The torque required at the driving road wheels of a vehicle is larger than the torque available at the engine flywheel. For example, the engine of a medium-sized family saloon may develop a torque of 100 N m and require a torque of 1500 N m at the driving wheels. This would require a torque multiplication of 15 times the engine torque. In order to operate the vehicle it is necessary to provide some means of multiplying engine torque. Use of the gears is the most commonly used method of torque multiplication on vehicles.

Example 7.9

Figure 7.13 shows a pair of gears and their action may be compared to the action of two simple levers.

The radius of each gear is related to the number of teeth on the gear. In this example the radius of the large gear may be taken as 40 mm and that of the small gear as 10 mm. The gear ratio = revolutions of input gear/revolutions of output gear.

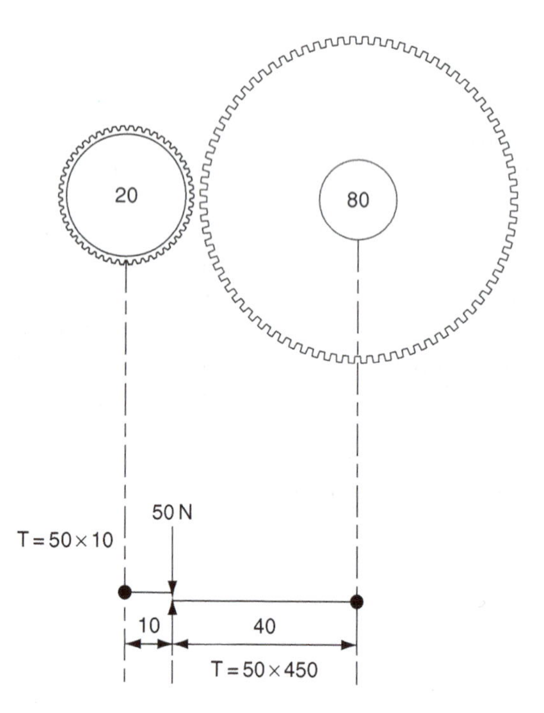

Fig. 7.13 Torque multiplication

In this case the small input gear must rotate four times to produce one revolution of the large output gear.

The gear ratio in this case is 4:1.

If the small gear is the input gear that carries a torque of $50\,N \times 10\,mm = 500\,N\ mm$; the torque on the large gear $= 50\,N \times 40\,mm = 2000\,N\ mm$.

This simple pair of gears provides a torque multiplication of 4.

Drivers and driven

In this simple gear train the small gear is said to be driving the large gear so the small gear is called the driver gear and the large one the driven gear.

This force of 2000 N is acting on the large gear at a radius of 20 cm. The torque on the large gear is now = Force × Radius, which means that:

$$\text{Torque on large gear} = 2000\,N \times 0.20\,m$$
$$= 400\,N\ m$$

The gear ratio multiplies torque and divides speed. In the case of the 4:1 ratio of the simple gear train shown in Figure 7.13, the output torque at the large gear is four times the input torque and the rotational speed of the large gear is $^1/_4$ of the rotational speed of the small gear.

7.9 Gear trains: calculating gear ratios

Spur gear ratios

Figure 7.14 shows part of a gear train. The input shaft pinion has 25 teeth; this drives a layshaft pinion with 35 teeth. The second layshaft pinion has 32 teeth and this drives the output shaft pinion, which has 28 teeth. The gear ratio is determined as follows.

(driven/driver) × (driven/driver);

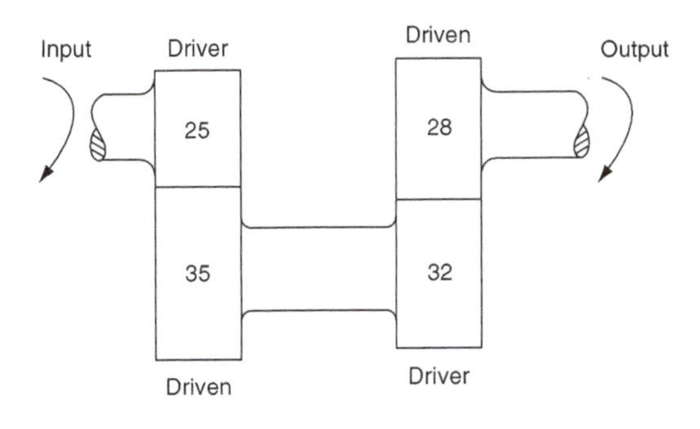

Fig. 7.14 Gear train

this is usually remembered as: $\dfrac{\text{driven}}{\text{drivers}}$

In this case, the ratio of the gear set shown

$$= \frac{35 \times 28}{25 \times 32} = 1.225$$

The normal type of manually operated gearbox utilises an input shaft that is driven by the engine; this shaft is known as the first motion shaft. The first motion shaft meshes with a gear on another shaft, which is known as the layshaft. This first step is normally a gear reduction. The motion is then transmitted from another gear on the layshaft to a meshing gear on the third shaft; this third shaft is known as the main shaft.

7.10 Couples

A couple is system of two equal but oppositely directed parallel forces. The perpendicular distance between the forces is called the arm. The turning moment of a couple is the product of one of the forces and the length of the arm. In Figure 7.15 the arm is d metres in length and the forces are F newtons; the turning moment = Fd N m.

7.11 Summary of main points

A moment of a force is the turning effect of the force.

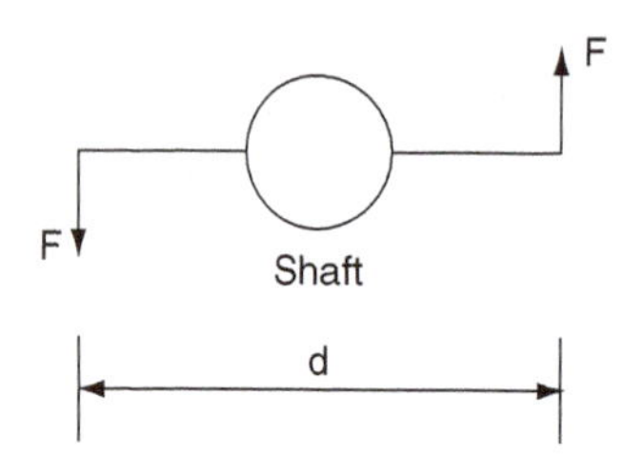

Fig. 7.15 Moment of a couple

Turning moment = force × perpendicular distance from the fulcrum to the line in which the force is acting.

Principle of moments: *If any number of forces, in one plane, act on a body free to rotate about a fixed point, and produce equlibrium, then the sum of the moments of the clockwise effect forces is equal to the sum of the moments of the anti-clockwise effect forces when the moments are taken about the fixed point, or fulcrum.*

Clockwise moments about fulcrum = anti-clockwise moments about fulcrum.

Torque = force × perpendicular distance from the fulcrum to the line in which the force is acting. T = F × R.

$$\text{Gear ratio} = \frac{\text{driven}}{\text{drivers}}$$

The gear ratio is a torque multiplier.

7.12 Exercises

7.1 You are given the details shown in Figure 7.16. The force F required to produce a thrust of 100 N at the clutch thrust bearing is:

(a) 300 N
(b) 33.3 N
(c) 27 N
(d) 333 N

7.2 A motor car of weight 1.8 tonnes has a wheelbase of 2.45 m. If the centre of gravity of the vehicle is located 1.3 m behind the front axle, calculate the loads on each axle.

7.3 A vehicle of gross weight 12.5 tonnes has a wheelbase of 3.32 m. When the vehicle is fully loaded, axle weights are checked and the front axle load is found to be 5.75 tonnes. Determine the distance of the centre of gravity from the centre line of the rear axle.

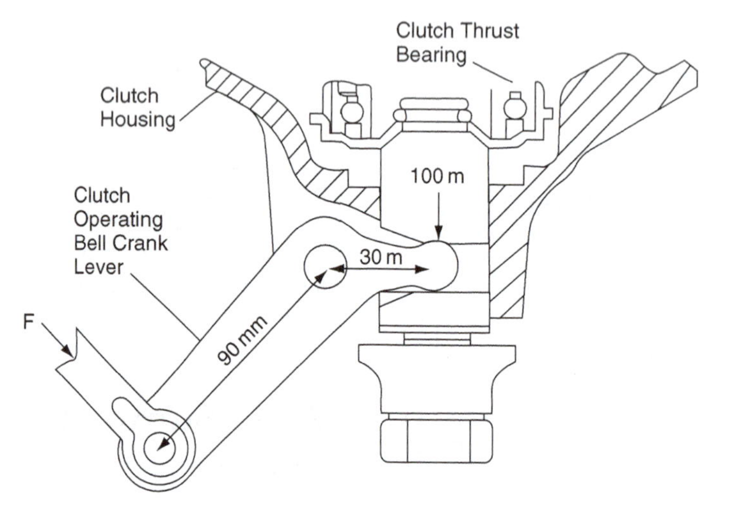

Fig. 7.16 Bell crank clutch mechanism (Exercise 7.1)

7.4 Calculate the reactions at the supports R_1 and R_2. (See Figure 7.17)

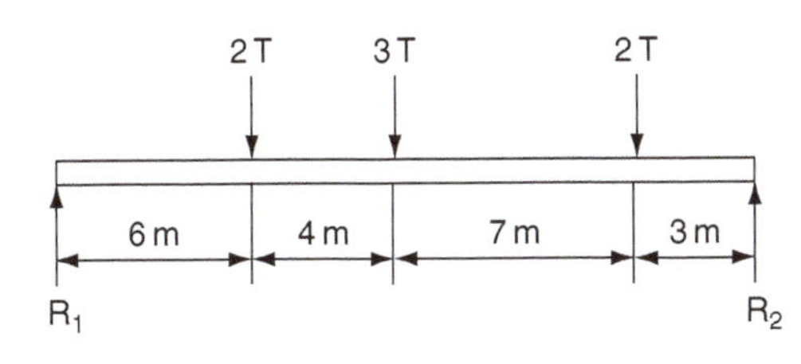

Fig. 7.17

7.5 Calculate R_1 and R_2. (See Figure 7.18)

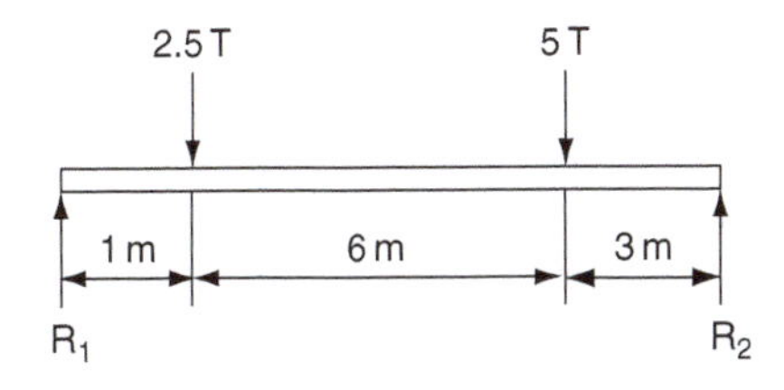

Fig. 7.18

7.6 The figure shows the outline of part of a suspension system. Calculate the spring force F. (See Figure 7.19)

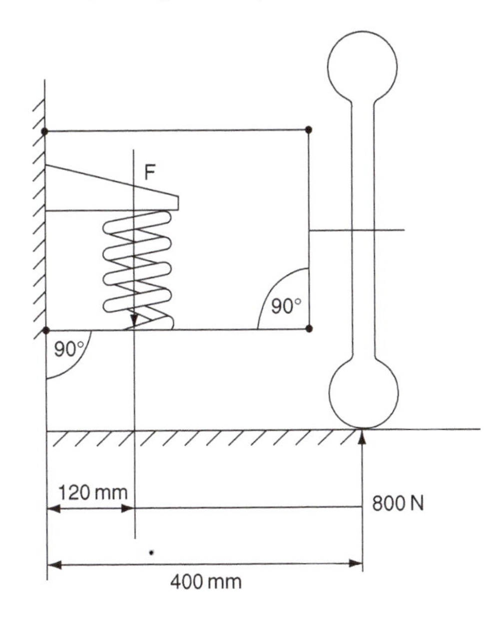

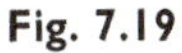

Fig. 7.19

7.7 The diagram represents a piston, connecting rod and crank mechanism. Calculate the torque on the crank. (See Figure 7.20)

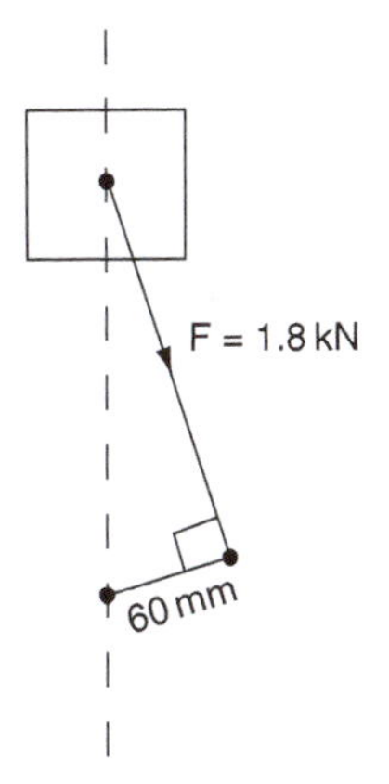

Fig. 7.20

7.8 Calculate (a) the gear ratio for the gear train shown in the diagram, and (b) the output torque for the input torque of 160 N m, assuming 100% efficiency. (See Figure 7.21)

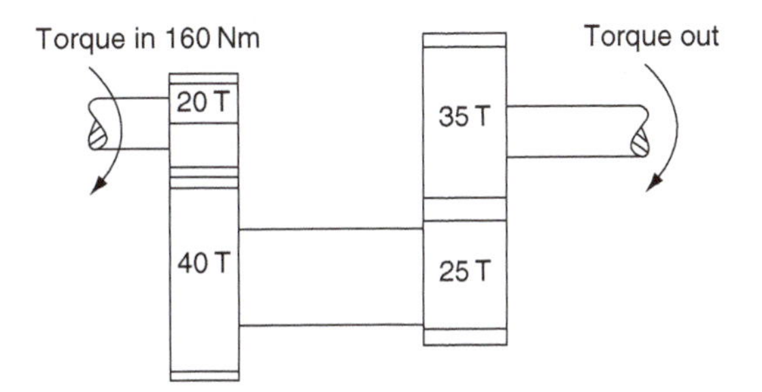

Fig. 7.21

7.9 The figure represents a hand brake lever. Calculate the force F. (See Figure 7.22)

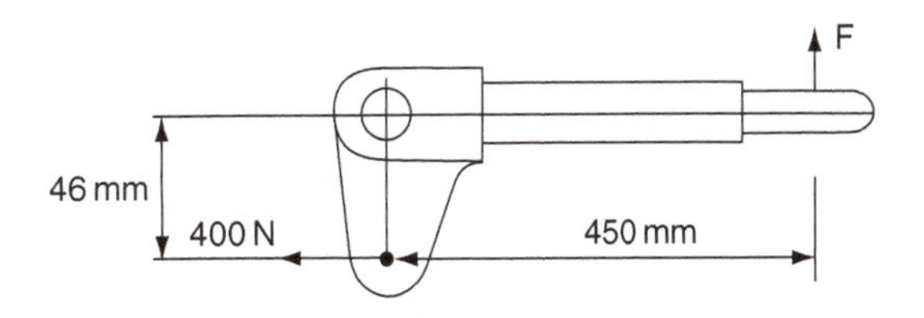

Fig. 7.22

7.10 Determine the values of R_r and R_f. (See Figure 7.23)

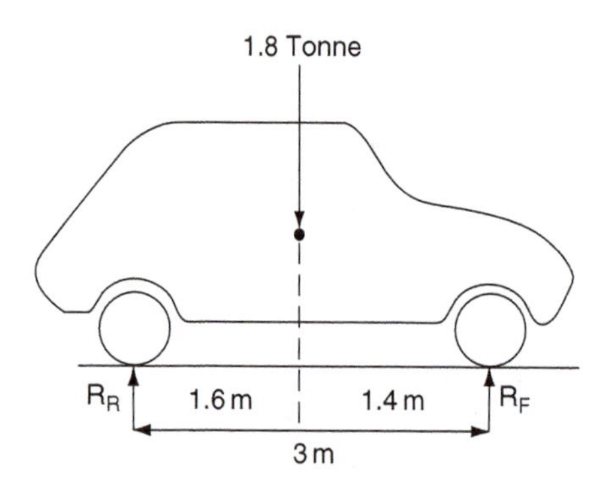

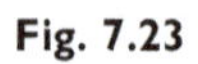

Fig. 7.23

8

Work energy, power and machines

8.1 Work

When an attempt is made to move a body, some resistance to movement occurs.

When lifting a body. as in Figure 8.1(a), the resistance arises from the force of gravity and when a body, such as the block shown in Figure 8.1(b), is moved along a horizontal surface the resistance to motion is caused by friction.

In both cases the force overcomes a resistance and causes the body to move and work is done.

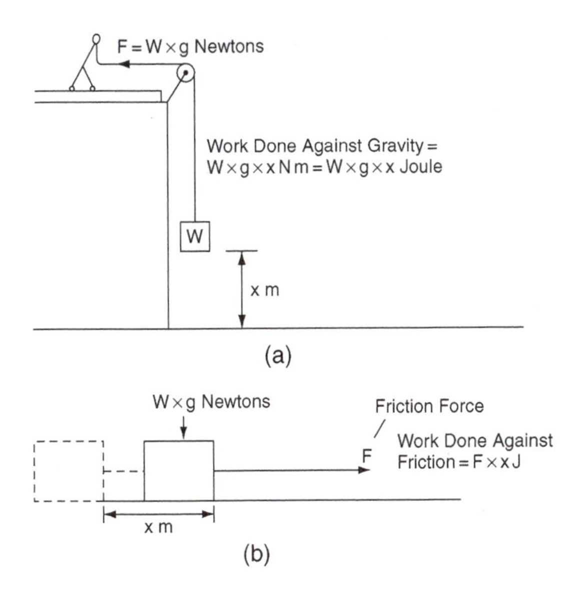

Fig. 8.1 Work = force × distance

The work done is found by multiplying the force by the distance moved.

Work done = force × distance moved in the direction of action of the force.

The distance is measured in metres and the force in newtons.

$$1\ \text{metre} \times 1\ \text{newton} = 1\ \text{m N} = \text{joule} \qquad \textbf{(8.1)}$$

Example 8.1

A vehicle is towed along a level surface for a distance of 200 m, by a force of 8 kN. Calculate the work done in moving the vehicle.

Solution

Work done = force × distance moved

$$= 8000\ \text{N} \times 200\ \text{m}$$

$$= 1\,600\,000\ \text{N m}$$

$$= 1\,600\,000\ \text{J, because}$$

$$1\ \text{N m} = 1\ \text{joule}$$

8.2 Power

Power is the rate of doing work, i.e. power $= \dfrac{\text{work done in joules}}{\text{time taken in seconds}}$

Power is measured in **watts**:

$$\text{Power P in watts} = \frac{\text{work done in joules}}{\text{time taken in seconds}} \qquad \textbf{(8.2)}$$

Because the watt is a small unit of power, the power of vehicle engines is normally quoted in kilowatts (kW).

Example 8.2

A vehicle is pushed along for a distance of 200 metres by a force of 220 newtons for a period of 10 seconds. Calculate the power that this represents.

Solution

$$\text{Work done} = \text{force} \times \text{distance}$$

$$= 220\,\text{N} \times 200\,\text{m}$$

$$= 44\,000\,\text{J}$$

$$= 44\,\text{kJ}$$

$$\text{Power P} = \text{Work done} \div \text{Time taken in seconds}$$

$$\text{P} = 44\,\text{kJ} \div 10\,\text{seconds}$$

$$\text{P} = 4.4\,\text{kW}$$

8.3 Work done by a torque

Figure 8.2 shows a force of F newtons acting at a radius of r metres. The force moves through an arc from P to Q. The distance moved, d, is an arc whose length is rθ metres, where r is the radius in metres and θ is the angle in radians. The work done by F in moving from P to Q = force × distance = F × rθ. But F × r = torque T.

$$\therefore \text{Work done by a torque} = \text{T}\theta \qquad \textbf{(8.3)}$$

In words, work done = torque × radians

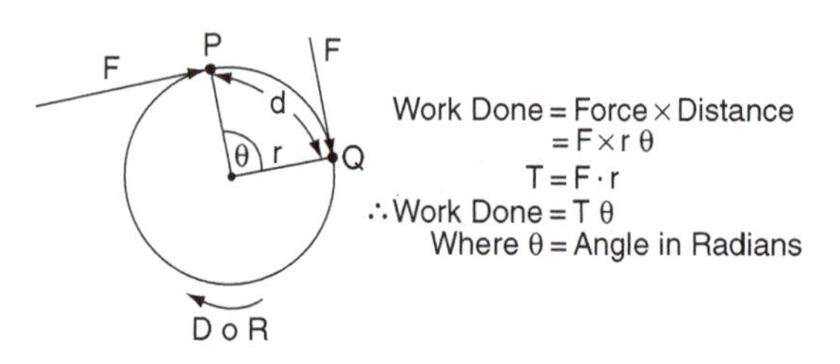

Fig. 8.2 Torque-work

Example 8.3

A torque of 200 N m is applied to a shaft. Calculate the total work done when the shaft rotates 1000 times.

Solution

As there are 2π radians in one revolution the work done per revolution is = 2π.T.

The work done in N revolutions = Work done in one revolution × N

$$\text{Total work done} = 2\pi.\text{T.N}$$

$$= 2\pi \times 200 \times 1000$$

$$= 1\,256\,800\,\text{N m}$$

$$= 1.2568\,\text{MJ}$$

8.4 Work done by a constantly varying force

Figure 8.3 shows a chain type block and tackle lifting gear. The weight being lifted is made up from the weight of chain plus the hook and the actual load. As the load is raised, the amount of chain associated with the load decreases.

Example 8.4

A chain-type lifting tackle is used to raise a load of 75 kg through a height of 2.5 metres. At the start of the lift the length of chain supporting the load is 12 m, and during the lift 5 m of chain are wound in. If the chain weighs 6 kg per metre and the pulley weighs 4 kg, determine the work done on the load in lifting it through the distance of 2.5 m. Take $g = 9.81\,\text{m/s}^2$.

Solution

The initial load

$$= \text{Chain} + \text{pulley} + \text{load}$$

$$= (12 \times 6 \times 9.81) + (4 \times 9.81) + (75 \times 9.81)$$

$$= 1481\,\text{N}$$

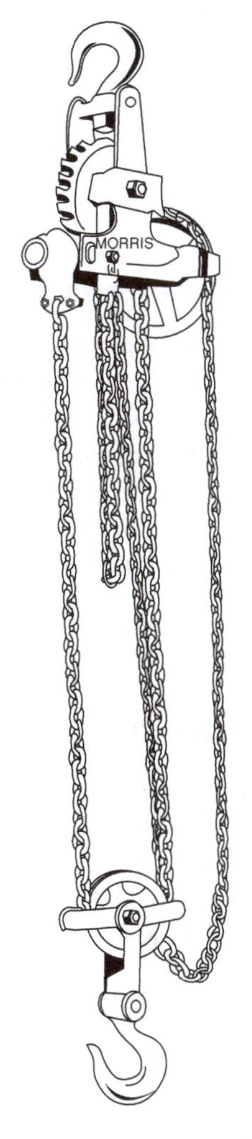

Fig. 8.3 Chain-type block and tackle lifting machine

The final load = $(7 \times 9.81 + 75 \times 9.81)$

$= 804.4\,\text{N}$

The distance moved by the load = 2.5 m

The work diagram in Figure 8.4(a) shows the operation.

The area under the work diagram = work done.

$$\text{Work done} = \frac{(1481 + 804.4)}{2} 2.5\,\text{J}$$

$$= 2857\,\text{J}$$

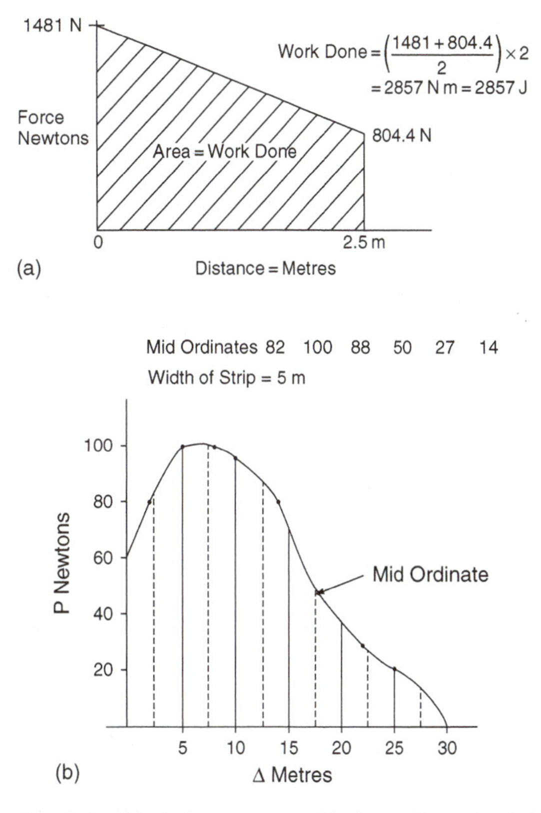

Fig. 8.4 Work diagram – variable force (Examples 8.4 and 8.5)

Mid-ordinate method for calculating work done

Example 8.5

A body has a variable pull of P newtons acting on it. Values of P and the distance D from rest are as shown in Table 8.1.

Plot a graph of P against D and use it to determine the amount of work done in moving the body through the distance of 30 m.

Table 8.1 Plotting a graph of P and D to find work done, using the mid-ordinate rule (Example 8.5)

D (m)	0	2	5	8	10	14	18	22	25	30
P (N)	60	80	100	100	96	80	48	30	20	0

Solution
Plot the graph as shown in Figure 8.4(b).

Mid ordinate rule

Divide the total base into a suitable number of equal parts and erect ordinates at the mid point of the base of each strip. Each strip is then considered to have an area equal to the width of one strip × mid ordinate length. Because each strip has the same width the total area is equal to the sum of the mid ordinates multiplied by the width of one strip.

In Example 8.5 the mid ordinates are: 82, 100, 88, 50, 27, 14

The width of each strip = 5 m

$$\text{The area under the curve} = 5(82+100+88+50+27+14)$$
$$= 5 \times 361\ \text{N m}$$
$$= 1.805\ \text{kJ}$$

Are under curve = work done = 1.805 kJ

8.5 Energy

Energy is the ability to do work. The unit of energy is the joule.

Potential energy is the energy that a body possesses by virtue of its height above some datum. Water, held back by a dam, in the reservoir of a hydro-electric power station whose turbines are at a lower level holds large reserves of potential energy.

Potential energy

Potential energy is the type of mechanical energy that a body possesses by virtue of its position; a typical example is water held in a reservoir which is used to drive water turbines for the generation of electricity. Another example is the energy that is stored in a spring brake that is held there until the brake is applied.

Chemical energy

The energy that drives the engine is derived from the chemical energy that is stored in the fuel. The chemical energy is converted into heat by the combustion process.

Conservation of energy

The principle of conservation of energy states that energy can neither be created nor destroyed. It can be changed from one form to another; for example, chemical energy can be transformed into electrical energy. This happens when energy is drawn from a battery to operate a starter motor; heat energy can be transformed into mechanical energy as happens when fuel is burned in an engine to drive the pistons; and so on.

Energy equation

Since energy is the capacity of a body to do work any change in the energy content of a body is accounted for by either (a) work done on the body or (b) work done by the body. In the form of an equation this becomes:

$$\text{Potential energy} + \text{kinetic energy} + \text{work done} = \text{constant}$$

Kinetic energy

Kinetic energy is the energy that an object possesses by virtue of its velocity.

Consider the motion of a body of mass m kg that is moved by a force of F newtons through a distance of l metres.

$$\text{Work done on the body} = F \times l \qquad \text{(i)}$$

$$\text{The acceleration produced, a,} = F/m \qquad \text{(ii)}$$

At end of the distance l the velocity v of the body may be obtained from the equation $v^2 = u^2 + 2al (u = 0)$

$\therefore l = v^2/2a$

Substituting for a from Equation (2) gives

$$l = \frac{v^2 m}{2F}$$

$$\text{Work done} = Fl = F.\frac{v^2 m}{2F} = \frac{mv^2}{2}$$

The work done on the body by the force F in moving through the distance l is $\frac{1}{2}mv^2$.

By the principle of conservation of energy, the work done is converted to converted to kinetic energy.

Kinetic energy is calculated by multiplying half the mass of the object by the square of the velocity of the object.

If the mass is m kg and the velocity is v m/s, the kinetic energy is calculated from the formula

$$\text{k.e.} = \tfrac{1}{2}mv^2 \qquad \textbf{(8.4)}$$

Example 8.6

A vehicle of mass = 1200 kg is travelling at 20 m/s. Calculate the kinetic energy of the vehicle.

Solution

Mass m = 1200 kg, velocity v = 20 m/s

Kinetic energy = half the mass × velocity squared

$$= 1200/2 \times 20 \times 20\,\text{J}$$

$$= 600 \times 400\,\text{J}$$

$$= 240\,000\,\text{J}$$

Example 8.7

A motor car with a mass of 1.2 tonnes is slowed down from a speed of 72 km/h to 45 km/h by application of the brakes. Calculate the change of kinetic energy that occurs during this process and state what happens to this energy.

Solution

Initial velocity u = 72 × 1000/3600 = 20 m/s; final velocity v = 45 × 1000/3600 = 12.5 m/s. Mass m = 1200 kg.

$$\text{Change of kinetic energy} = \tfrac{1}{2} - 1200(20^2 - 12.5^2)$$

$$= 600(400 - 156.25)$$

$$= 146.25\,\text{kJ}$$

The kinetic energy is converted into heat through friction at the brakes and is then dissipated through the atmosphere.

Energy of a falling body

Consider an object of mass m kilogram that is on a ledge at height h metres above the ground. If the object is pushed off the ledge it will fall to the ground and as it falls the potential energy is reduced because the height above the ground level is reduced. However, the velocity of the object increases as falls and this results in an increase in kinetic energy.

As no external work is done by the object, the total energy which it possesses remains constant.

Potential energy + kinetic energy = constant

$$\text{Total energy before falling} = \text{potential} + \text{kinetic}$$

$$= mgh + 0$$

$$= mgh \text{ joules}$$

Since no work is done by the object while falling h metres, the total energy on reaching the ground must equal mgh joules. At the instant that the object reaches the ground the only energy that it possesses is kinetic.

Total energy on reaching the ground

$$= \text{potential} + \text{kinetic}$$

$$= 0 + \tfrac{1}{2}mv^2$$

Because no work has been done,

$$mgh = \tfrac{1}{2}mv^2$$

which gives $v = \sqrt{2gh}$.

Kinetic energy of rotation

When a body such as an engine flywheel is rotating, the whole of the rotating, mass (weight) is considered to be concentrated at a certain radius that is known as the radius of gyration.

Consider a flywheel (see Figure 8.5) of mass m kg and radius of gyration K metres revolving at N rev/s.

The linear velocity v of the concentrated mass at radius K is: $v = 2\pi KN$ m/s

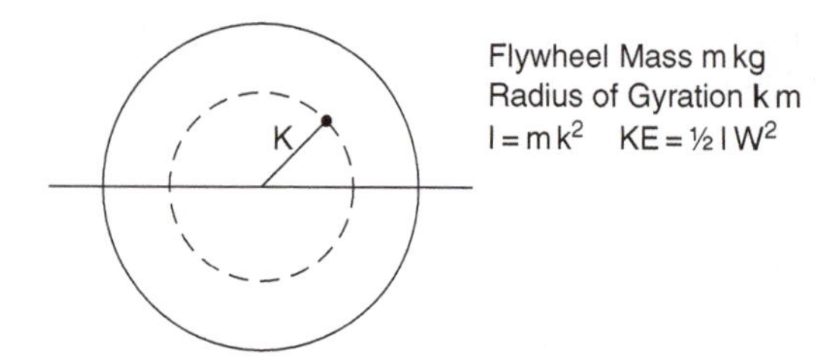

Fig. 8.5 Kinetic energy of flywheel

The kinetic energy $= \frac{1}{2} m v^2 = \frac{1}{2} m(2\pi KN)^2$

$2\pi N =$ angular velocity ω, and $mK^2 = I$ the moment of inertia of the flywheel.

This gives another expression for kinetic energy of a flywheel, or similar rotating disc, namely:

$$\text{k.e. of flywheel} = \tfrac{1}{2} I\omega^2 \qquad \textbf{(8.5)}$$

Where I = moment of inertia and ω = angular velocity in radians per second.

Example 8.8

An energy-saving railcar is fitted with a system that transfers braking energy into a rotating flywheel. When acceleration is required, the energy from the rotating flywheel is transferred to the railcar's propulsion system. The flywheel of the energy-saving system has a mass of 0.5 tonne and a radius of gyration of 0.36 m. Determine the kinetic energy of the flywheel when it is rotating at 1200 rev/min.

Solution

Kinetic energy of flywheel $= \frac{1}{2} I\omega^2$,

$$\omega = 2\pi N/60 = 125\,\text{rad/s}$$

$$I = m K^2$$

$$= 500(0.36 \times 0.36)$$

$$= 65\,\text{kg}\,\text{m}^2$$

$$\text{k.e.} = \tfrac{1}{2} \times 65 \times 125 \times 125$$

$$= 507.8\,\text{kJ}$$

8.6 Machines

A machine is a device that receives energy in some available form and converts it to a useable form. For example, a person may wish to lift a weight of 2 tonnes; this is not possible unaided but the use of a jack or a hoist permits the person to achieve their objective. A lever such as a pry bar is an example of a simple machine. A relatively small manual force can be converted into a large force in order to lift or move an object.

Figure 8.6 shows a hydraulic trolley jack. A load of 2 tonnes (196 kN) is being lifted by an effort of 150 N. In order to perform this operation the operator will make several pumps of the handle, which means that the effort will move much further than the load.

Machines exist in many different forms but the following rules apply to them all.

Mechanical advantage

The mechanical advantage of a machine is the ratio of the load to the effort:

$$\text{mechanical advantage MA} = \frac{\text{load}}{\text{effort}}$$

Lifting machines such as jacks and cranes have a large mechanical advantage so that relatively small manual forces can be used to raise heavy weights.

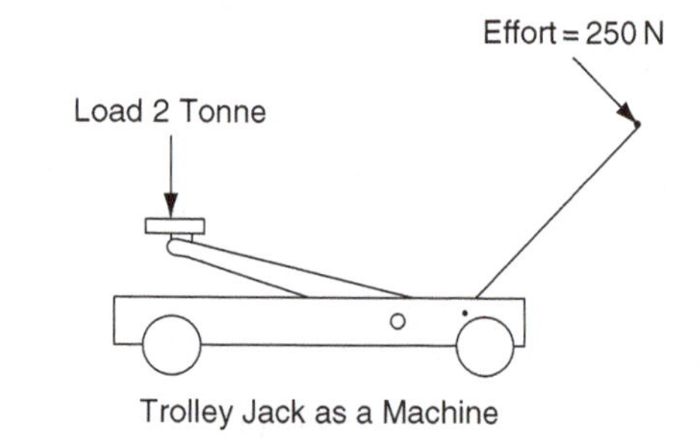

Fig. 8.6 Trolley jack (Example 8.9)

Velocity ratio (movement ratio)

The velocity ratio, VR, of a machine, or movement ratio, MR, is the ratio of the distance moved by the effort to the distance moved by the load:

$$\text{movement ratio MR} = \frac{\text{distance moved by effort}}{\text{distance moved by load}}$$

Efficiency of a machine

The efficiency of a machine = (energy output/energy input) in the same time and this also

$$= \frac{\text{MA}}{\text{VR}} \times 100\%$$

Example 8.9

In the trolley jack example shown in Figure 8.6 an effort of 250 newtons is lifting a load of 2 tonnes. In lifting the load through a distance of 15 cm the operator performs 40 pumping strokes of the handle each of which is 50 cm long. Calculate the mechanical advantage, velocity ratio and efficiency of the jack.

Solution

$$\text{MA} = \frac{\text{load}}{\text{effort}}$$

$$= \frac{2 \times 1000 \times 9.81}{250}$$

$$\text{MA} = 78.5$$

$$\text{MR} = \frac{\text{distance moved by effort}}{\text{distance moved by load}}$$

$$= \frac{40 \times 50}{15}$$

$$= 133.3$$

$$\text{The efficiency} = \frac{\text{energy output}}{\text{energy input}} \times 100$$

$$= (\text{MA}/\text{VR}) \times 100$$

$$= (78.5/133.3) \times 100$$

$$= 59\%$$

Work done against friction

In the example of the jack, the energy input = effort × distance moved = 250 × 40 × 0.5 = 5000 J. In the same time, the energy output = load × distance raised = 2000 × 9.81 × 0.15 = 2943 J.

The difference between the input energy of 5000 J and the output energy of 2943 J is a loss of 2057 J. This loss of energy is accounted for by work done against friction in the machine.

Example 8.10

When using a simple floor crane, the operator inputs 25 000 joules to lift a load of 1 tonne through a height of 1.2 metres. Taking g = 9.81 m/s^2 calculate the efficiency of the crane and determine the energy loss due to friction.

Solution

$$\text{efficiency} = \frac{\text{energy output}}{\text{energy input}}$$

$$\text{energy output} = \text{work done in lifting the load}$$

$$= 1000 \times 9.81 \times 1.2$$

$$= 11\,772\ \text{J}$$

$$\therefore \text{efficiency} = \frac{11\,772}{25\,000}$$

$$= 0.471$$

$$= 47.1\%$$

$$\text{energy loss due to friction} = 25\,000 - 11\,772$$

$$= 13\,228\ \text{J}$$

A steering mechanism as a machine

The force required to steer a vehicle is often considerably larger than a driver can comfortably exert. The steering mechanism is a machine that allows the driver of a vehicle to operate the steering without having to exert a large force at the steering wheel. The rack and pinion steering

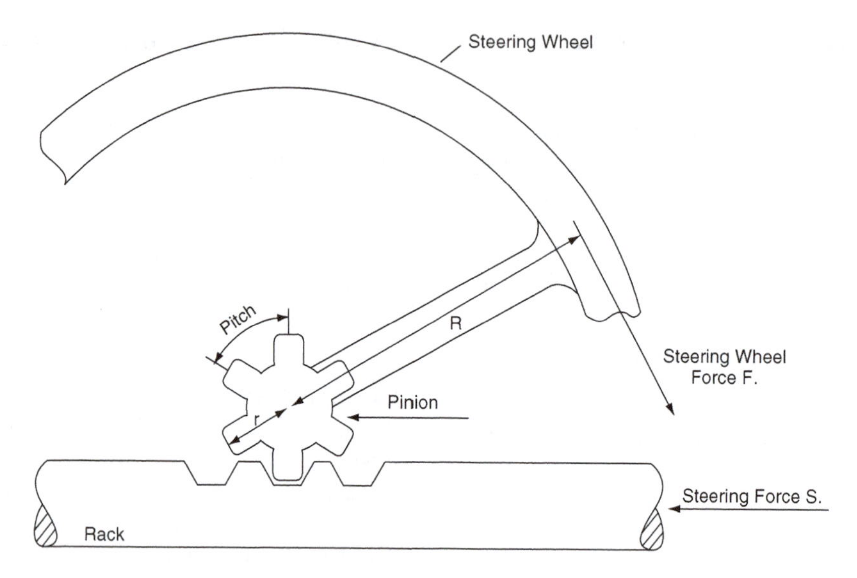

Fig. 8.7 Rack and pinion as a machine

mechanism that is widely used on light vehicles is a convenient example of such a machine (see Figure 8.7).

The steering wheel radius = R
The tangential force at the rim of the steering wheel = F (this is the effort)
The number of teeth on the pinion = Nt
The circular pitch of the pinion gear = p
The force produced at the rack = S (this is the load)

The mechanical advantage MA $= \frac{\text{Load}}{\text{Effort}} = \frac{S}{F}$

Consider one revolution of the steering wheel.

The circular pitch of the gear is the distance between the teeth as measured on the circumference of the pitch circle – one revolution of the pinion is a circular distance of Nt × p, where Nt = number of teeth on the pinion. This is the distance that the rack moves when the steering wheel makes one revolution.

$$\text{The movement ratio MR} = \frac{\text{distance moved by effort}}{\text{distance moved by load}}$$

$$= \frac{2\pi R}{\text{Nt.p.}}$$

Example 8.11

A rack and pinion steering system has 5 teeth of 10 mm pitch and a steering wheel of 320 mm diameter. Calculate:

(a) the movement ratio
(b) the mechanical efficiency of the steering gear if a tangential force of 30 N at the rim of the steering wheel produces a force of 560 N on the rack.

Solution

(a) $\text{MR} = 2\pi R$

$$= \frac{2 \times \pi \times 160}{\text{Nt.p.}}$$

$$= \frac{10\,005.4}{5 \times 10}$$

$$= 20.1$$

(b) $\text{Mech. effcy} = \frac{\text{MA}}{\text{MR}}$

$$\text{MA} = \frac{\text{load}}{\text{effort}} = \frac{560}{30} = 18.7$$

$$\therefore \text{Mech. effcy} = \frac{18.7}{20.1} = 0.93 = 93\%$$

Example 8.12

A worm and sector steering gear of the type shown in Figure 8.8 has a 4-start worm and a sector that is a sector of a 48-tooth worm wheel. The steering wheel has a diameter of 400 mm and the drop arm has a length of 150 mm. Two tangential forces each of 15 N applied at the rim of the steering wheel produce a force in the drag link of 450 N, which acts at right angles to the drop arm. Calculate:

(a) the movement ratio, MR, of the steering system;
(b) the mechanical advantage, MA, of the system;
(c) the mechanical efficiency of the system.

Solution

(a) $\text{MR} = \dfrac{\text{distance moved by effort}}{\text{distance moved by load}}$

The worm is 4-start so the worm wheel (sector) rotates by four teeth when the steering wheel makes one revolution.

The distance moved by effort at the steering wheel during one revolution $= 2\pi \times 200 = 400\pi$

Distance moved by load $= \dfrac{4}{48} \times 2\pi \times 150 = 25\pi$

$\therefore \text{MR} = \dfrac{400\pi}{25\pi} = 16$

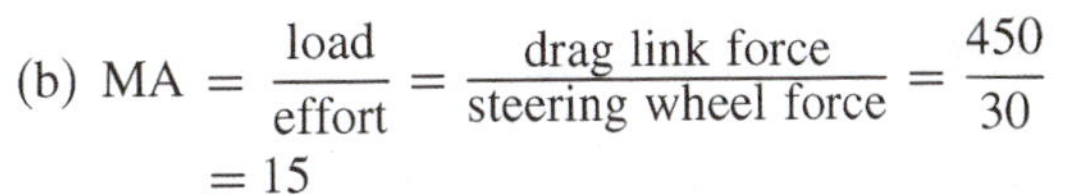

(b) $\text{MA} = \dfrac{\text{load}}{\text{effort}} = \dfrac{\text{drag link force}}{\text{steering wheel force}} = \dfrac{450}{30} = 15$

(c) $\text{Efficiency} = \dfrac{\text{MA}}{\text{MR}} \times 100 = \dfrac{15}{16} \times 100 = 94\%$

8.7 Summary of formulae

Work done = force × distance. Units are joules

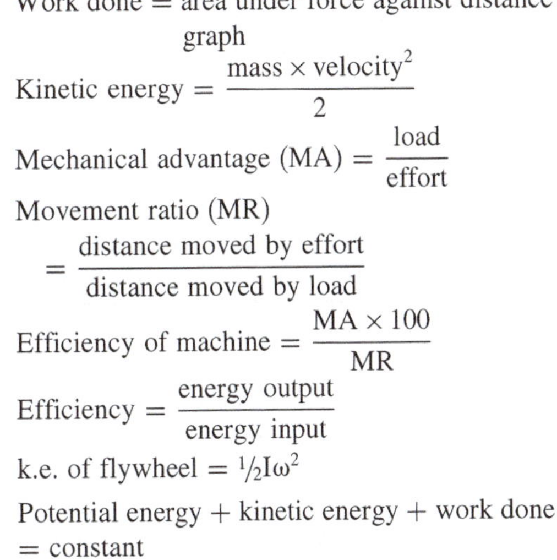

Energy is ability to do work, units are joules

Work done = area under force against distance graph

$\text{Kinetic energy} = \dfrac{\text{mass} \times \text{velocity}^2}{2}$

$\text{Mechanical advantage (MA)} = \dfrac{\text{load}}{\text{effort}}$

Movement ratio (MR) $= \dfrac{\text{distance moved by effort}}{\text{distance moved by load}}$

$\text{Efficiency of machine} = \dfrac{\text{MA} \times 100}{\text{MR}}$

$\text{Efficiency} = \dfrac{\text{energy output}}{\text{energy input}}$

k.e. of flywheel $= \frac{1}{2} I\omega^2$

Potential energy + kinetic energy + work done = constant

Work done by torque $= T\theta$, where T = torque in N m and θ = angle in radians

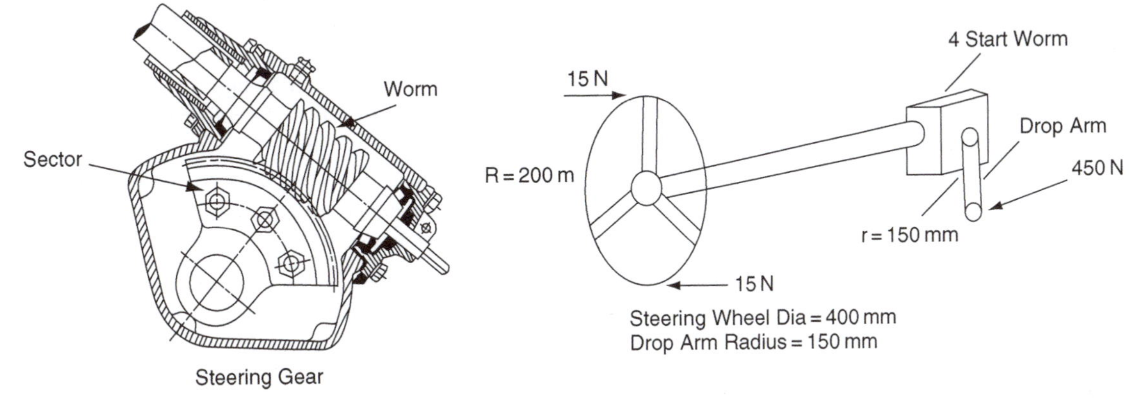

Fig. 8.8 Worm and sector steering gear (Example 8.12)

8.8 Exercises

8.1 In climbing an incline 200 metres long a truck exerts a steady pull of 1 kN on a trailer. Calculate the total work done by the truck.

8.2 During the power stroke of a certain diesel engine that has a stroke of 160 mm, the force acting on the piston varies throughout the length of the power stroke as shown in Table 8.2; F = force on piston, D = distance the piston moves from TDC. Use the mid-ordinate rule to calculate the work done in kilojoules during this process.

Table 8.2 Calculation using mid-ordinate rule (Exercise 8.2)

F (kN)	75	80	70	60	55	43	28	18	5
D (mm)	0	20	40	60	80	100	120	140	160

8.3 An engine develops a torque of 400 N m at a crankshaft speed of 1200 rev/min. Calculate:
(a) the work done per revolution;
(b) the power in kW.

8.4 An vehicle engine develops a torque of 180 N m at a speed of 3000 rev/min. Calculate:
(a) the engine power at this speed;
(b) the torque at the gearbox output shaft when the vehicle is in third gear, which is 1.2:1, and the gearbox efficiency is 96%.

8.5 The wheel, tyre and hub assembly for a certain truck has a combined weight of 200 kg. If the radius of gyration of the assembly 0.15 m, calculate:
(a) the kinetic energy of the assembly when it is rotating at 5 rev/s;
(b) the work done by, and the power of, the brake when the wheel is brought to rest from this speed in a period of 1.5 seconds.

8.6 A certain rear axle final drive has gear ratio of 6:1. At a propeller shaft speed of 1800 rev/min the input torque to the final drive is 420 N m. Calculate:
(a) the speed of the crown wheel in rev/min;
(b) the torque at the crown wheel when the final drive efficiency is 95%.

8.7 An articulated vehicle of 40 tonnes gross weight and travelling at 72 km/h is brought to a standstill by the brakes in a distance of 180 m. Calculate the average braking force.

8.8 A hydraulic system on a tipper truck lifts a load of 2 tonnes through a distance of 0.125 m while the effort moves through a distance of 20 m. Calculate:
(a) the movement ratio;
(b) the actual effort required if the efficiency of the lifting system is 75%. Take $g = 10\,m/s^2$.

8.9 An effort of 180 N will lift a load of 3860 N in a certain lifting machine. If the efficiency of the machine is 80% what is its movement ratio?

8.10 In a test on a hoisting machine it is found that the effort moves 0.6 m while the load rises by 40 mm. Find the effort required to lift a load of 1 tonne assuming an ideal frictionless machine. Take $g = 10\,m/s^2$

8.11 A trolley jack requires an effort of 200 newtons to lift a load of 2 tonnes. In lifting the load through a distance of 15 cm the operator performs 40 pumping strokes of the handle each of which is 50 cm long. Calculate the mechanical advantage, movement ratio and efficiency of the jack. Take $g = 10\,m/s^2$.

8.12 A vehicle is towed along a level surface for a distance of 400 m, by a force of 8 kN. Calculate the work done in moving the vehicle.

8.13 A vehicle is pushed along at uniform speed for a distance of 200 metres by a force of 120 newtons for a period of 10 seconds. Calculate the power that this represents.

8.14 A vehicle hoist raises a vehicle weighing 1.8 tonnes through a distance of 2 metres in 15 seconds. If the overall efficiencyof the hoist is 80%, calculate the power of the lift motor. Take $g = 10\,m/s^2$.

9
Friction

9.1 Introduction

Surfaces which appear to be quite smooth are, when minutely examined, normally found to have surface irregularities and imperfections. When two such surfaces are made to slide, one over the other, these surface imperfections cause resistance to motion. This resistance to motion is called **friction**.

In Figure 9.1(a) the surface imperfections are enlarged. The peaks of the imperfections slot into valleys and as one surface moves over the other the force causing movement attempts to break the peaks off; this causes the resistance to movement that is known as friction and also results in wear and deposition of particles in surrounding parts of any mechanism. Figure 9.1(b) shows the same surfaces separated by a layer of oil. The layer of oil separates the surfaces so the one may slide over the other with reduced resistance to movement; lubrication also reduces wear.

Coefficient of friction

The apparatus shown in Figure 9.2 is used to determine the coefficient of friction. The block W is loaded with suitable weights. Other weights are then placed in the small weight pan and the amount is increased until W moves steadily across the flat surface on which it is placed. The weight of W (in newtons) represents the force pressing the block on to the flat surface and the weight in the pan represents the force F (in newtons) that

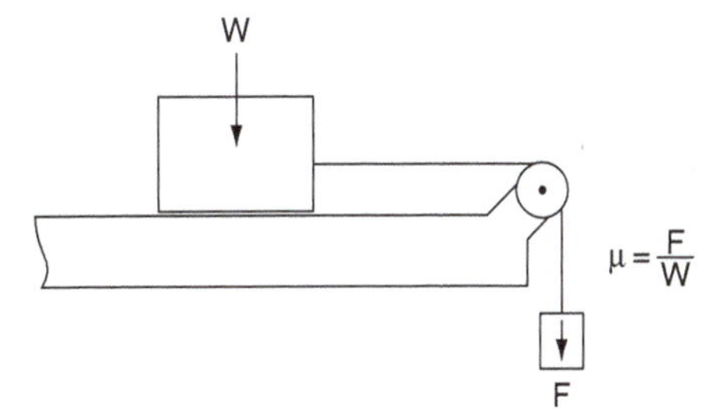

Fig. 9.2 Friction experiment

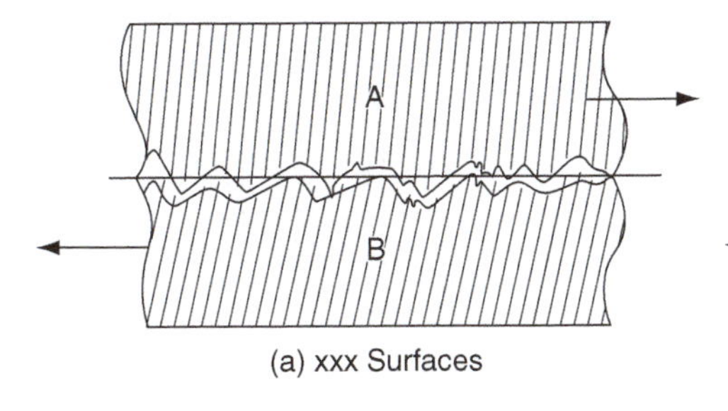

(a) xxx Surfaces

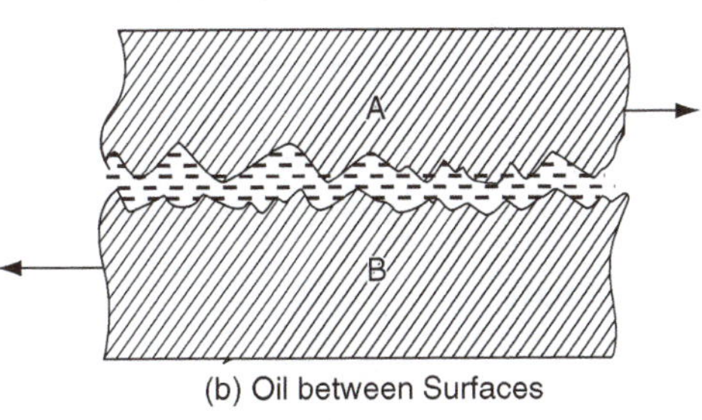

(b) Oil between Surfaces

Fig. 9.1 Friction

causes the block W to move steadily along the flat surface.

The ratio of F to W is called the coefficient of friction and it is denoted by the symbol μ.

$$\text{Coefficient of friction } \mu = \frac{F}{W} \qquad \textbf{(9.1)}$$

where F is the force to move the block and W is the force pressing the surfaces together.

Example 9.1

In an experiment to determine the coefficient of friction between brake lining material and steel, a force of 60 newtons steadily moves a block lined with the brake lining material across a steel surface. The block weighs 100 newtons. Calculate the coefficient of friction.

$$\mu = F/W \quad F = 60\,\text{newtons} \quad W = 100\,\text{newtons}$$

$$\mu = 60/100$$

$$\mu = 0.6$$

Values of μ for some materials are shown in Table 9.1.

Table 9.1 Some approximate values of μ

Phosphor bronze on steel – dry	0.35
Phosphor bronze on steel – oiled	0.15
Steel on steel – dry	0.55
Steel on steel – oiled	0.1
Unworn tyre on dry tarmac	0.8
Unworn tyre on wet tarmac	0.7 (falls rapidly as speed increases)
Brake lining material on steel	0.3

Static friction

The force required to cause initial movement against friction is greater than the force required to maintain uniform motion.

Sliding friction

The sliding force of friction is directly proportional to the force pressing the surfaces together – this force is slightly less than the static friction force.

9.2 Making use of friction

Clutch

Torque transmitted by a plate clutch

In the single plate clutch shown in Figure 9.3 the force provided by the springs is applied uniformly over both sides of the friction surfaces of the clutch plate. The friction force (F) = μW, where W = total spring force.

This force acts at the mean radius of the clutch linings = $R_1 + R_2/2$, where R_1 and R_2 are the inner and outer radii of the clutch linings respectively. The maximum torque that the clutch can transmit = friction force × mean radius of clutch linings.

∴ Clutch torque $T = \mu W \cdot \dfrac{R_1 + R_2}{2}$, but this is for one side of the clutch lining.

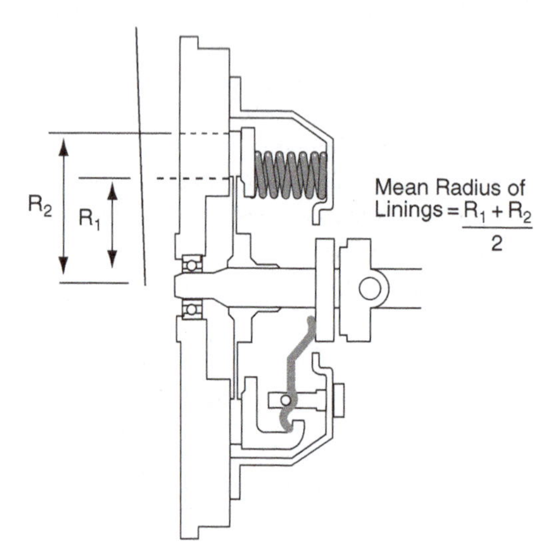

Fig. 9.3 Single plate clutch

There are two sides to the clutch plate so the maximum torque is

$$T = 2\left(\mu W \cdot \frac{R_1 + R_2}{2}\right)$$

In general the formula for torque transmitted by a plate clutch is

$$T = n \cdot \mu W \tfrac{1}{2}(R_1 + R_2)\ \text{N m} \qquad \textbf{(9.2)}$$

where n = number of friction surfaces.

Example 9.2

Figure 9.4 shows a twin plate clutch. The linings have an inner radius of 250 mm and an outer radius of 320 mm. The total spring force is 4 kN and the coefficient of friction of the linings and the pressure plate and flywheel is 0.35. Calculate the maximum torque that this clutch can transmit.

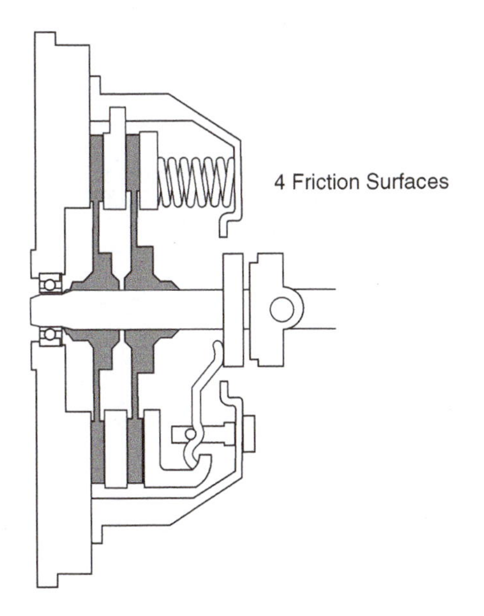

Fig. 9.4 Twin plate clutch (Example 9.2)

Solution

$T = n \cdot \mu\ W\ \tfrac{1}{2}(R_1 + R_2)$

Mean radius of linings $= \tfrac{1}{2}(0.25 = 0.32) = 0.285$ m

Number of friction surfaces n = 4

$T = 4 \times 0.35 \times 4000 \times 0.285$

$T = 1596$ N m

$T = 1.6$ kNm

Power transmitted by a clutch

$\text{Power} = \dfrac{(2\pi TN)}{1000}$ kW, where T is torque in N m, N = revolutions per second.

Example 9.3

Calculate the maximum power transmitted by a single plate clutch at speed of 3600 rev/min if the coefficient of friction is 0.4 and the linings have a radii of 160 mm inner and 190 mm outer. The total spring force is 2.5 kN.

Solution

Maximum clutch torque $= \mu W n\ \tfrac{1}{2}(R_1 + R_2)$

Rev/min = 3600; ∴ rev/sec = 3600/60 = 60

Mean radius $= \tfrac{1}{2}(0.16 + 0.19) = 0.175$ m

W = 2500 N

$T = 0.4 \times 2500 \times 2 \times 0.175$ N m = 350 N m

$$\text{Power} = (2\pi TN)/1000$$

$$= (2 \times 3.142 \times 350 \times 60)/1000$$

Power = 132 kW

Belt drive

Belt drive systems rely on friction between the belt and the pulley for their ability to transmit power (Figure 9.5). The friction force derives from the force exerted on the pulleys by the difference in tension between the tight and slack sides of the belt, $T_1 - T_2$. In a V-belt drive the friction force is provided by the sloping sides of the pulley grooves and this serves to increase the effectiveness of these drives.

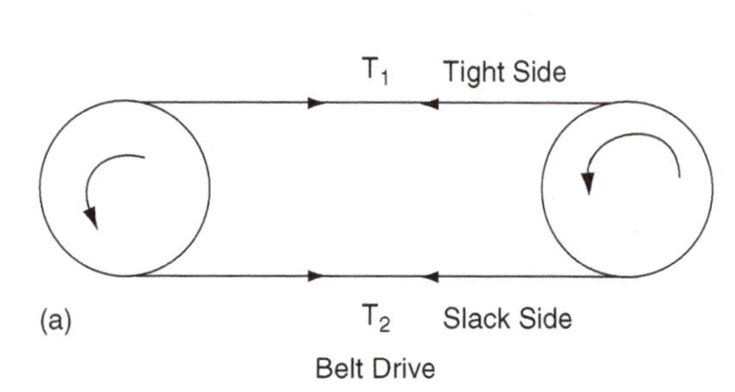

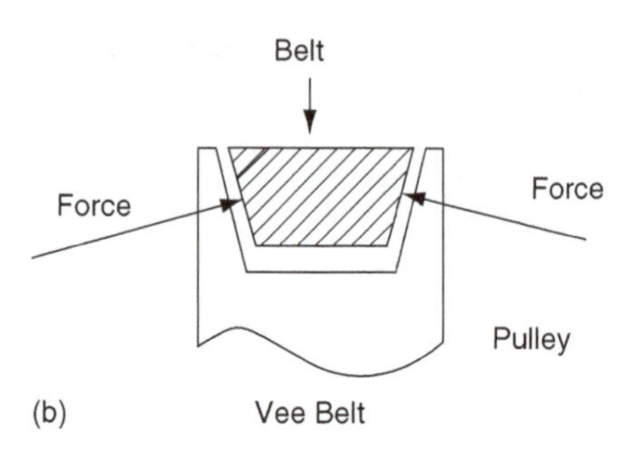

Fig. 9.5 Belt drive

Torque transmitted by belt drive

In both cases the effective torque is $T = (T_1 - T_2)\ r$, where T is the torque in N m, T_1 and T_2 are the belt tensions in newtons, and r = radius of pulley in metres.

Power of belt drive

Power = $(2\pi T N)/1000$

$$\text{Power} = \frac{2\pi(T_1 - T_2)\, r\, N}{1000}\ \text{kW, where N = rev/s} \qquad \textbf{(9.3)}$$

Example 9.4

A pulley of a certain belt drive system has a diameter of 45 mm and it rotates at 1800 rev/min. If the belt tension is 180 N on the tight side and 45 N on the slack side, calculate the power that is being transmitted by the belt drive.

Solution

Power = $(2\pi T N)/1000$

$= [2\pi(T_1 - T_2)\, r\, N]/1000$,

where N = rev/s

$(T_1 - T_2) = (180 - 45) = 135\ \text{N}$

$r = 0.045/2\text{m},\ N = 1800/60 = 30$

Power = $(2\pi \times 135 \times 0.0225 \times 30)/1000$

$= 0.573\ \text{kW}$

Speed ratio of belt drive

In the belt drive shown in Figure 9.6 the small diameter wheel is driving the larger one. One revolution of the small wheel produces $\pi \times 15 = 47.12$ millimetres of movement of the belt.

This produces the same amount of movement around the circumference of the large wheel.

The circumference of the large wheel = $\pi \times 30 = 94.24$ mm.

A movement of 47.12 mm on the circumference of the large wheel represents an angular movement of $(47.12/94.24) \times 360° = 180° = \frac{1}{2}$ revolution.

In general, speed ratio of belt drive system

$$= N_l/N_s$$
$$= D_s/D_l \qquad \textbf{(9.4)}$$

Where N_l = rev/min of large wheel, N_s = rev/min of small wheel, D_s = diameter of small wheel and D_l = diameter of the large wheel.

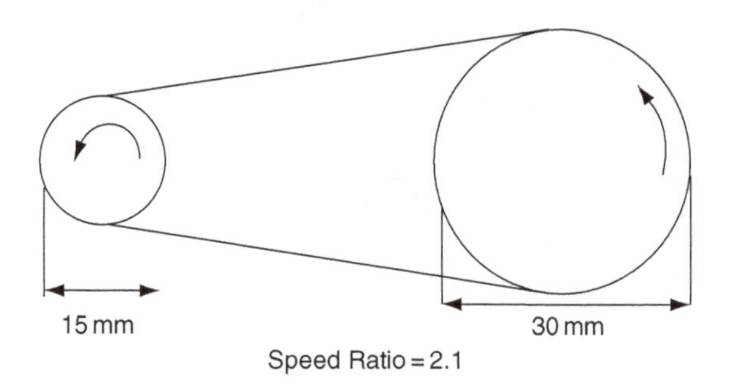

Fig. 9.6 Speed ratio – belt drive

9.3 Brakes

Drum brake – basic principle

Figure 9.7 shows the basic principle of a drum brake with one leading shoe and one trailing shoe.

The actuating force P is provided by a hydraulic cylinder or cam. The shoes are pivoted at A_1 and A_2.

The action of force P creates a friction force between the rotating brake drum and the brake lining.

The friction force on the leading shoe is μF and it acts tangentially to the drum. On the trailing shoe the friction force is μF_1.

The pivots are distance y from the centre of the drum and the actuator force P is at distance X from the pivots. The drum radius is R.

Action of the leading shoe

Taking moments about pivot A_1, $Px = Fy - \mu FR$

$$F = Px/(y - \mu R) \qquad \textbf{(9.5)}$$

Action of the trailing shoe

Taking moments about pivot A_2, $Px = F_1 y + \mu F_1 R$

$$F_1 = Px/(y + \mu R) \qquad \textbf{(9.6)}$$

The effective braking force on the leading shoe is greater than that on the trailing shoe. It should be noted that this effect is dependent on the direction of rotation of the brake drum, which is the reason why two leading shoe brakes are less effective when reversing a vehicle.

Example 9.5

In a certain drum brake the distance x as shown in Figure 9.7(b) is 12 cm, the distance y is 24 cm and the drum radius R = 14 cm. If the actuating force P is 800 newtons and the coefficient of friction $\mu = 0.4$ calculate the effective friction forces on each of the brake shoes.

Solution

Normal force on leading shoe

$$F = \frac{Px}{(y - \mu R)}$$

$$= \frac{800 \times 12}{(24 - 0.4 \times 14)}$$

$$F = \frac{9600}{(24 - 5.6)}$$

$$F = 522\,\text{N}$$

Normal force on trailing shoe

$$F_1 = \frac{Px}{(y + \mu R)}$$

$$F_1 = 9600/(24 + 5.6)$$

$$F_1 = 324\,\text{N}$$

Disc brake

The disc brake shown in Figure 9.8 utilises opposed cylinders to apply force to the friction pads and these pads apply a clamping force to the disc. In Figure 9.8 the following data apply:

hydraulic pressure is P N/m^2; area of the pistons is A m^2; the coefficient of friction between the pads and the disc is μ; the effective radius of the brake is R m. The hydraulic force on each pad = P × A newtons.

The friction force at each pad = μPA newtons

The braking torque

$= 2R\mu PA$ newton metres (two pads) **(9.7)**

Example 9.6

In a disc brake of the type shown in Figure 9.8 the hydraulic pistons are 50 mm in diameter and the effective radius of the brake disc is 150 mm. Calculate the braking torque when the hydraulic pressure is 20 bar and the coefficient of friction is 0.4.

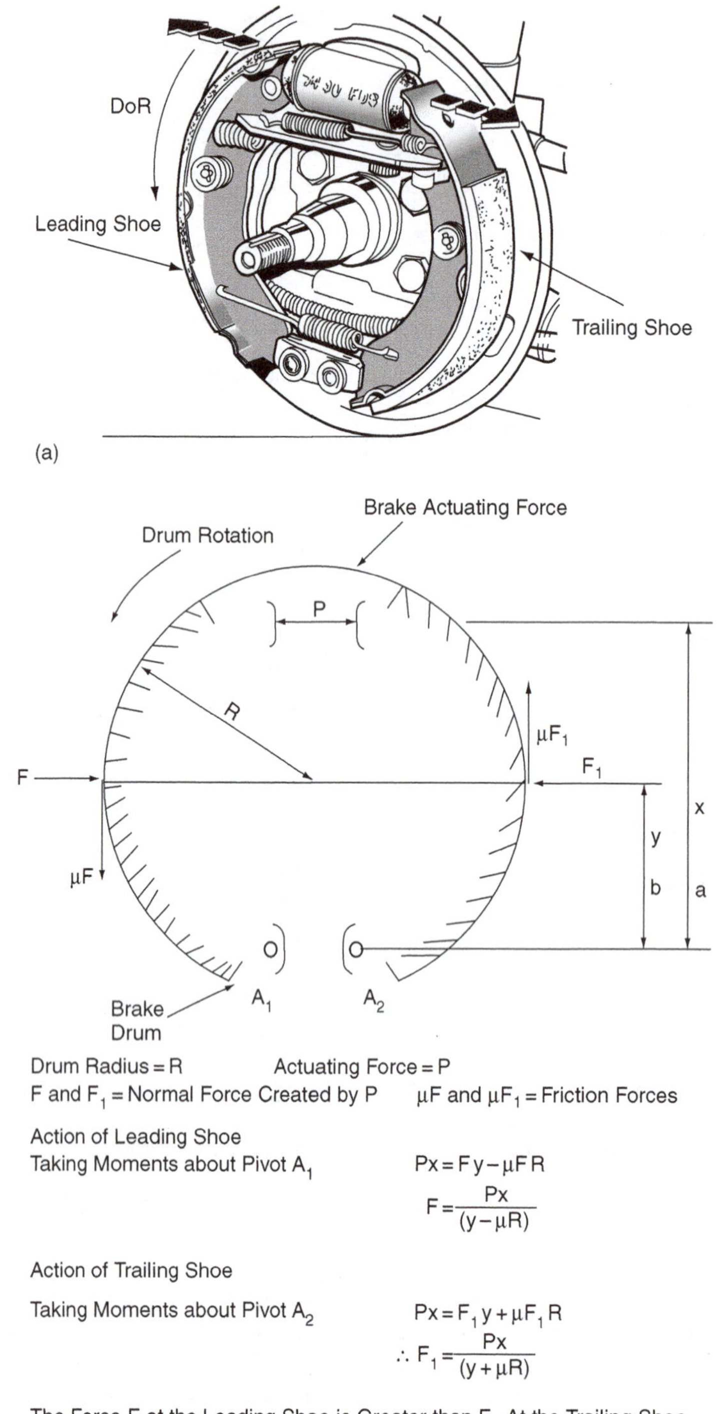

Fig. 9.7 Drum brake

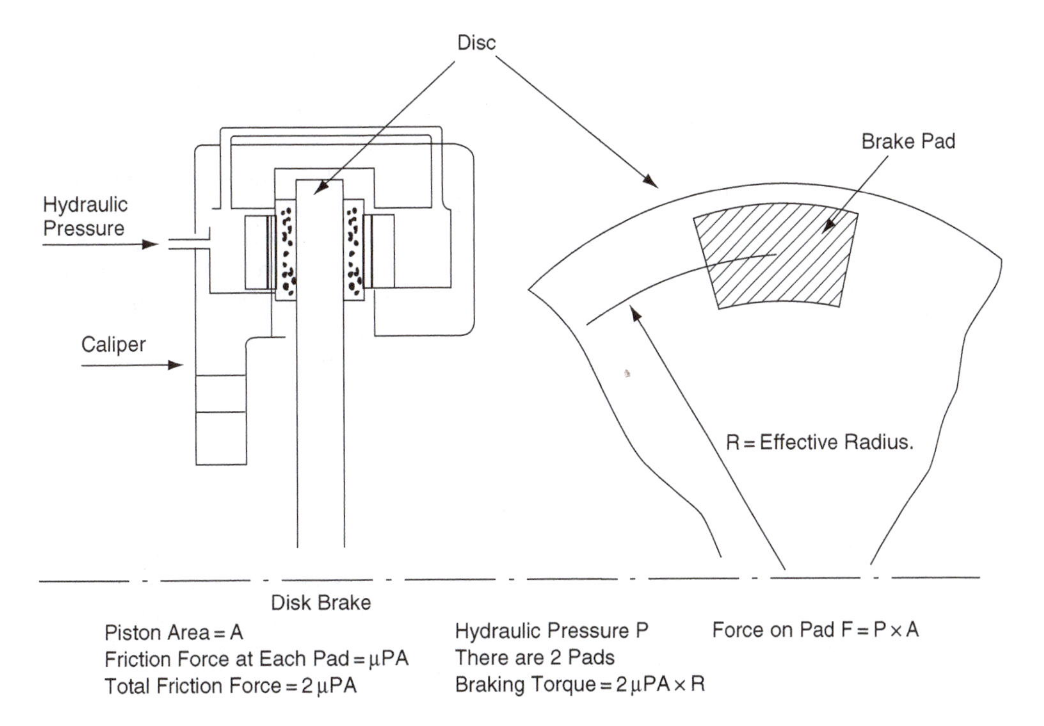

Fig. 9.8 Disc brake

Solution

Braking torque = $2R\mu PA$

$P = 20 \times 100\,000$

$P = 2\,000\,000\,N$

$A = \pi d^2/4$

$A = 1.96 \times 10^{-3}\,m^2$

$R = 0.15\,m$

Braking torque $= 2 \times 0.15 \times 0.4 \times 2\,000\,000 \times (1.96/1000) = 470\,N\,m$.

Tyres

Tyres rely on friction between the tyre tread and the driving surface to provide the grip for driving purposes. The coefficient of friction for tyres is affected by the rubber compounds used in the tyre construction, the tread pattern, the tyre dimensions (low profile) the type and condition of the driving surface and the speed of the vehicle etc. The graph in Figure 9.9 gives an indication of tyre friction on different surfaces.

Anti-lock braking systems (ABS) make use of the fact that greater friction is obtained when the amount of slip is kept at a low level. Traction control systems are also designed to control slip and thus maximise tyre friction.

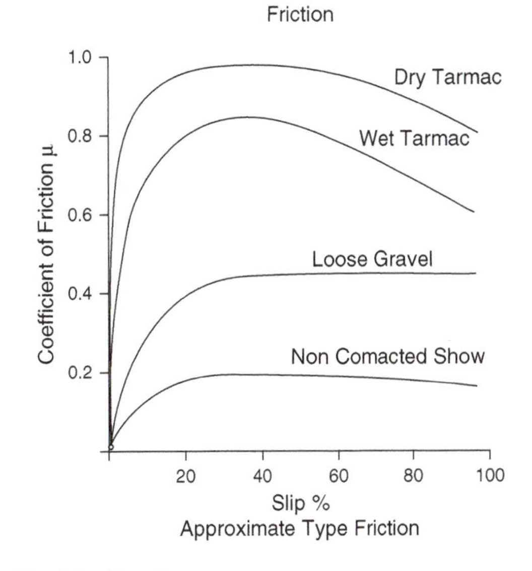

Fig. 9.9 Tyre friction

Braking efficiency

The power of a braking system is determined ultimately by the friction between the tyres and the driving surface. The concept of braking efficiency is based on the assumption that the coefficient of friction between tyre and driving surface can never be greater than 1. Figure 9.10 shows a vehicle with its wheels locked by the brakes being pulled along horizontally by a force that is equal to the weight of the vehicle. The retardation achieved under such conditions is equal to gravitational acceleration g ($9.81\,m/s^2$) and this is taken to be 100% braking efficiency. If the actual retardation of a vehicle under braking is a m/s^2 the braking efficiency is $a/g \times 100\%$.

$$\text{Braking efficiency} = (a/g) \times 100\% \qquad \textbf{(9.8)}$$

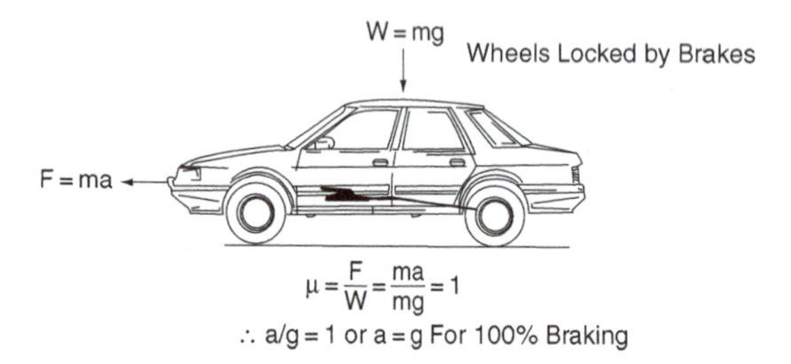

Fig. 9.10 Braking efficiency

Example 9.7

The maximum braking deceleration of a certain vehicle is $7.2\,m/s^2$. Calculate the braking efficiency. Take $g = 9.81\,m/s^2$.

Solution

Braking efficiency $= (a/g) \times 100\%$
Braking efficiency $= (7.2/9.81) \times 100$
Braking efficiency $= 73\%$

Static brake tests

In the UK, brake efficiency tests are performed on machines that measure the horizontal braking forces at each wheel. These forces are then added together to give a total horizontal braking force for the vehicle; this total force is then divided by the weight of the vehicle to give the braking efficiency of the vehicle under test station conditions.

Figure 9.11 shows the basic principle of a dynamometer that is used to determine braking forces on a vehicle. The vehicle is driven on to the rollers, one for each side of the axle, and the front and rear axles are tested separately. An electric motor drives the test rollers. The rollers are driven by the armature shaft and the casing of the electric motor is free to rotate on its axis but is restrained from doing so by a torque arm and spring balance. The torque exerted by the motor and roller on the wheel and tyre is equal and opposite to the torque measured at the torque arm of the electric motor.

In the diagram, F_B = braking force, r = radius of roller, F = force on torque arm of the electric motor, and R = radius of torque arm.

Torque at point of contact between tyre and roller = torque at the electric motor:

$$\therefore F_B \cdot r = F \cdot R$$

$$F_B = FR/r$$

In practice, the force at the torque arm is measured by a transducer and the reading is transmitted to gauges on a console and the braking forces for the left-hand and right-hand side of the vehicle can be compared. The total braking force (front + rear) is divided by the weight of the vehicle to give the braking efficiency.

This test relies on the following principles of mechanics.

Retarding force $F = M{\cdot}a$,

where F = braking force,

M = mass of vehicle and

a = retardation of vehicle.

$$F = M.a$$

$$\therefore a = F/m$$

Braking efficiency $= (a/g) \times 100\%$

Substituting for F/m for a in this equation gives

$$\text{braking efficiency} = \frac{F \cdot 100}{Mg} \qquad \textbf{(9.9)}$$

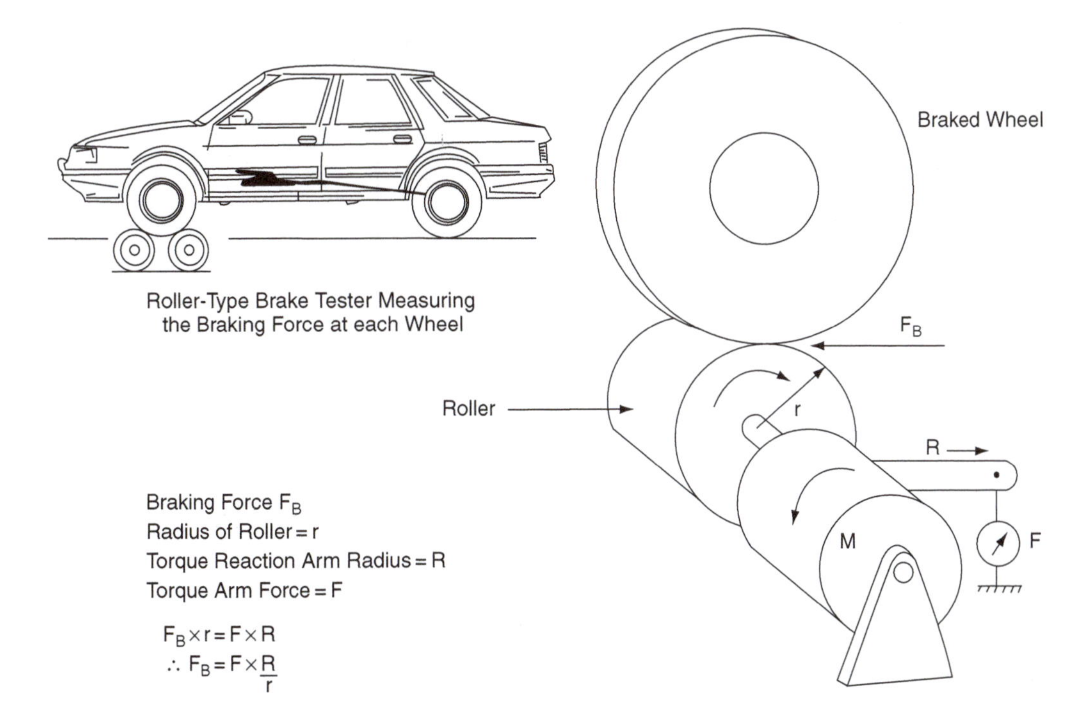

Fig. 9.11 A static brake test

Where Mg = weight of vehicle in newtons $\therefore$ braking efficiency = (F/M) × 100, where M = weight of vehicle in kg × 9.81 and F = total braking force in newtons.

Example 9.8

In a test-station brake-test on a vehicle, the front brake forces total 2200 N and the rear brake forces total 1400 N. If the vehicle has a mass (weight) of 1.2 tonne, calculate the braking efficiency. Take $g = 9.81\,m/s^2$.

Solution

Braking efficiency BE = (total braking force ÷ weight of vehicle in newtons) × 100

BE = (3600/1200 × 9.81) × 100

BE = (3600/11772) × 100

BE = 31%

9.4 Angle of friction

If a block of weight W is placed on an adjustable inclined plane such that the angle of the incline can be altered from zero degrees to the angle at which the block will slide at uniform velocity down the slope, the forces acting are as shown in Figure 9.12. The angle at which this occurs is known as the **angle of friction**.

The forces acting are

force acting down the slope = $W \sin \phi$

force pressing block on to slope = $W \cos \phi$

the coefficient of friction μ

$$= \frac{\text{force pressing surfaces together}}{\text{force moving the block}}$$

$$\therefore \mu = \frac{W\sin\phi}{W\cos\phi} = \tan\phi \qquad \textbf{(9.10)}$$

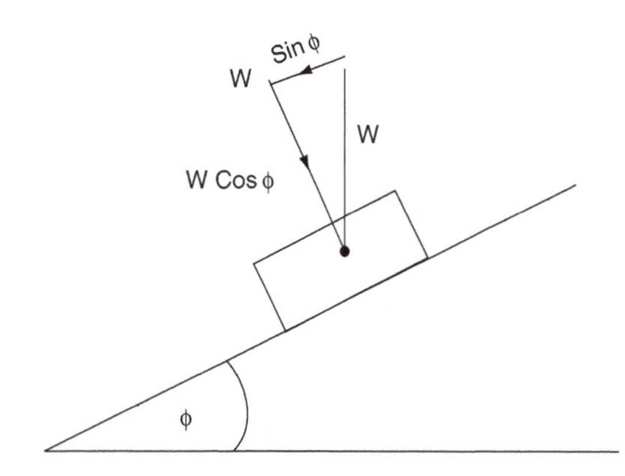

Force Down Slope to Move Block Against Friction = W sin ϕ
Force Pressing Block on to Slope = W cos ϕ

Coefficient of Frition $\mu = \dfrac{W \sin\phi}{W \cos\phi} = \tan\phi$

Fig. 9.12 Angle of friction

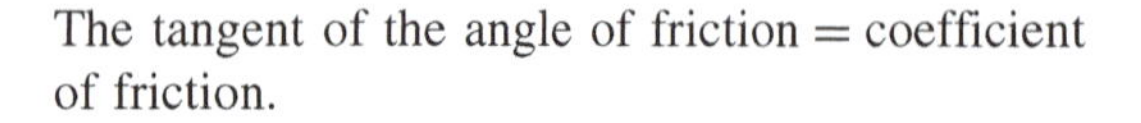

The tangent of the angle of friction = coefficient of friction.

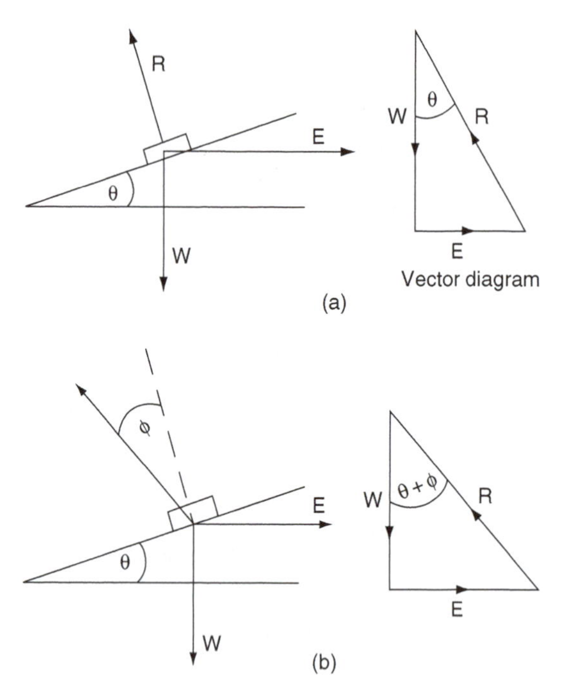

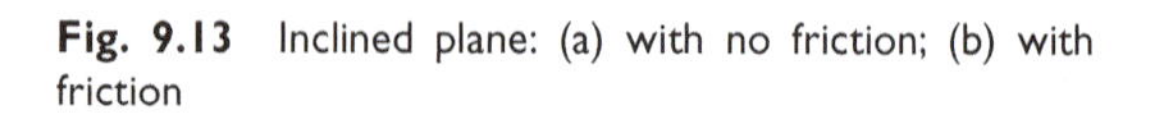

Fig. 9.13 Inclined plane: (a) with no friction; (b) with friction

9.5 Inclined plane

The inclined plane is, in effect, a simple machine. When a body of weight W newtons is moved up an incline by a force E newtons, there is a mechanical advantage of E/W and a relatively small force can lift a heavy load.

Inclined plane without friction

Consider the inclined plane shown in Figure 9.13(a).

(a) with the force E acting horizontally

$$\text{mechanical advantage} = W/E = 1/\tan\theta$$

(b) with the force E acting parallel to the incline

$$\text{mechanical advantage} = W/E = 1/\sin\theta$$

Inclined plane with friction

Figure 9.13(b) shows a diagram of the forces that are acting on the inclined plane when friction is present. The reaction R is now acting at an angle of ϕ to the normal to the plane, where ϕ is the angle of friction.

From the force diagram, $E = W\tan(\theta + \phi)$

The mechanical advantage

$$= W/E = 1/\tan(\theta + \phi) \qquad \textbf{(9.11)}$$

9.6 Screw thread

A screw thread may be treated as an inclined plane where the height of the plane is equal to the pitch of the thread p and the base is equal to the mean circumference of the thread = πd, where d is the mean diameter of the thread.

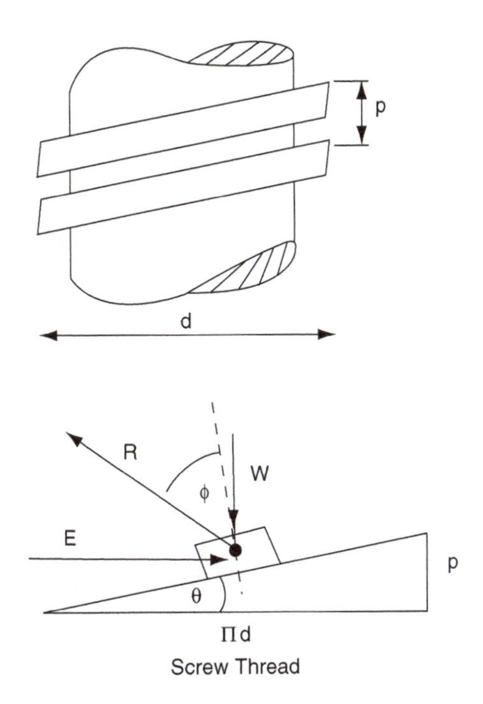

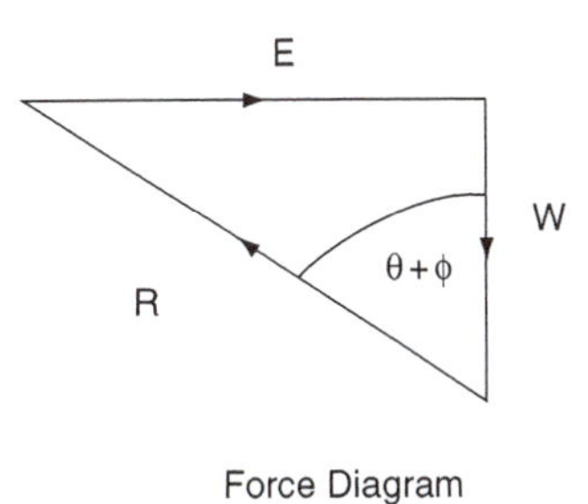

Fig. 9.14 Screw thread as inclined plane

The forces acting on screw thread are the same as those on the inclined plane with friction. The vector diagram for the forces acting on the thread are shown in Figure 9.14(b). The horizontal force $E = W\tan(\theta + \phi)$, where θ = helix angle of thread and ϕ is the angle of friction.

The torque required to tighten the nut against the axial force $W = d/2.E$

$$\text{Torque} = W\frac{d}{2}\tan(\theta + \phi) \qquad \textbf{(9.12)}$$

because $E = W\tan(\theta + \phi)$

V-thread

For a V-thread the normal force R between the nut and the thread is increased since the vertical component of R must be equal to W.

From the force diagram in Figure 9.15, $R = W/\cos\alpha$.

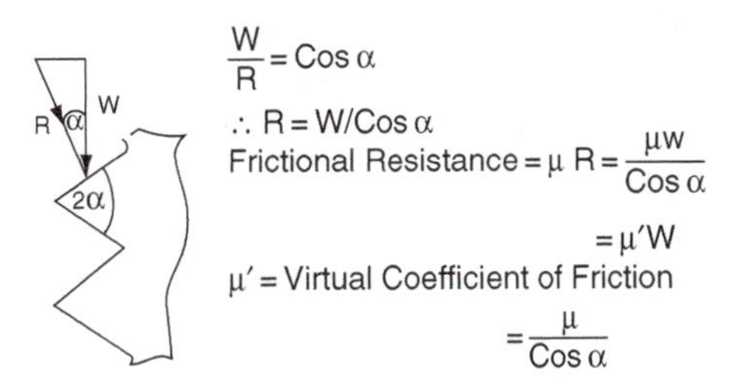

Fig. 9.15 V-thread

$$\therefore \text{frictional resistance} = \mu R$$
$$= \mu W/\cos\alpha$$
$$= \mu' W$$

Where $\mu' = \mu/\cos\alpha$

α is half the thread angle.

The V-thread is therefore equivalent to a square thread with a coefficient of friction μ' instead of μ.

The angle of friction that corresponds to

$$\mu' \text{ is } \phi' = \tan^{-1}\mu'. \qquad \textbf{(9.13)}$$

Example 9.9

A screw jack has a thread of 45 mm mean diameter and a pitch of 7 mm. The thread is of V-form with an included angle of 60° and a coefficient of friction $\mu = 0.15$. Calculate the torque required to raise a load of 1 tonne. Take $g = 9.81\,m/s^2$.

Solution

The formula required here is the same as that for the square thread except that the angle of friction is ϕ' instead of ϕ.

$$\text{Torque} = W\frac{d}{2}\tan(\theta+\phi),$$

$$\text{because } E = W\tan(\theta+\phi)$$

$$\mu' = \mu/\cos\alpha = 0.15/\cos 30°$$

$$= 0.15/0.8660 = 0.173$$

$$\phi' = \tan^{-1} 0.173 = 9.82°$$

$$\tan\theta = \text{pitch/mean circumference.}$$

$$\text{Mean circumference} = 0.045\,\pi = 0.143\,m$$

$$\text{Pitch} = 0.007\,m$$

$$\tan\theta = 0.007/0.143 = 0.05;$$

$$\therefore \text{angle } \theta = 2.9°$$

$$\therefore \text{Torque} = 1000 \times 9.81 \times 0.045/2 \times \tan(2.9° + 9.82°)$$

$$= 221 \times \tan 12.72°$$

$$= 221 \times 0.225$$

$$= 50\,N\,m$$

9.7 Friction in a journal bearing

Figure 9.16 represents a shaft rotating inside a plain bearing. The radius of the shaft is r and the load acting is W. If the coefficient of a friction is μ, the frictional force opposing motion is μW and the resulting torque $T = \mu Wr$.

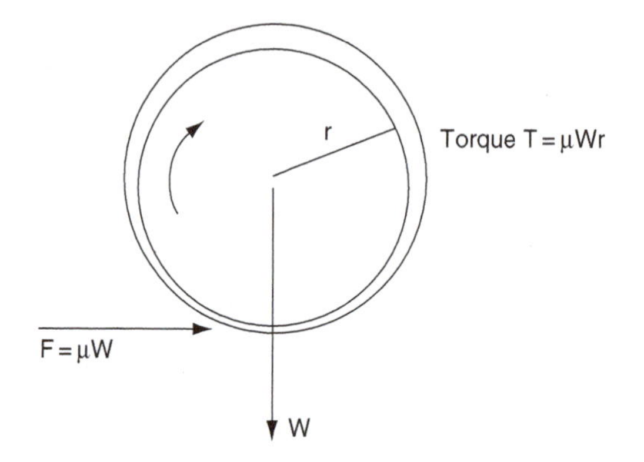

Fig. 9.16 Friction in a journal bearing

$$\text{power lost in friction} = \frac{2\pi TN}{1000}\,kW.$$

Where N = shaft speed in rev/s.

$$\therefore \text{power lost in friction} = \frac{2\pi\mu Wr}{1000}\,kW \quad \textbf{(9.14)}$$

Example 9.10

A 75 mm crankshaft main bearing journal rotates at 3000 rev/min while carrying a normal load W = 20 kN. If the coefficient of friction in the bearing = 0.02, calculate:

(a) the friction torque on the crankshaft;
(b) the power lost in friction at this engine speed; and
(c) state what happens to the energy absorbed in friction.

Solution

(a) Friction force $= \mu W = 0.02 \times 20\,000 = 400\,N$

Friction torque $T = \mu Wr = 400 \times \frac{0.075}{2} = 15\,N\,m$

(b) Friction power

$$= \frac{2\pi TN}{1000}\,kW, \text{ where } N = \text{shaft speed in rev/s}$$

$$= \frac{2\pi \times 15 \times 3000/60}{1000}\,kW$$

$$= 4.71\,kW$$

(c) The 4.71 kJ of heat energy is transferred to the lubricating oil and surroundings.

9.8 Summary of formulae

Laws of static friction

1. Friction always opposes motion.
2. The force of friction depends on the materials and the properties of the surfaces in contact.
3. When motion is about to take place, the magnitude of the limiting friction is directly proportional to force pressing the surfaces together, that is:

Force of friction = coefficient of friction × force pressing surfaces together

4. Friction is independent of surface area.

Coefficient of friction

$$\mu = \frac{\text{friction force}}{\text{force pressing surfaces together}}$$

Friction torque = friction force × radius at which it acts

9.9 Exercises

9.1 A metal casting weighing 250 kg is dragged along a concrete floor surface by a force of 1.2 kN. Determine the coefficient of friction between the casting and the concrete floor surface. Take $g = 10\,m/s^2$

9.2 A shaft of 50 mm diameter runs in a plain bearing while carrying a normal load of 500 N. The coefficient of friction between the bearing and the shaft is 0.08. Calculate the friction torque.

9.3 What effect does lubrication have on the friction of dry surfaces? Describe the principles of lubrication and explain the advantages of pressure lubrication for bearings.

9.4 A vehicle of mass 1.4 tonnes is pulled across a horizontal surface with the wheels locked, by a horizontal force 8 kN. Determine the coefficient of friction between the tyres and the surface. Take $g = 10\,m/s^2$.

9.5 Figure 9.17 shows a V-belt drive system of the type that is used in some Volvo vehicles. The effective diameter of the small pulley is changed during operation by a centrifugal device and this provides a continuously variable speed ratio. At a certain speed the small pulley has a radius of 60 mm. If the power transmitted is 45 kW at a pulley speed of 3000 rev/min, calculate the difference in tension between the tight and slack sides of the drive belt.

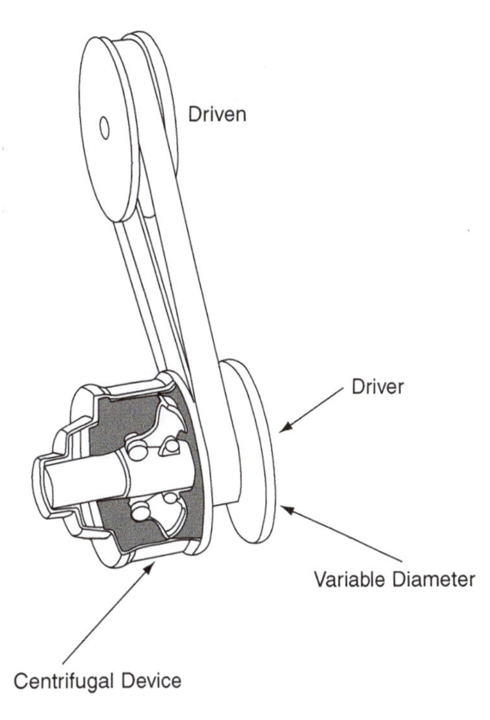

Fig. 9.17 Belt drive variable ratio

9.6 The wheel-fixing studs on a certain truck have a mean thread diameter of 20 mm and the pitch of the thread is 2 mm. Determine the torque required to produce a clamping force W of 2.6 kN when the coefficient of friction μ' is 0.18.

9.7 The recommended tightening torque for a hub nut on a front wheel drive axle shaft

is 205 N m with a lightly oiled thread. The thread has a mean diameter of 24 mm, a pitch of 2 mm and a thread angle of 60°. When oiled as specified the thread has a coefficient of friction of 0.1, but when the thread is dry the coefficient of friction is 0.15. Calculate:
(a) the axial force produced when the thread is oiled;
(b) the axial force produced when the thread is dry; and
(c) state the effect that the reduced axial force will have on wheel-bearing pre-load and wheel security.

9.8 A certain twin plate clutch has a mean diameter of 0.20 m. If the coefficient of friction of the linings and pressure plate/flywheel is 0.4, determine the minimum spring force that will be required to transmit 150 kW at an engine speed of 3000 rev/min.

9.9 If the coefficient of friction in a 75 mm diameter main bearing is 0.035, determine the power absorbed in friction when the load on the shaft is 500 N and the engine speed is 1800 rev/min.

9.10 A disc brake has two opposing pads that are operated by hydraulic pistons which are 50 mm in diameter. If $\mu = 0.4$ and the pads act at an effective radius of 0.140 m, calculate:
(a) the brake torque when the hydraulic pressure applied to the pistons is 20 bar;
(b) the braking force between the tyre and the road if the wheel and tyre diameter is 0.65 m.

9.11 During a static brake test the following braking force readings are obtained:

off-side front 2.1 kN
near-side front 2.05 kN
off-side rear 1.4 kN
near-side rear 1.38 kN.

(a) If the weight of the vehicle is 1.2 tonne, calculate the braking efficiency; take $g = 9.81 m/s^2$;
(b) Check the MOT testers manual to establish whether or not the side-to-side imbalance shown is acceptable.

10
Velocity and acceleration, speed

10.1 Speed and velocity

Speed is the rate at which distance is covered and speed can be specified by its size alone. Velocity is a more specific quantity because it possesses direction as well as size; velocity is said to be a vector quantity.

$$\text{velocity} = \frac{\text{distance covered}}{\text{time taken}}$$

10.2 Acceleration

Acceleration occurs whenever the velocity of an object changes. When the velocity of an object decreases it is said to have decelerated.

$$\text{acceleration} = \frac{\text{change in velocity}}{\text{time taken}}$$

10.3 Velocity–time graph

Uniform velocity

When an object travels at constant velocity of v m/s for a given period of time, t seconds, a graph of velocity against time produces a straight line, as shown in Figure 10.1.

The area enclosed by this graph is $v \times t$. As v is in metres/second and t is in seconds, the units of area cancel down to give metres.

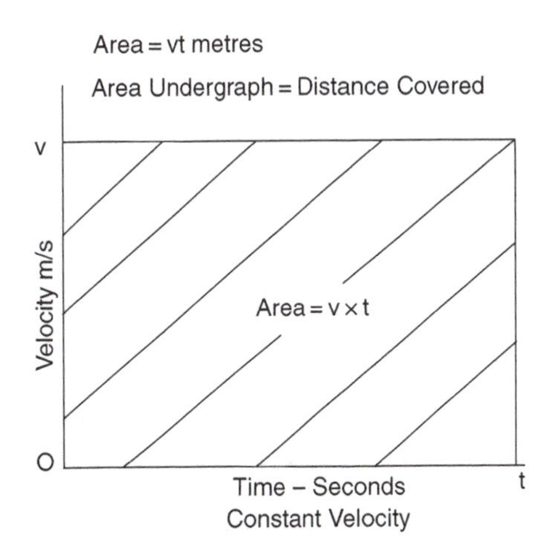

Fig. 10.1 Constant velocity

This leads to the useful fact that the area enclosed by a velocity–time graph is equal to the distance covered.

$$\text{distance} = \text{uniform velocity} \times \text{time in seconds}$$

Uniform acceleration

When an object is accelerated uniformly from an initial velocity of u m/s to a final velocity of v m/s the graph produced is a straight line of the type shown in Figure 10.2.

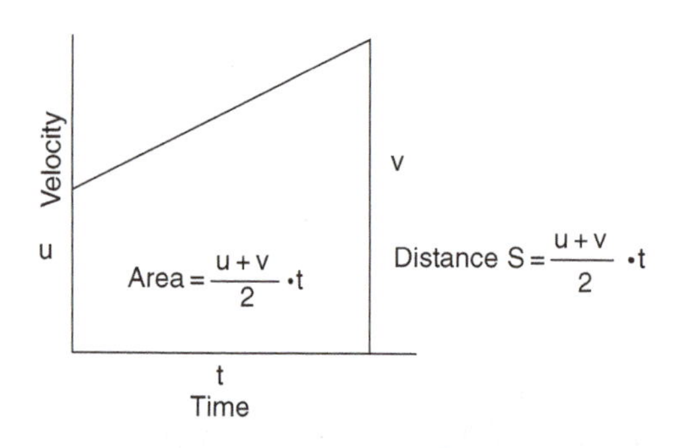

Fig. 10.2 Constant acceleration

The area enclosed by this graph is $\frac{(u+v)}{2} \cdot t$

But $\frac{u+v}{2}$ is the average velocity

The distance covered = average velocity × time in seconds.

These basic principles of velocity and acceleration are used to develop a set of formulae that are known as equations of motion.

10.4 Equations of motion and their application to vehicle technology

$$v = u + at$$

$$s = \frac{(u+v)t}{2}$$

$$v^2 = u^2 + 2as$$

$$s = ut + \frac{at^2}{2}$$

In these equations: u = initial velocity in m/s; v = final velocity in m/s; a = acceleration in m/s^2; t = time in seconds; s = distance.

To convert km/h to m/s: multiply by 1000 and divide by 3600.

Example 10.1

A vehicle travels at a constant speed of 50 km/h for 3 minutes; it then accelerates uniformly to a speed of 80 km/h in a period of 2 minutes. Draw a speed against time graph and determine the distance covered by the vehicle in the 5-minute period.

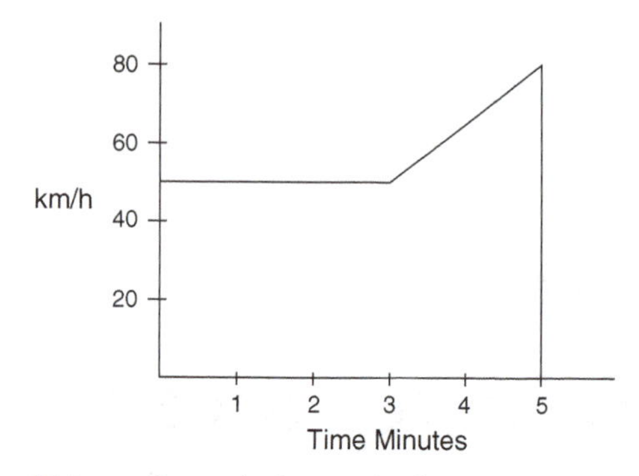

Distance Covered = Area under Graph

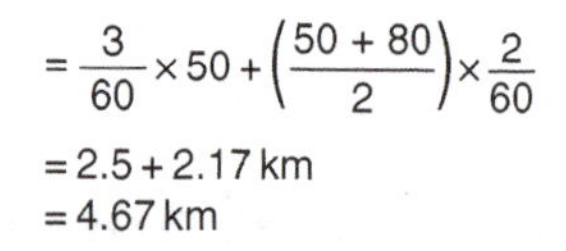

Fig. 10.3 Example 10.1

Solution (Figure 10.3)

Distance covered in first 3 minutes = $3/60 \times 50$ km = 2.5 km

Distance covered in next 2 minutes = $\frac{50+80}{2} \cdot 2/60 = 2.17$ km

Total distance in 5 minutes = 4.67 km

Example 10.2

A vehicle travelling at 72 km/h is braked uniformly to a speed of 28 km/h in a time of 10 seconds. Draw a speed–time graph and determine the distance that the vehicle travels while it is slowing down.

Solution (Figure 10.4)

Distance covered while slowing down = area under speed–time graph

$$\text{Distance covered} = \frac{72+28}{2} \cdot \frac{1000}{3600} \times 10 = 139\,\text{m}$$

Example 10.3

A vehicle travelling at 60 km/h is braked to a standstill by a retardation of 24 m/s^2. Use the

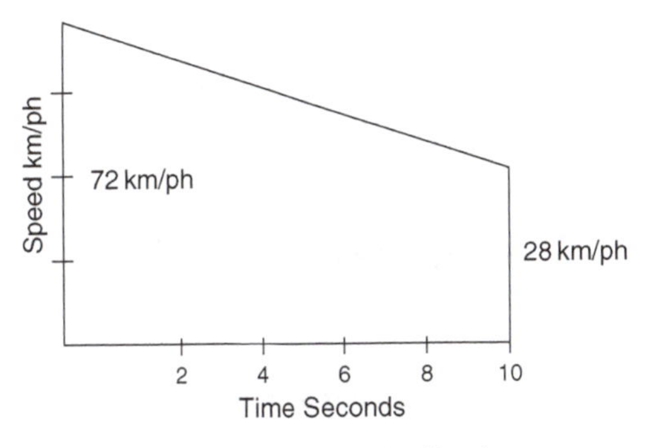

Fig. 10.4 Example 10.2

equation $v = u + at$ to determine the time taken to stop the vehicle.

Solution

$u = 60 \times 1000/3600 = 16.7\,\text{m/s}$
$a = -24\,\text{m/s}^2$
Final velocity $v = 0$
$v = u + at$

$$\therefore t = \frac{(v-u)}{a}$$

$$t = \frac{0 - 16.7}{-24}$$

$t = 0.7\,\text{s}$

10.5 Force, mass and acceleration

Newton's laws of motion

The basic laws used in the study of motion are those that were first formulated by Sir Isaac Newton in the seventeenth century. These laws are as follows.

The ***law of inertia*** states that a body will remain in a state of rest or uniform motion in a straight line until it is compelled to change that state by being acted upon by an external force.

The ***law of change of momentum*** states that the rate of change of momentum is equal to the applied force.

The ***law of reaction*** states that every action has an equal and opposite reaction.

The basic unit of force is the newton; 1 newton is the magnitude of the force that will produce an acceleration of 1 m/s^2 when applied to a mass of 1 kg.

The relationship between force, mass and acceleration is given in the following equation.

$$F = ma$$

where F = force in newtons, m = mass in kg, a = acceleration in m/s^2.

10.6 Relation between mass and weight

The weight of an object is the amount of force that is exerted on the object by the gravity of the earth. If an object falls freely it will accelerate at the gravitational acceleration of 9.81 m/s^2. The gravitational acceleration is known as g.

The weight of an object = mass in kilograms × gravitational constant.

This is normally written as weight = $m \times g$, or mg.

In simple calculations the approximation of $g = 10\,\text{m/s}^2$ is sometimes used.

10.7 Inertia

Inertia is the sluggishness or resistance that a body offers to starting from rest or to a change of velocity.

Example 10.4

A motorcycle and its rider increase speed from 30 km/h to 80 km/h in a time of 5 seconds.

Calculate the acceleration in m/s^2. If the rider weighs 52 kg what force will be felt by the rider's body?

Solution

Initial velocity $u = 30 \times 1000/3600 = 8.33\,m/s$
Final velocity $v = 80 \times 1000/3600 = 22.22\,m/s$
$t = 5$ seconds
$v = u + at$

$$\therefore a = \frac{v-u}{t} = \frac{(22.22 - 8.33)}{5}$$

$$= 13.89/5 = 2.78\,m/s^2$$

Force felt by rider $= m \times a$
$= 52 \times 2.78$
$= 144.6\,N$

10.8 Motion under gravity

When a body is acted on by the attractive force of the earth and drawn towards the earth's centre the body is said to be moving under gravity. It can be shown that all bodies that are small in comparison with the earth, whatever their mass, are influenced in practically the same manner by this force of gravity. The acceleration that is caused by the force of gravity varies slightly across the earth's surface. A figure for gravitational acceleration that is used as an average is $\mathbf{g = 9.8\,m/s^2}$ and this is known as the gravitational constant.

Example 10.5

A vehicle falls off the edge of a viaduct 50 m high. How long will it take to reach the ground?

Solution

The initial velocity $u = 0\,m/s$
The acceleration $a = g = 9.81\,m/s^2$
The distance $s = 50$ metres
Using the equation $s = ut + \frac{at^2}{2}$

$$50 = 0 \times t + \frac{9.81\,t^2}{2}$$

$$100 = 9.81\,t^2$$

$$t = \sqrt{(100/9.81)}$$

$$t = 3.2 \text{ seconds}$$

Time taken to reach the ground = 3.2 seconds

10.9 Angular (circular) motion

Revolutions per minute is a term that is commonly used when quoting speeds of rotation of objects such as shafts. In engineering science, revolutions per minute is replaced by radians per second.

A radian is 57.3 degrees and there are 2π radians in one revolution.

To convert rev/min to radians per second, a speed in rev/min is multiplied by 2π and divided by 60.

For example a shaft rotating at 2400 rev/min has an angular velocity ω:

$$\omega = \frac{2400 \times 2 \times 3.142}{60} = 251.4 \text{ radians per second.}$$

The symbol ω (omega) is used to denote angular velocity.

10.10 Equations of angular motion

The symbol θ (theta) is used to denote total angle turned through. It is analogous to s in the linear motion equations. The symbol α (alpha) is used to denote angular acceleration in rad/s^2; it is analogous to a in the linear motion equations. ω_1 is analogous to u, and $= \omega_2$ is analogous to v in the linear equations of motion.

The equations of angular motion are:

$$\theta = \frac{\omega_1 + \omega_2}{2} \cdot t$$

$$\omega_2 = \omega_1 + \alpha t$$

$$\theta = \omega_1 \cdot t + {}^1/_2 \alpha t^2$$

$$(\omega_2)^2 = (\omega_1)^2 + 2\alpha\theta$$

Example 10.6

The speed of an engine increases from 1200 rev/min to 3600 rev/min in 0.2 seconds. Calculate the angular acceleration of the crankshaft in radians/s^2.

Solution

Use $\omega_2 = \omega_1 + \alpha t$

Transposing gives $\alpha = (\omega_2 - \omega_1)/t$

$$\omega_2 = \frac{3600 \times 2\pi}{60} = 377\,\text{rad/s}$$

$$\omega_1 = \frac{(1200/60) \times 2\pi}{60} = 125.7\,\text{rad/s}$$

$$\alpha = (\omega_2 - \omega_1)/t$$

$$\therefore \alpha = (377 - 125.7)/0.2$$

$$\alpha = 1257\,\text{rad/s}^2$$

10.11 Relation between angular and linear velocity

Figure 10.5(a) shows an object at point P that is rotating at velocity v m/s in an anti-clockwise direction at a radius r, about point O.

Point P moves round the circumference of a circle to point Q in a time of t seconds.

The angle turned through is θ radians.

The arc from P to Q has a length of $r\theta$ metres and this is the distance that P moves in t seconds.

$$\text{linear velocity} = \frac{\text{distance moved}}{\text{time in seconds}} = \frac{r\theta}{t},$$

but $\frac{\theta}{t} = \omega$ the angular velocity

This gives linear velocity $\boldsymbol{v = \omega r}$.

10.12 Centripetal acceleration

Velocity has both magnitude and direction and, by definition, any change of velocity is accompanied by an acceleration. When an object is moving in a circular path, as shown in Figure 10.5(a) the velocity is constantly changing. The vector diagram in Figure 10.5(b) shows the change of velocity U that occurs when P moves to Q.

Let the change of velocity be u m/s

The acceleration of P $= \frac{\text{change of velocity}}{\text{time taken}} = \frac{u}{t}\ \text{m/s}^2$

If t and θ are made very small, the sector OPQ may be regarded as a triangle; this triangle is similar to the triangle formed by the vector diagram ACB.

The distance moved by P $= vt$

From this $\frac{AC}{AB} = \frac{r}{vt}$

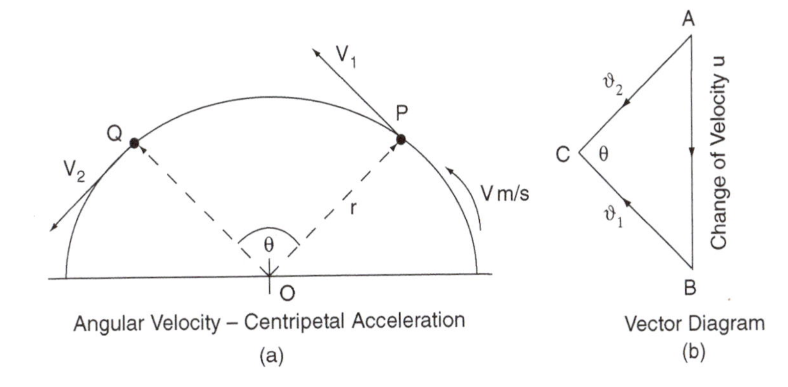

Fig. 10.5 Centripetal acceleration

But AC = v, and AB = u,

$$\therefore \frac{v}{u} = \frac{r}{vt}$$

Cross-multiplying gives

$$\frac{v^2}{r} = \frac{u}{t}$$

But $\frac{u}{\text{t}}$ = acceleration of P

$$\text{Acceleration} = \frac{u}{t} = \frac{v^2}{r}\ \text{m/s}^2$$

This acceleration is called centripetal acceleration and it gives rise to centripetal force, the reaction to which is called centrifugal force.

In terms of angular acceleration, the centripetal acceleration = $\omega^2 r$, because $v = \omega r$

$$\text{Centripetal force} = \text{mass} \times \text{acceleration}$$

$$= m\frac{v^2}{r} = m\omega^2 r$$

Centrifugal force is the reaction to centripetal force; it is therefore equal in magnitude but opposite in direction.

Example 10.7

A truck of mass = 8 tonnes is travelling at 48 km/h round a curve in the road that has a radius of 100 m. Calculate the centrifugal force (CF) acting at the centre of gravity of the truck.

Solution

$$\text{Centripetal force} = \frac{mv^2}{r}$$

$$\text{v} = 48 \times 1000/3600 = 13.33\ \text{m/s}$$

$$\text{CF} = \frac{2000 \times 13.33 \times 13.33}{100} = 200 \times 177.7$$

$$= 35547\ \text{N} = 35.55\ \text{kN}$$

$$= 200 \times 177.7$$

$$= 35547\ \text{N}$$

$$= 35.55\ \text{kN}$$

Centrifugal force acting at the centre of gravity = 33.55 kN

Example 10.8

A diesel pump governor weight has a mass of 0.4 kg. The centre of gravity of the weight is 25 mm from the centre of the pump shaft. If the pump shaft rotates the governor at 1800 rev/min determine the force exerted by the governor weight.

Solution

Centrifugal force = $m\omega^2 r$

$$\omega = \frac{\text{rev/min}}{60} \times 2\pi = 1800/60 \times 6.248$$

$$= 187.4\ \text{rad/s}$$

$$\text{CF} = 0.4 \times 187.4 \times 187.4 \times 25/1000$$

Centrifugal force at the governor weight = 351.2 N

Accelerating torque

The torque required to accelerate a rotating mass is analogous to the force required to accelerate an object in linear motion. For example, the torque required to accelerate a flywheel is dependent on the mass of the flywheel and its radius of gyration.

If the mass of the flywheel is m kg and the radius of gyration is k, the torque T required to accelerate the flywheel at α radians/s^2 is

$$T = mk^2\alpha$$

But $mk^2 = I$, where I is the moment of inertia of the flywheel,

$$\therefore T = I\alpha \text{ (compare this with } F = m.a)$$

Example 10.9

A flywheel has a mass of 6 kg and a radius of gyration of 100 mm. Calculate the torque exerted on the flywheel when the engine accelerates from 1200 rev/min to 4800 rev/min in a period of 0.5 s.

Solution

Moment of inertia of the flywheel $I = m.k^2$

The radius of gyration $k = 0.1$ m

$\therefore I = 6 \times 0.1 \times 0.1 = 0.06\ \text{kg m}^2$

Angular acceleration

$$\alpha = \frac{\text{change of angular velocity}}{\text{time taken}}$$

$$\alpha = \frac{(4800 - 1200) \times 2\pi}{(60 \times 0.5)} = 754$$

$$\alpha = 754\,\text{rad/s}$$

$$T = I\alpha$$

$$T = 0.06 \times 754$$

Torque $T = 45.2\,\text{N m}$

10.13 Exercises

10.1 Convert the following into m/s: 10 km/h; 100 km/h; 56 km/h; 64 mile/h.

10.2 At a given instant a car is moving with a velocity of 5 m/s and a uniform acceleration of 1.2 m/s^2.What will the velocity be after 10 s?

10.3 A bullet is travelling with a velocity of 300 m/s, hits a target, and penetrates it by a distance of 100 mm. Assuming retardation is constant, (a) determine the time taken for the target to stop the bullet, and (b) determine the retardation of the bullet.

10.4 The maximum retardation that the brakes of a certain truck can produce is 7 m/s^2. If the truck is travelling at 80 km/h, what is the shortest distance in which the truck can be brought to a standstill?

10.5 With what initial velocity must a body be travelling if it is brought to rest in a distance of 35 m with a uniform retardation of 2.4 m/s^2?

10.6 Convert the following speeds to metres per second (take 0.625 mile = 1 km): (a) 30 mile/h; (b) 90 km/h; (c) 205 km/h.

10.7 A truck passes a certain kilometre post with a velocity of 20 m/s and the next kilometre post with a velocity of 35 m/s. What is the average velocity of the truck? How many seconds will it take to cover the distance between the kilometre posts? What is the magnitude of the acceleration of the truck?

10.8 A motorist travelling at steady speed of 18 km/h in a car experiences engine failure. From the point at which the engine cuts out, the motorist decides to coast and the vehicle retards uniformly. In the first 30 seconds the vehicle covers 110 metres and continues to slow down and then stops after a further 30 seconds. Sketch the speed–time graph and then determine the distance.

10.9 A car travelling at 36 km/h is decelerated uniformly at 3.5 m/s^2 until the speed reaches 21 km/h. Determine the time taken to reduce the speed and the distance covered during the retardation period.

10.10 A vehicle starts from rest and accelerates at 8 m/s^2 until the speed is 72 km/h. It continues at this speed until the brakes are applied to stop the vehicle in a distance of 60 m. If the total time from start to stop is 25 seconds, calculate the distance covered while the vehicle is travelling at 72 km/h.

10.11 A vehicle accelerates at 8 m/s^2 for 10 seconds, from a standing start. Calculate the velocity of the vehicle at the end of the 10 second period.

10.12 An engine runs at 3600 rev/min. Calculate its angular velocity in radians/s.

10.13 The speed of an engine increases from 2000 rev/min to 4500 rev/min in a time of 0.5 seconds. Determine the angular acceleration in radians/s^2.

10.14 Calculate the angular velocity in rad/s of a wheel that has of rolling diameter of 0.8 m when the vehicle to which the wheel is fitted is travelling at 72 km/h.

10.15 An out-of-balance weight of 30 grams acts at a radius of 0.3 m on a tyre. The tyre has a rolling circumference of 2.5 m. Calculate the force acting on the weight when the vehicle is travelling at 72 km/h.

10.16 The driver of a certain formula 1 car takes a corner of radius 175 m at a speed of 160 km/h. If the driver weighs 68 kg calculate the force exerted on the driver's body.

11 Vehicle dynamics

11.1 Load transfer under acceleration

When a vehicle accelerates, the force causing acceleration is opposed by a force of equal magnitude acting horizontally through the centre of gravity of the vehicle. The opposing force is known as the inertia force and because it acts at height of h metres above the datum (ground level) it exerts a turning moment that tends to lift the front of the vehicle. The lifting effect reduces the load on the front axle and increases the load on the rear axle by the same amount.

There is an apparent transfer of load from the front to the rear axle.

The principle of moments can be used to determine the amount of load transfer.

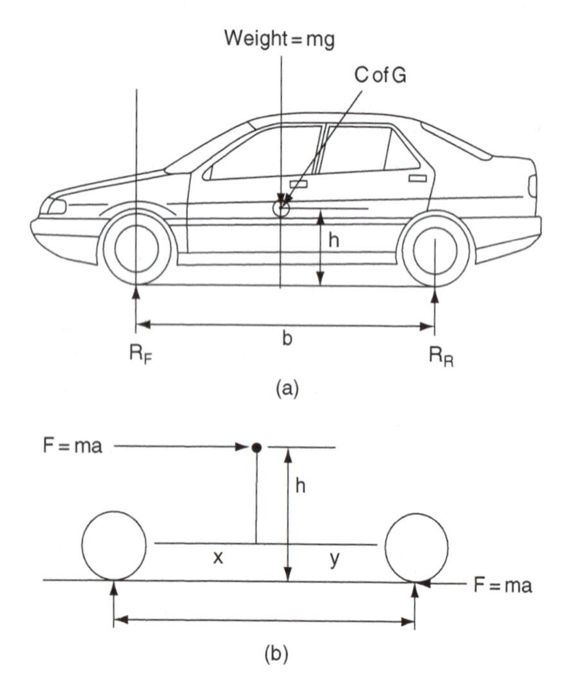

Fig. 11.1 Load transfer

11.2 Static reactions

Figure 11.1(a) shows a vehicle of mass m kilograms.

The centre of gravity is h metres above ground level.

The wheel base = b metres.

The centre of gravity is x metres behind the front axle and y metres in front of the rear axle.

The static reaction forces are R_f kg at the front axle and R_r kg at the rear axle.

By considering the vehicle as a skeleton beam, as shown in Figure 11.1(b):

$$R_f = mg\frac{y}{b} \quad \textbf{(11.1)}$$

$$R_r = mg\frac{x}{b} \quad \textbf{(11.2)}$$

11.3 Vehicle under acceleration

Assume the vehicle to be accelerating at a m/s^2

The inertia force $F = ma$

The front reaction R_f is now the dynamic reaction.

Take moments about the point of contact of the rear tyre on the road:

$$\text{CWM} = \text{ACWM}$$

$$R_f b + Fh = mgy$$

$$R_f b = mgy - Fh$$

$$R_f = \frac{mgy}{b} - \frac{Fh}{b} \qquad \textbf{(11.3)}$$

In this expression the term Fh/b is the apparent load transfer. Under acceleration the apparent load transfer Fh/b is effectively placed on the rear axle and this causes the front end of the vehicle to rise; under braking conditions the apparent load transfer is effectively placed on the front axle and this causes the front end of the vehicle to dip.

Example 11.1 (see Figure 11.2)

A vehicle of mass = 1200 kg has a wheelbase of 2.5 m. The centre of gravity of the vehicle is at a height of 0.65 m above ground level and 1.2 m behind the front axle.

If the vehicle accelerates at 5 m/s^2 calculate:

(a) the amount of load transfer;
(b) the dynamic loads on the front and rear axles.

Take $g = 9.81$ m/s^2.

Solution

First determine load transfer $\frac{Fh}{b}$

$$F = ma$$

$$F = 1200 \times 5$$

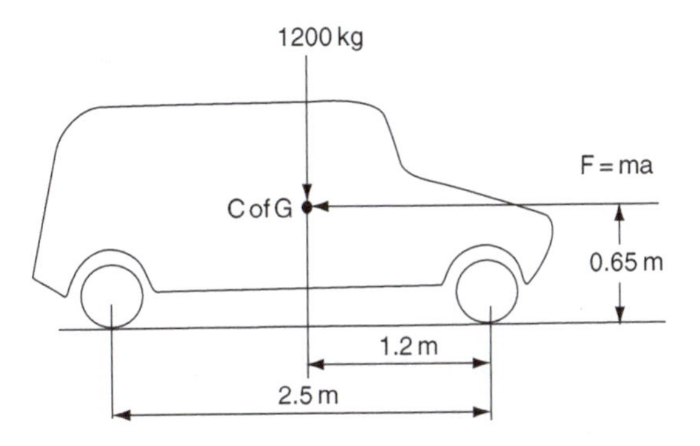

Fig. 11.2 Calculation of load (Example 11.1)

Force acting at centre of gravity $F = 6000$ N

$$\text{Load transfer} = \frac{Fh}{b} = \frac{6000 \times 0.65}{2.5} = 1560\,\text{N}$$

Next find the static reactions R_f and R_r

Take moments about rear axle

$$\text{CWM} = \text{ACWM}$$

$$(1200 \times 9.81) \times 1.3 = R_f \times 2.5$$

$$R_f = \frac{1200 \times 9.81 \times 1.3}{2.5}$$

$$R_f = 6121\,\text{N}$$

To find R_r, use total downward force = total upward force

$$1200 \times 9.81 = R_f + R_r$$

$$R_r = 11\,772 - 6121$$

$$R_r = 5651\,\text{N}$$

The dynamic loads are:

Front axle dynamic load

= static load − load transfer

= 6121 − 1560

= 4561 N

Rear axle dynamic load = static load + load transfer

$= 5651 + 1560$

$= 7211\,\text{N}$

11.4 Vehicle acceleration – effect of load transfer

Front wheel drive

The maximum tractive effort Te that can be applied is governed by the coefficient of friction μ between the tyre and the driving surface.

The coefficient of friction

$$\mu = \frac{\text{tractive effort}}{\text{force pressing the surfaces together}}$$

This means that

$$\text{Te} = \mu \times \text{force pressing surfaces together}$$

In this case the force pressing the surfaces together is the vertical reaction at the point of contact of the tyre on the driving surface which, in the case of front wheel drive, is:

From Equation (11.3), static reaction at the front wheels – load transfer $= R_F - Fh/b$

Maximum tractive effort for front wheel drive:

$$\text{Te} = \mu(R_F - Fh/b) \tag{11.4}$$

Example 11.2

A vehicle of weight = 1.8 tonnes has wheelbase of 2.6 m and its centre of gravity is 0.7 m above ground level and midway between the axles. If the coefficient of friction between the tyres and the road is 0.7, calculate the maximum acceleration that is possible on a level surface:

(a) when the vehicle has front wheel drive;
(b) when the vehicle dimensions are the same but rear wheel drive is used.

Take $g = 9.81\,\text{m/s}^2$.

Solution: (a) front wheel drive

The most direct way to tackle this problem is to equate the propelling force (tractive effort Te) to the maximum force that can be applied at the front wheels.

As shown in Equation (11.4) the maximum force (tractive effort) is dependent on axle load and friction.

From Equation (11.4),

$\text{Te} = \mu(R_F - Fh/b)$

The tractive effort Te is equal and opposite to the inertia force that acts at the centre of gravity of the vehicle.

If the maximum possible acceleration is a m/s^2, the inertia force $F = ma$, and this is the F in the load transfer term, Fh/b.

By this reasoning we may say that:

$\text{Te} = ma = \mu(R_F - Fh/b)$

By writing ma in place of F inside the bracket, this expression becomes

$$ma = \mu R_F - \mu mah/b \tag{11.5}$$

The numerical values of m, R_F, μ and h and b are known and these may now be substituted for the corresponding symbols. Mass of vehicle $m = 1800\,\text{kg}$; coefficient of friction $\mu = 0.7$; height of centre of gravity $h = 0.7\,\text{m}$; wheelbase $b = 2.6\,\text{m}$.

The centre of gravity is midway between the axles so the weight of the vehicle is evenly divided between the two axles, which means that

$$R_f = \frac{1800 \times 9.81}{2} = 8829\,\text{N}$$

Substituting these values in the equation $ma = \mu R_F - \mu m\,ah/b$ gives

$$1800\,a = 0.7 \times 8829 - \frac{0.7 \times 1800 \times 0.7\,a}{2.6}$$

$$1800\,a = 6180 - 339\,a$$

$$1800\,a + 339\,a = 6180$$

$$2139\,a = 6180$$

$$a = \frac{6180}{2139}$$

$$a = 2.89\,\text{m/s}^2$$

Max. acceleration with front wheel drive

$$= 2.89\,\text{m/s}^2$$

Maximum acceleration – rear wheel drive

Solution (b) rear wheel drive

When considering the maximum acceleration possible in rear wheel drive it is important to remember that the load transfer is added to the rear axle static load.

In this case, maximum tractive effort

$$\text{Te} = \mu(R_R + Fh/b) \qquad \textbf{(11.6)}$$

By using the same method as that used to determine the max acceleration in front wheel drive the maximum acceleration of the vehicle when driven by the rear wheels may be determined as follows:

$$ma = \mu R_R + \mu mah/b$$

Note that the load transfer is now added to the rear axle.

This gives $1800a = 0.7 \times 8829$

$$+ \frac{0.7 \times 1800 \times 0.7 \times a}{2.6}$$

$$1800a - 339a = 6180$$

$$1461a = 6180$$

$$a = \frac{6180}{1461}$$

$$a = 4.23\,\text{m/s}^2$$

The maximum acceleration of the vehicle in rear wheel drive $= 4.23\,\text{m/s}^2$

Four wheel drive – fixed

The drive is now transmitted through all four points of contact of the tyres on the driving surface; the force pressing down on the surface is the weight of the vehicle mass × gravitational constant mg. The maximum tractive effort $\text{Te} = \mu mg$.

Four wheel drive – with third differential

The purpose of the third differential is to balance the driving effort between the front and rear axles. The simple differential transmits equal torque front to rear. This means that the maximum torque, and therefore tractive effort Te, is limited by the amount of torque at the axle with the least frictional grip. This problem is overcome by the use of a differential lock. However, the simple differential lock affects the handling of road vehicles. To manage vehicle handling of road-going four-wheel drive vehicles, various forms of differential locks and electronic control are used.

11.5 Accelerating force – tractive effort

Tractive effort is the force that is required to propel the vehicle against all resistance to motion.

11.6 Tractive resistance

The resistance to motion is known as the tractive resistance and it consists of several elements that vary according to operating conditions (Figure 11.3). The principal elements of tractive resistance are:

Rolling resistance – this accounts for approximately 40% of the total resistance to motion for a medium size family saloon at a speed of 90 km/h (56 mile/h) on a level road. A large proportion of rolling resistance is attributable to tyres and the type of road surface.

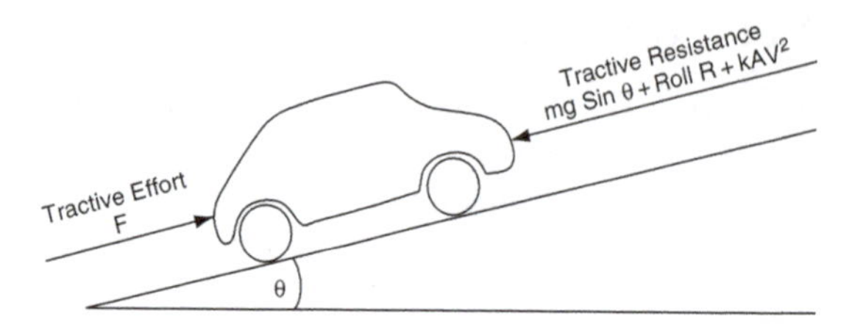

Fig. 11.3 Tractive resistance

Wind, or air, resistance – wind resistance is dependent on vehicle design and it varies as the square of vehicle speed. For a family saloon travelling at 90 km/h on a level road, wind resistance accounts for approximately 60% of the total resistance to motion. The wind resistance is calculated by using the formula $Rw = kAv^2$, where k is a constant that is derived from the aerodynamics of the vehicle, A = the effective area of the vehicle facing the wind, and v is the vehicle's velocity.

Gradient resistance – when a vehicle is climbing a slope (gradient), extra tractive effort is required to overcome the effect of gravity. This extra tractive effort $F = mg \sin\theta$, where m = mass of vehicle in kg, g = gravitational constant, and θ = angle of the gradient in degrees.

Inertia – when accelerating additional tractive effort is required to overcome inertia. The extra tractive effort for acceleration $F = ma$, where m is the mass of the vehicle in kg and a = acceleration in m/s^2.

11.7 Power required to propel vehicle

$$\text{power of a vehicle } P = \frac{\text{work done}}{\text{time taken in seconds}} = \frac{\text{force} \times \text{distance}}{\text{time taken in seconds}}$$

$$\text{But velocity } v = \frac{\text{distance moved}}{\text{time taken in seconds}}$$

This means that power of vehicle P can be expressed as $P = Fv$, where F = propelling force in newtons and v = velocity in m/s.

$$\text{power of vehicle} = \text{force} \times \text{velocity}$$

$$P = Fv$$

Power available

The power available at any given road speed is the power that is delivered to the driving wheels. The power available at the driving wheels is equal to engine power at an engine speed that relates to the road speed multiplied by the overall transmission efficiency.

The power curve at the driving wheels has the same shape as the engine power curve. When the power available and the power required to overcome resistance are plotted on the same graph, as shown in Figure 11.4, a number of features are revealed. The following two of these features are of interest at this point:

1. The point at which the two curves meet represents the maximum speed that the vehicle can reach on a level road.

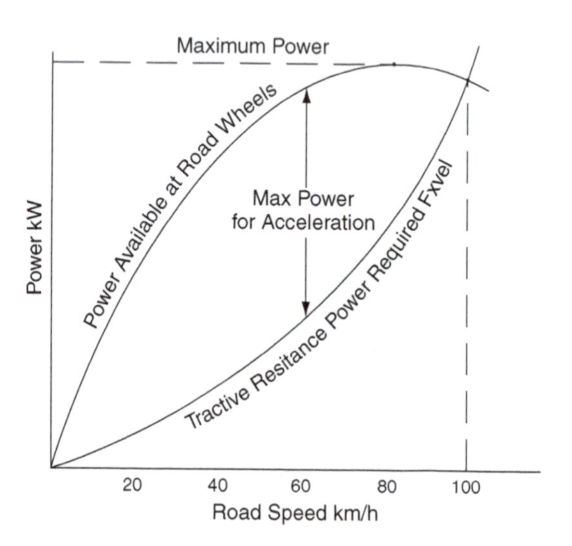

Fig. 11.4 Graph of power required vs. power available

2. The maximum vertical distance between the two curves is the point where the power available is greater than the power required by the largest margin and this is the point at which the maximum amount of power is available for acceleration.

Example 11.3

The figures given in Table 11.1 relate to a motor vehicle travelling on a level road.

Plot this data on a base of road speed and use the graph (Figure 11.5) to determine:

(a) the maximum amount of power available for acceleration;
(b) the velocity in m/s at which this acceleration occurs;
(c) the value of the acceleration if the vehicle weighs 1.2 tonne.

Solution

The following steps are recommended as the approach to the solution of this question.

1. Plot the graph using suitable scales.
2. Count the squares at the point where the curves are furthest apart.
3. Use the scale to convert the number of squares to kW, in this case 18.2 kW.
4. Read off the speed at which maximum power for acceleration is available.

Table 11.1 Vehicle travelling on level road (Example 11.3)

Road speed (km/h)	**15**	**30**	**45**	**60**	**75**	**90**
Power required to overcome resistance to motion (kW)	1.12	2.8	5.4	9.5	15.6	24.25
Power available at road wheels (kW)	4.9	12.6	20.3	27.6	33.3	36.25

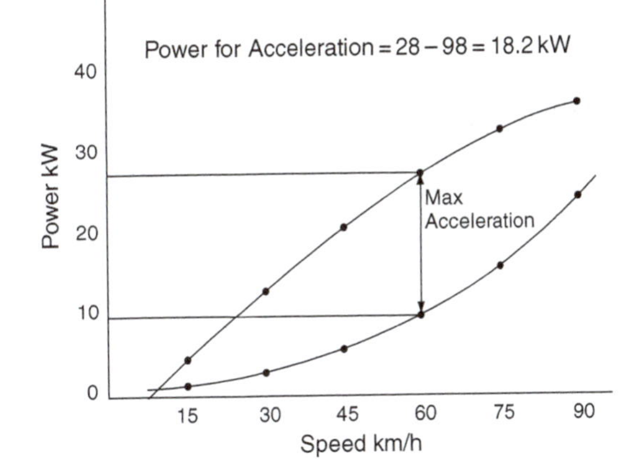

Fig. 11.5 Graph of power vs. speed (Example 11.3)

5. Convert the speed in km/h to m/s; in this case

$$60\,\text{km/h} = \frac{60 \times 1000}{3600} = 16.7\,\text{m/s}$$

6. Use power in kW = force × velocity to find the force available for acceleration.

In this case the power available for acceleration = 18.2 kW; the velocity at which it occurs is 16.7 m/s.

$$P = Fv$$

where P is power in kW and v = velocity in m/s; F = force in newtons

Transposing gives:

$$F = \frac{P}{v} = \frac{18.2 \times 1000}{16.7} = 1090\,\text{N}$$

7. Use $F = ma$ to find the acceleration.

Transposing gives $a = \dfrac{F}{m} = \dfrac{1090}{1200} = 0.91\,\text{m/s}^2$

Answers

(a) maximum power available for acceleration = 18.2 kW
(b) the velocity = 16.7 m/s
(c) the acceleration = $0.91\,\text{m/s}^2$

11.8 Forces on a vehicle on a gradient – gradient resistance

When a vehicle is climbing a gradient, additional force is required to overcome the effect of gravity. This additional force is known as the gradient resistance. In order to determine the amount of gradient resistance it is necessary to resolve the forces acting on the vehicle. This is achieved with the aid of a force diagram, as shown in Figure 11.6(b).

The weight of the vehicle mg acts vertically downwards through the centre of gravity of the vehicle.

The component of the weight that attempts to make the vehicle run back down the gradient also acts through the centre of gravity of the vehicle. This is the gradient resistance and it is the force that must be overcome if the vehicle is to move at uniform speed up the gradient.

The force diagram shows that the gradient resistance $= mg \sin\theta$, where θ is the angle of the gradient in degrees, m is the mass of the vehicle in kg and g is the gravitational constant.

$$\text{Gradient resistance} = mg \sin\theta \qquad \textbf{(11.7)}$$

The other force that acts normal (at right angles) to the gradient has a magnitude of $mg \cos\theta$ and this is used in other calculations.

Example 11.4

Calculate the force (tractive effort) that is required to overcome the gradient resistance of a vehicle weighing 3.5 tonnes on a gradient of 18°. Take $g = 9.8\,\text{m/s}^2$.

Solution

Force, F, required to overcome gradient resistance $= mg \sin\theta$

$$F = 3500 \times 9.8 \times 0.309\,\text{N}$$
$$= 10\,599\,\text{N}$$
$$= 10.6\,\text{kN}$$
$$F = \text{tractive effort} = 10.6\,\text{kN}$$

11.9 Gradeability

Gradients are often measured in percentages, as shown on signs at the roadside. A gradient of 10% means that the average amount that the road rises is 10 m for every 100 m covered horizontally. The ability of a vehicle to ascend a gradient is normally

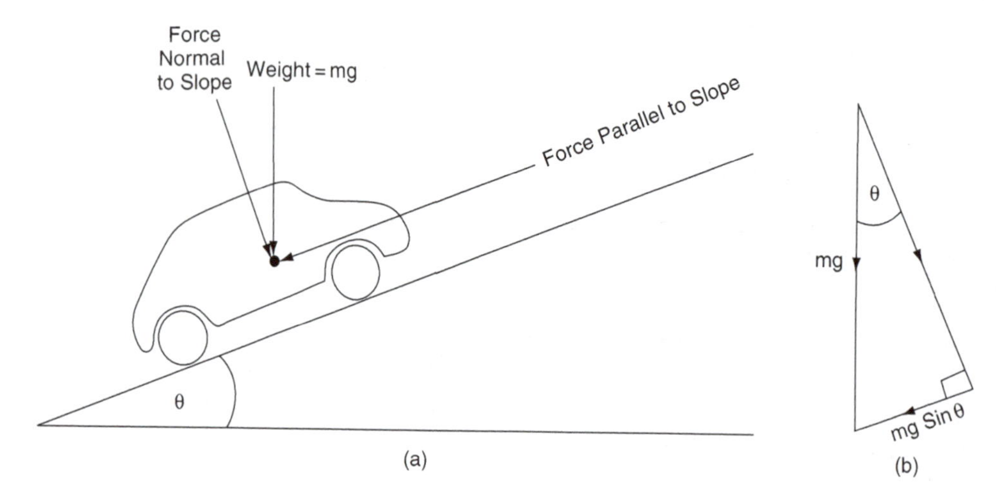

Fig. 11.6 Forces acting on vehicle on a gradient

expressed as a percentage. Table 11.2 shows the gradeabilty figures for a large vehicle.

The actual method of determining gradeability appears not to be standardised. However, the figure is vitally important to road hauliers and a transport engineering technician may well be called upon to assess any claims that may be made about the performance of a particular vehicle. The following example shows how this assessment of the grade ability may be made.

Example 11.5

The figures in Table 11.2 state that the maximum restart gradient is 15% for the 6.98 litre engine vehicle with an axle ratio of 4.62. Other details state that the gross vehicle weight (GVW) is 12.5 tonnes. The maximum torque for the 6.98 litre naturally aspirated engine is 340 Nm at 1650 rev/min. The tyres are 10R 22.5, which have a loaded rolling radius of 0.494 m.

Given that the gearbox ratio in the lowest forward gear is 8.97:1 and that the overall transmission efficiency is 93% in this gear, calculate the maximum theoretical gradient that can be climbed from a standing start, assuming that the maximum torque can be applied to the driving wheels. Compare this figure with the 15% quoted. Take $g = 9.81\,\text{m/s}^2$.

Solution

The torque available at the driving wheels T = engine torque × overall gear ratio × transmission efficiency.

$$T = 340 \times 4.62 \times 8.97 \times 0.93$$

$$= 13104\,\text{Nm}$$

The rolling radius of the tyres = 0.494 m

The wheel torque $T = F \times r$

$$F = \frac{T}{r}$$

$$= \frac{13\,104}{0.494}$$

$$F = 26526\,\text{N}$$

This force F is the theoretical maximum force available to overcome the gradient resistance.

The gradient resistance is also $= mg \sin\theta$.

$$\text{Force } F = mg \sin\theta$$

Angle θ is required so the next step is to transpose this equation to give

$$\sin\theta = \frac{F}{mg}$$

$$\sin\theta = \frac{26\,526}{12\,500 \times 9.81}$$

$$\sin\theta = 0.216$$

Using the $\sin^{-1}$ function on the calculator gives angle $\theta = 12.5°$.

The tangent of 12.5° is 0.22, and this corresponds to a gradient of 22%.

The quoted figure of 15% for the restart grade ability is thus well within the maximum theoretical figure.

Table 11.2 Gradeability

Engine type	Gearbox type	Axle ratio	Max. geared speed (km/h)	Max. climb gradient	Max. restart gradient
6.98 l	5 speed	4.62	105	16%	15%
		5.14	94	18%	17%
		5.57	87	19%	18%
		6.14	79	21%	20%
6.98 l turbo	6 speed	4.62	100	26%	25%

11.10 Vehicle power on a gradient

In order to proceed at uniform speed up a gradient the vehicle must have sufficient power to overcome all resistances opposing motion. These amount to the sum of wind resistance, rolling resistance and gradient resistance. These forces are shown in Figure 11.3.

The total resistance to motion on the gradient $= mg\ \sin\theta + \text{roll resistance} + kAv^2$

Example 11.6

Determine the power required to drive a vehicle weighing 1.5 tonnes up a gradient of 8° at a steady speed of 72 km/h. The air resistance is $0.08\,\boldsymbol{v}^2$ where $\boldsymbol{v}$ is the speed of the vehicle in km/h and the rolling resistance is 250 N/tonne of vehicle weight.

Take $g = 9.81\text{m/s}^2$.

Solution

$$\text{gradient resistance} = mg \sin\theta$$
$$= 1500 \times 9.81 \times \sin 8°$$
$$= 1500 \times 9.81 \times 0.14$$
$$= 2060\,\text{N}$$

$$\text{rolling resistance} = 1.5 \times 250$$
$$= 375\,\text{N}$$

$$\text{air resistance} = 0.08 \times 72 \times 72$$
$$= 415\,\text{N}$$

$$\text{total resistance} = 2060 + 375 + 415$$
$$= 2850\,\text{N}$$

The total force F_t (tractive effort) required to drive vehicle up the gradient is equal and opposite to the resisting force.

$$F_t = 2850\,\text{N}$$

Power required P = tractive effort × velocity of vehicle in m/s

$$P = 2850 \times 72 \times \frac{1000}{3600}$$
$$P = 2850 \times 20$$
$$= 57\,000\,\text{W}$$
$$= 57\,\text{kW}$$

Power required to drive the vehicle up the gradient at $72\,\text{km/h} = 57\,\text{kw}$.

11.11 Vehicle on a curved track

When a vehicle is moving round a curve, such as a corner on the road, a number of factors affect the speed at which the vehicle can proceed. The two factors that are considered here are: (1) maximum speed before overturning occurs; (2) maximum speed before skidding occurs.

Overturning speed

Figure 11.7(a) shows a vehicle proceeding around a left hand curve, away from the viewer, at a velocity of $v\,\text{m/s}$. Figure 11.7(b) shows the same vehicle represented by a skeleton framework.

Let m = mass of vehicle in kg; g = gravitational constant; h = height of centre of gravity of the vehicle above ground level; d = track width in metres.

The forces acting on the vehicle are:

mg – the weight of the vehicle pressing down through the centre of gravity;

R_A and R_B – the normal reactions at the centre of a wheel's point of contact on the road;

Centrifugal force $= \dfrac{mv^2}{r}$, acting horizontally through the centre of gravity.

v = velocity of vehicle in m/s and r = radius of turn in metres.

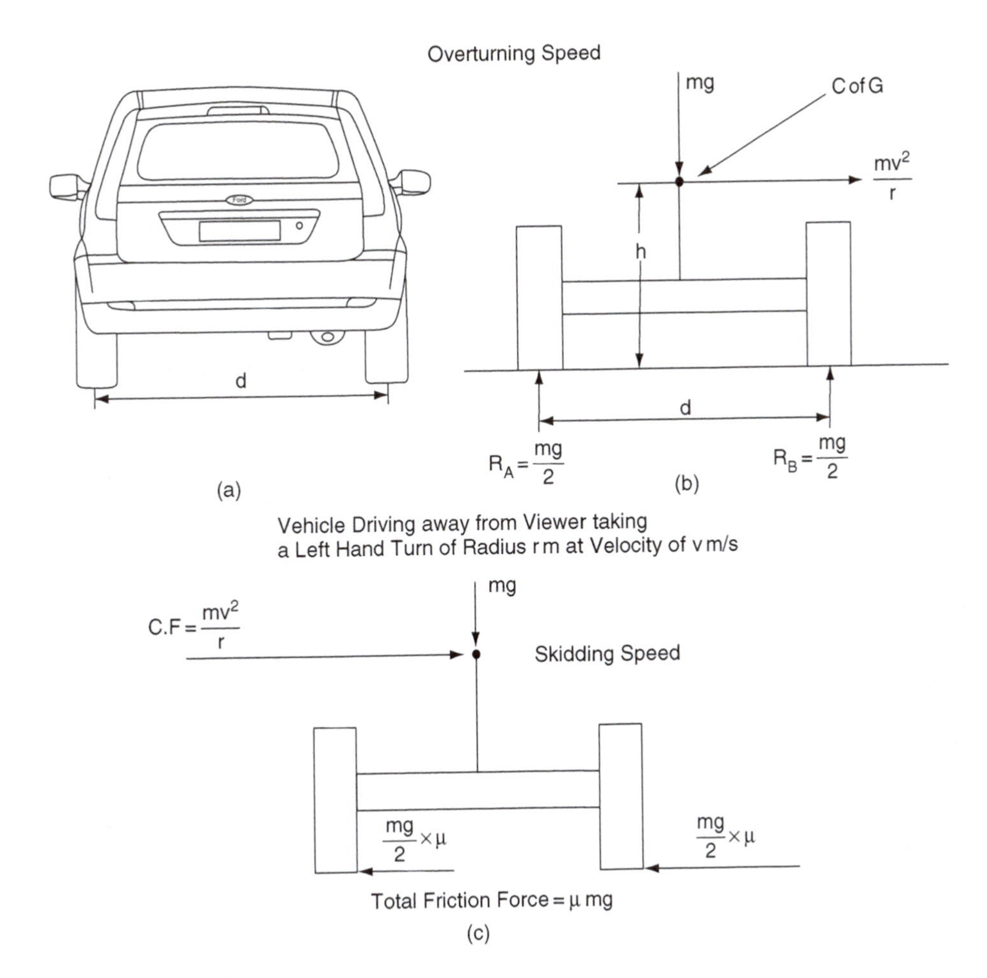

Fig. 11.7 Overturning speed

At the point of overturning, the wheel on the inside of the bend starts to lift and there is no force between the tyre and road. Under these conditions the overturning effect is equal and opposite to the righting effect.

Take moments about the contact point of the outer wheel:

Overturning moment = righting moment

$$\frac{mv^2 \cdot h}{r} = \frac{mg \cdot d}{2}$$

$$v^2 = \frac{grd}{2h}$$

The overturning velocity $v = \sqrt{(grd/2h)}$ **(11.8)**

Example 11.7

In the vehicle shown in Figure 11.7(b) the track width $d = 1.44\,\text{m}$ and the centre of gravity is 0.8 m above ground level. Determine the overturning speed on a curve of 80 m radius. Take $g = 9.81\,\text{m/s}^2$.

Solution

$h = 0.8\,\text{m}$; $d = 1.44\,\text{m}$; $r = 80\,\text{m}$; $g = 9.81\,\text{m/s}^2$

$$v = \sqrt{(grd/2h)}$$

$$= \sqrt{\{(9.81 \times 80 \times 1.44)/2 \times 0.8\}}$$

$$= \sqrt{706.3}$$

$$v = 26.58\,\text{m/s} = 95.7\,\text{km/h}$$

Skidding speed

Consider the vehicle in Figure 11.7(c) to be travelling in the same direction at v m/s on the same curved track where the coefficient of friction between the tyres and the track $= \mu$.

The friction force that is resisting sideways motion $= \mu mg$ newtons.

The force pushing the vehicle sideways = centrifugal force $= \frac{mv^2}{r}$ newtons.

At the point of skidding the centrifugal force is equal and opposite to the friction force.

$$\frac{mv^2}{r} = \mu mg$$

$$v^2 = \mu gr$$

Skidding velocity $$v = \sqrt{\mu gr} \quad \textbf{(11.9)}$$

Example 11.8

Determine the skidding speed of a vehicle on a curve of 80 m radius if the coefficient of friction between the tyre and the road is 0.7. Take $g = 9.81\,\text{m/s}^2$.

Solution

$g = 9.81\,\text{m/s}^2$; $\mu = 0.7$; $r = 80\,\text{m}$

$$v = \sqrt{\mu gr}$$

$$v = \sqrt{0.7 \times 9.81 \times 80}$$

$$= \sqrt{549.4}$$

$$v = 23.43\,\text{m/s} = 84.4\,\text{km/h}$$

11.12 Summary of formulae

Force F = mass × acceleration

Load transfer $= \frac{Fh}{b}$

$$R_f = \frac{mgy}{b} - \frac{Fh}{b}$$

Maximum tractive effort for front wheel drive $\text{Te} = \mu(R_F - Fh/b)$

Maximum tractive effort rear wheel drive $\text{Te} = \mu(R_R + Fh/b)$

Gradient resistance $= mg \sin\theta$

Overturning velocity $v = \sqrt{(grd/2h)}$

Skidding velocity $v = \sqrt{\mu gr}$

11.13 Exercises

11.1 Table 11.3 shows how the air resistance against a certain heavy vehicle varies with road speed. Plot a graph of these figures on a base of road speed. From the graph determine the air resistance at a speed of 100 km/h and use this figure to calculate the power absorbed in overcoming wind resistance at 100 km/h.

11.2 A vehicle of gross vehicle weight of 12 tonnes is propelled by an engine that produces a power of 120 kW at an engine speed of 4500 rev/min. This engine speed corresponds to a road speed of 90 km/h and the tractive resistance at this speed is 2320 N.

If the overall efficiency of the transmission is 90% calculate:

(a) the power available at the driving wheels;

(b) the maximum possible acceleration at this speed.

11.3 A vehicle with a wheelbase of 3.2 m has its centre of gravity located midway between the axles at a height of 0.75 m. If the vehicle weight is 2 tonnes, calculate the load

Table 11.3 Wind resistance at various speeds (Exercise 11.1)

Vehicle speed (km/h)	48	56	64	72	80	88	97	105
Wind force (N)	510	690	899	1145	1280	1710	2012	2363

transfer that occurs when the vehicle accelerates at $5\,m/s^2$.

11.4 The figures in Table 11.4 relate to a small car that is operating in top gear on a level road.

Plot the curves of power required and power available and determine:

(a) the maximum possible speed of the car on the level;

(b) the maximum power available for acceleration.

Table 11.4 Calculation of vehicle speed and power (Exercise 11.4)

Road speed (km/h)	20	40	60	70	80
Power required to overcome tractive resistance (kW)	2.5	10	27	40	60
Power developed by engine (kW)	18	38	49	48	36

11.5 A vehicle of mass 1.6 tonnes accelerates uniformly from 20 km/h to 90 km/h in 5 seconds. Determine:

(a) the acceleration;

(b) the force acting on the vehicle to produce this acceleration.

11.6 A vehicle weighing 1.4 tonnes is driven at uniform speed of 90 km/h along a level road while the engine is developing 25 kW. If the overall transmission efficiency is 80% calculate the tractive resistance to motion of the vehicle.

11.7 A vehicle weighing 1800 kg is driven at uniform speed up an incline of 5%.

Calculate:

(a) the force required to overcome gradient resistance;

(b) the power at the driving wheels if wind and rolling resistances amount to 300 N.

11.8 A vehicle has its centre of gravity at a height of 0.7 m. If the track width is 1.3 m, calculate the overturning velocity on a bend of 100 m radius. Take $g = 9.81\,m/s^2$.

11.9 A truck is travelling round a curve of 50 m radius on a horizontal track. At a speed of 60 km/h the vehicle is just on the point of skidding. Determine the coefficient of friction between the tyres and the track. Take $g = 9.81\,m/s^2$.

11.10 A vehicle of weight 1.6 tonnes has a wheelbase of 2.6 m and its centre of gravity is 0.7 m above ground level and midway between the axles. If the coefficient of friction between the tyres and the road is 0.7, calculate the maximum acceleration that is possible on a level surface when the vehicle has front wheel drive. Take $g = 9.81\,m/s^2$.

11.11 A vehicle weighing 1.3 tonnes has its centre of gravity at a height of 0.6 m above ground level. The wheelbase of the vehicle is 3.4 m and the centre of gravity is 1.6 m behind the centre of the front axle. When the brakes are fully applied the vehicle stops in a distance of 20 m from a speed of 50 km/h. Determine the loads on each axle under these conditions. Take $g = 9.81\,m/s^2$.

11.12 A vehicle of mass 16 tonnes is brought to rest with a braking efficiency of 80%. Taking $g = 9.81\,m/s^2$, calculate the braking force.

12
Balancing and vibrations

12.1 Introduction

Many vehicle systems can be adversely affected by vibrations that arise from rotation and reciprocation of moving parts. Two prominent examples are wheels and tyres and the moving parts of engines. In both of these cases care in maintenance and repair operations must be taken to ensure: (1) in the case of wheels and tyres, that the assembly is correctly balanced; and (2) in the case of engines, that parts are correctly fitted to ensure that balance is maintained. To develop an appreciation of this topic it is useful to understand some basic principles of mechanical balancing.

12.2 Balance of rotating masses acting in the same plane (coplanar)

Figure 12.1 shows a mass attached to a shaft by means of a stiff rod. When the shaft rotates, the mass at the end of the rod is made to move in a circular path and this causes a centrifugal force that acts on the shaft bearings.

To counteract this centrifugal force a balance mass is introduced into the system, as shown in Figure 12.2.

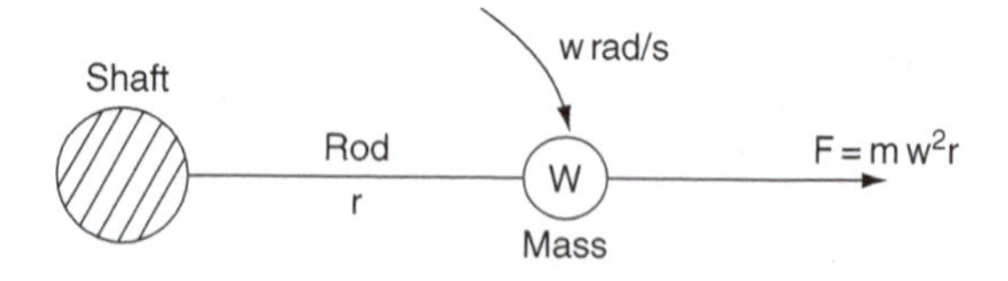

Fig. 12.1 Centrifugal force on rotating mass

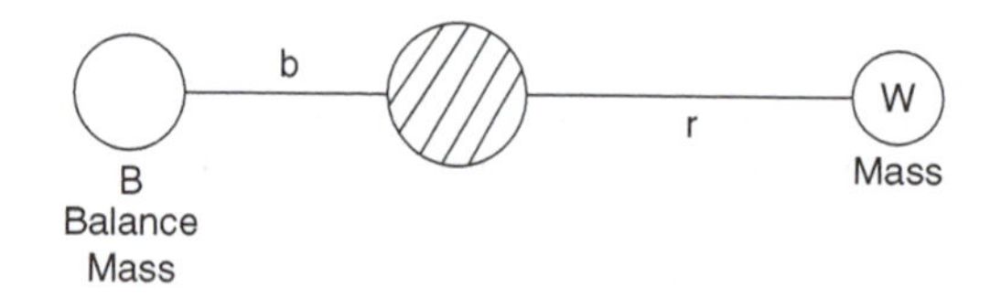

Fig. 12.2 Balance mass

Let B = mass of balance weight; b = radius at which balance weight is attached; W = out-of-balance mass; r = radius of out-of-balance mass; ω = angular velocity in rad/s.

For balance, the force on the out-of-balance weight must be equal to the force created by the balance weight:

$$B\omega^2 b = W\omega^2 r \qquad \textbf{(12.1)}$$

Because the products Bb and Wr are directly proportional to the centrifugal forces, this equation may be written as

$$Bb = Wr \qquad \textbf{(12.2)}$$

Bb = Wr is also the condition for static balance.

When the forces are acting in the same plane the condition for dynamic balance is the same as that for static balance.

When a system is **dynamically balanced it is always in static balance**.

12.3 Balancing of a number of forces acting in the same plane of revolution (coplanar forces)

Figure 12.3 shows a thin disc which has three out-of-balance masses. W_1, W_2 and W_3 are the out-of-balance masses acting at radii r_1, r_2 and r_3, respectively. A balance mass B acting at a radius b is to be introduced to balance the system. The problem is to determine the position and size of this balancing weight.

The procedure for finding the equilibrant of a number of forces that was introduced in Chapter 5 can be used.

In this case the centrifugal forces acting on the out-of-balance masses are proportional to the product Wr and it is sufficient to draw a force diagram, where the Wr products represent the forces.

It is helpful to tabulate the data in the form of a table (see Table 12.1) before proceeding to draw the force diagram to a suitable scale.

The force diagram is then drawn to a suitable scale, as shown in Figure 12.4. The equilibrant represents the magnitude and position of the product Bb from which the radius b and the mass B may be determined.

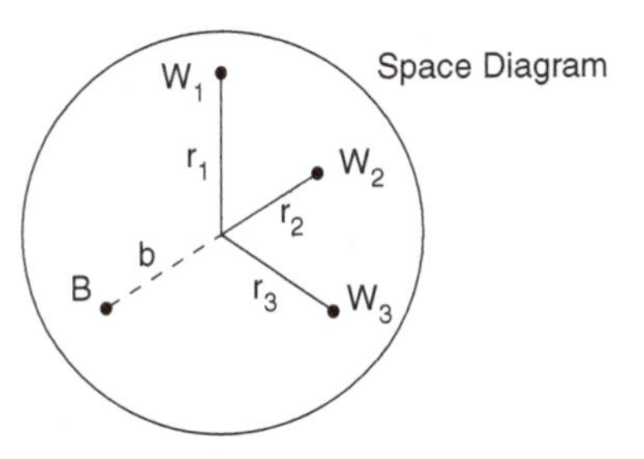

Fig. 12.3 Three unbalanced forces in the same plane

Table 12.1 Data re centrifugal forces and out-of-balance masses

Mass	Radius	Product
W_1	r_1	W_1r_1
W_2	r_2	W_2r_2
W_3	r_3	W_3r_3
B	b	Bb

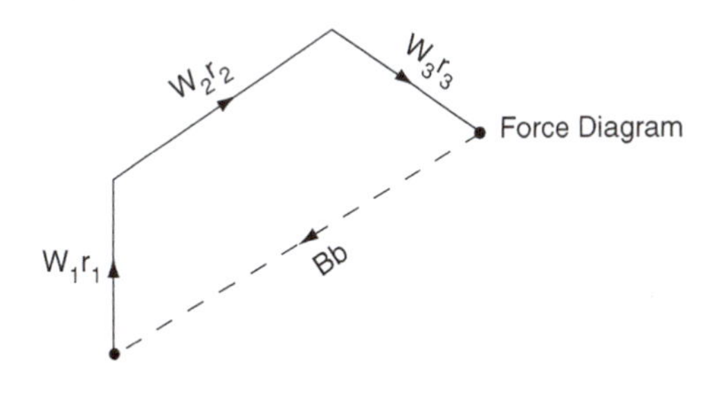

Fig. 12.4 Single-plane balancing

Example 12.1

As a result of an inefficient repair operation, two of the eight bolts that secure a clutch to a flywheel are replaced by incorrect bolts that have an excess weight of 5 grams each. The incorrect bolts are placed 135° apart at a radius of 100 mm. Construct a force diagram and use it to determine the magnitude of the out-of-balance force that will arise at an engine speed of 6000 rev/min as a result of the use of these bolts.

Solution

The procedure adopted here is to construct a mass × radius diagram, as shown in Figure 12.5.

The resultant vector from S to F on the diagram is the out-of-balance effect.

Scaling this vector shows it to be 4 cm in length, which is 400 g mm, or 0.0004 kg m.

This out-of-balance effect is the product of mr.

Centrifugal force $= mr\omega^2$

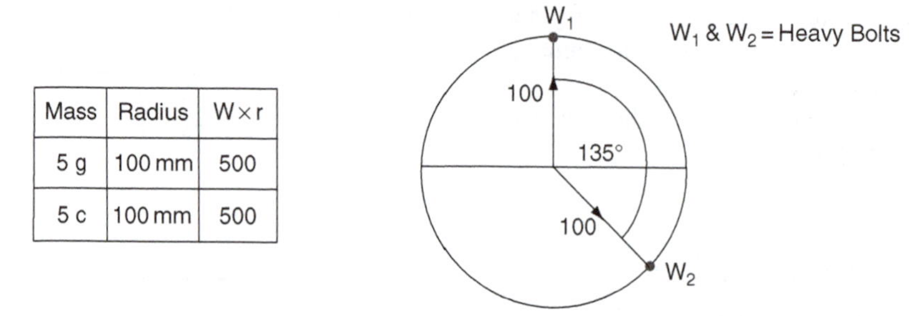

Mass	Radius	W × r
5 g	100 mm	500
5 c	100 mm	500

Wr - Vector Diagram Scale 100 gm mm = 1 cm

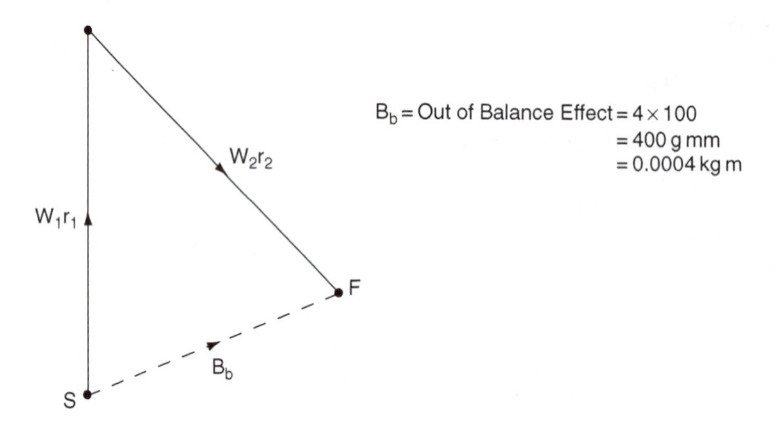

Fig. 12.5 Balancing diagram

The flywheel speed = 6000 rev/min

$$= \frac{6000 \times 2\pi}{60} = 314.2\,\text{rad/s}$$

The out of balance force = 0.0004×314.2^2

$$= \mathbf{39.5\,N}$$

12.4 Wheel and tyre balance

A road wheel and tyre assembly may be out of balance owing to slight imperfections in the tyre or wheel. Any imbalance in the wheel and tyre assembly will give rise to disturbing forces as the wheel rotates. In the vertical plane an out-of-balance force will attempt to move the wheel up and down and the effect may be to cause vibrations in the suspension system. If, as is usually the case, the out-of-balance force acts at some distance from the plane of rotation of the wheel and tyre, the effect is to cause the wheel to wobble, as shown in Figure 12.6. This wobble affects the steering and at critical speeds the effect can be dangerous.

In order to correct this wobble, the wheel and tyre are rotated on a dynamic test machine known as a wheel balancer. The wheel balancer is calibrated so that the place on the rim of the wheel where a balance weight of a specified amount is secured. Because a rotating assembly is in static balance when dynamic balance has been established, a single wheel-balancing operation ensures that the wheel and tyre are balanced for all conditions.

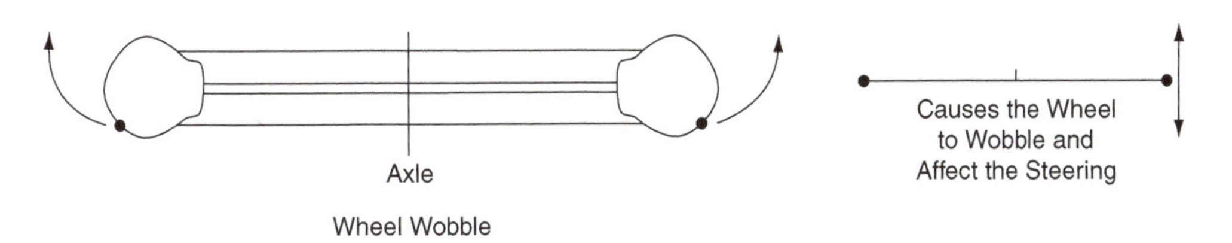

Fig. 12.6 Wheel and tyre wobble

12.5 Engine balance

To achieve engine balance the following two types of balancing have to be considered:

1. reciprocating parts, mainly pistons and part of the connecting rods;
2. rotating parts, mainly flywheels, crankshafts and part of the connecting rods, and other components such as camshafts.

12.6 Balance in a single-cylinder engine

Figure 12.7 shows the piston, connecting rod and crank assembly of a single-cylinder engine. The inertia force on the piston arises from the acceleration and deceleration of the piston as it moves along the stroke. At top dead centre (TDC) and bottom dead centre (BDC) the piston acceleration and inertia force are at maximum values. This inertia force is transmitted to the crankshaft and the engine frame, with the result that the engine would move up and down if some method were not used to prevent this from happening; i.e. the engine must be balanced to overcome this effect.

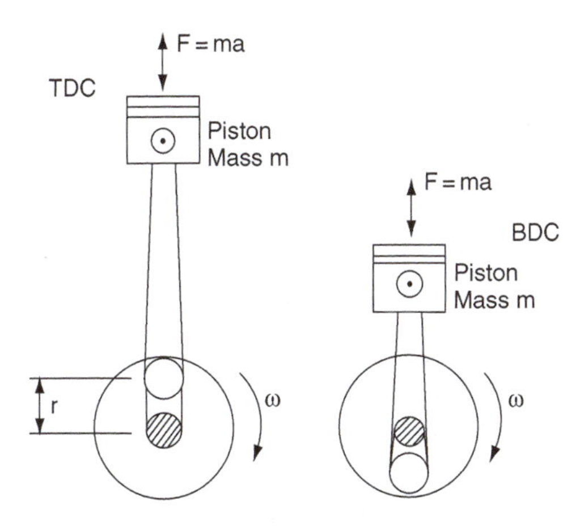

Fig. 12.7 Forces acting on reciprocating parts

The method employed to balance a single-cylinder engine involves the use of a balance mass (weight) that is attached to the crank so that it rotates with it, as shown in Figure 12.8(a). The balance weight can be designed to exactly balance the inertia force on the piston; however, the solution is not that simple. Figure 12.8(b) shows the situation that arises when the crank has turned through 90° from the TDC position. It is now evident that the force on the balance weight is acting at right angles to the line of stroke and is exerting a force at right angles that no longer balances the inertia force on the piston.

This shows that it is not possible to provide complete balance of a reciprocating mass, such as the piston, by the use of a rotating mass. The force on the reciprocating parts varies in magnitude but is always directed along the line of the piston stroke. The force on the rotating mass is constant but varies in its direction, which is the point made in Figure 12.8.

In practice, a compromise is reached that aims to achieve the maximum amount of balance. This compromise varies according to engine design but it is usually achieved by including one third of the weight of the connecting rod and small end in the weight of the piston and then providing a

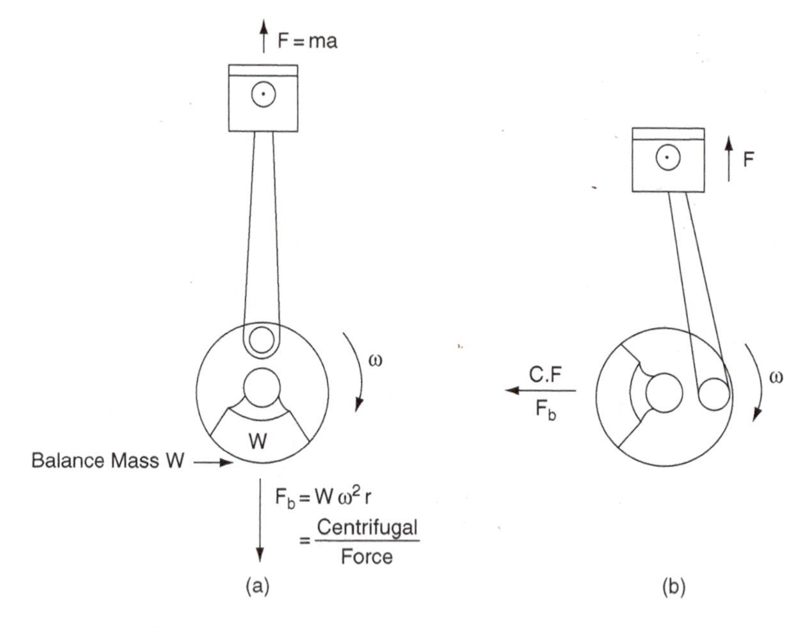

Fig. 12.8 Effect of balance weight (mass)

balance weight that is equivalent to 50% of the piston weight.

12.7 Primary and secondary forces

The acceleration of the piston is at its maximum at each end of the stroke, as shown in Figure 12.7.

Analysis of piston and crank motion shows that at any point during the stroke a very close approximation of the piston acceleration is given by the formula

$$a = \omega^2 r\{(\cos\theta + (\cos 2\theta)/n\} \qquad \textbf{(12.3)}$$

where ω = angular velocity of the crank, θ = angle that the crank has turned through, and n = connecting rod to crank ratio.

$$n = \frac{\text{connecting rod length}}{\text{radius of crank throw}}$$

The force on the piston is

$$Fp = m\omega^2 r\{(\cos\theta) + (\cos 2\theta)/n\} \qquad \textbf{(12.4)}$$

For engine balance purposes the force Fp is normally considered in two parts, called the **primary force** and the secondary force respectively.

The **primary** force is given by the first part of the equation, e.g. primary force

$$Fp = m\omega^2 r(\cos\theta) \qquad \textbf{(12.5)}$$

The **secondary** force is given by the second part of the equation, e.g. secondary force

$$Fs = m\omega^2 r(\cos 2\theta)/n \qquad \textbf{(12.6)}$$

Graph of primary and secondary forces

Figure 12.9 shows how the primary and secondary forces vary during one complete revolution of the crankshaft; the plus and minus signs indicate the direction of the force along the line of the piston stroke.

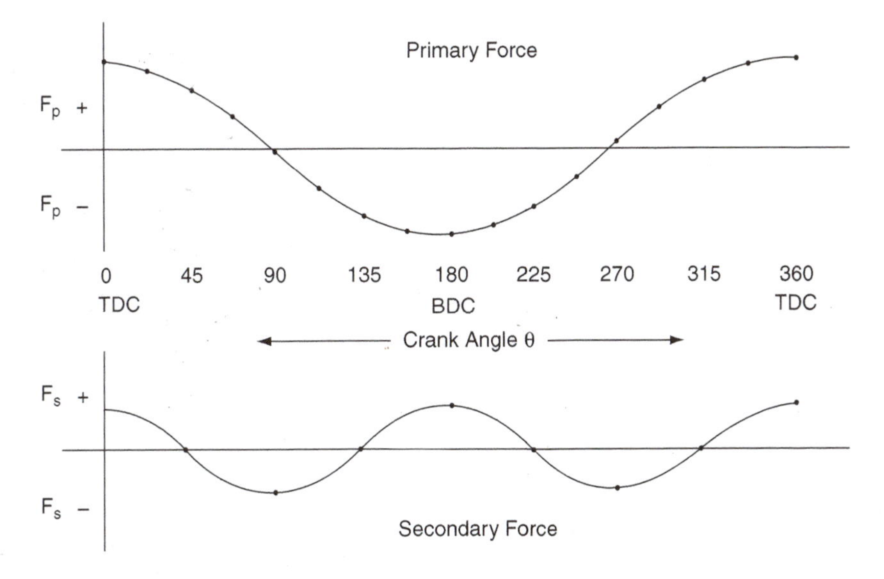

Fig. 12.9 Primary and secondary force variation

Significant points in Figure 12.9

The graph shows that the primary force is at a maximum at TDC and BDC, i.e. **twice per revolution of the crankshaft**.

The graph also shows that the secondary force is at a maximum at TDC, at 90° after TDC, at BDC and at 270° of crank rotation , i.e. **four times per revolution of the crankshaft**.

While a reasonable balance of the primary forces in a single-cylinder engine can be achieved by the use of balance weights, as shown in Figure 12.8, a solution to the problem of the secondary forces is more difficult to achieve. One method that is used to counter the effect of secondary forces makes use of rotating weights that are driven from the crankshaft at twice engine speed, as described in the following section.

12.8 Secondary force balancer

The balance weights are attached to shafts that are driven from the crankshaft. One shaft rotates in a clockwise direction, the other rotates in an anti-clockwise direction; both shafts are driven at twice engine speed.

The rotation of the balance weights is timed so that they are effective at the points required, i.e. TDC, and 90 and 135 degrees after TDC. The diagrams in Figure 12.10 show how the weights are effective when required and how the forces they generate cancel each other out when the secondary force is zero.

Harmonics

Vibrations caused by unbalanced forces in an engine occur at a certain engine speeds; the lowest speed at which the vibrations are a problem is called the **fundamental frequency**. Further vibrations will occur at higher speeds that are multiples of the fundamental frequency; these higher speeds are called **harmonics** and the harmonic balancer is designed to counteract the effects of the vibrations at these speeds. That is why the secondary force balancer is often referred to as a harmonic balancer.

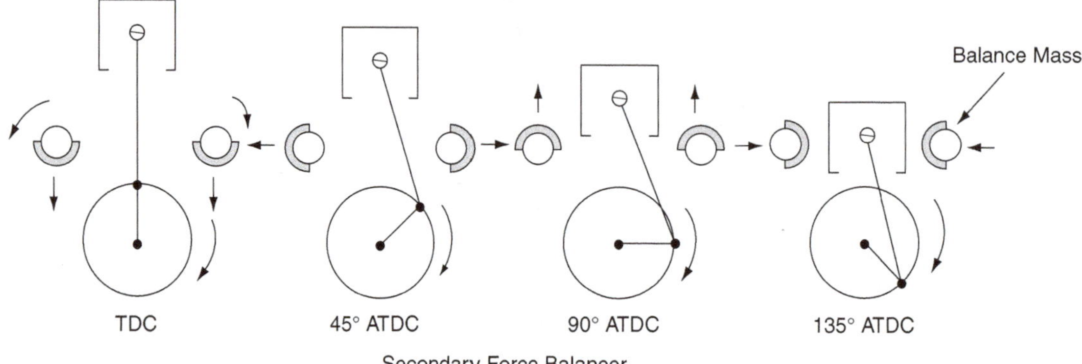

Fig. 12.10 Secondary force balancer

12.9 Balance of rotating parts of the single cylinder engine

For the purposes of this explanation, the rotating parts considered are the crank pin and the crankshaft, and part of the mass of the connecting rod. The normal procedure for balancing these parts is to attach balance masses on the opposite side to the crank pin, as shown in Figure 12.11.

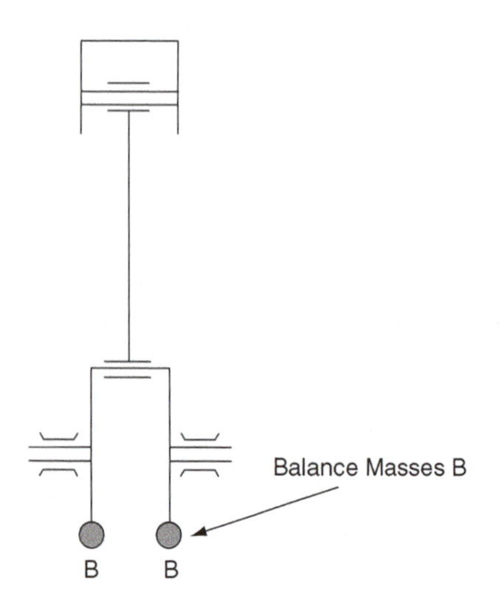

Fig. 12.11 Single-cylinder balance masses

Example 12.2

A single-cylinder engine is to be balanced by placing masses on the crank, as shown in Figure 12.11. The piston and gudgeon pin have a mass of 0.450 kg, the connecting rod mass is 0.65 kg, and the crank pin mass is 0.14 kg. The crank pin and part of the connecting rod act at an effective radius of 80 mm from the centre of the crank and the balance masses are to be placed at an equal radius on the opposite side of the crank. To achieve a satisfactory degree of balance, 50% of the mass of the reciprocating parts plus the mass of the rotating parts is added in the form of balance masses. For balancing purposes, 70% of the connecting rod mass is taken to be a rotating mass. Determine the mass of the two balance masses that act at the same radius as the crank pin.

Solution

$$\text{Mass of rotating parts} = 0.14 + (0.7 \times 0.65) = 0.595\,\text{kg}$$

$$50\%\ \text{of reciprocating parts} = \frac{0.45 + (0.3 \times 0.65)}{2} = 0.326\,\text{kg}$$

Total mass to be added as balance masses $= 0.595 + 0.326 = 0.921\,\text{kg}$

$$\text{Each balance mass} = \frac{0.921}{2} = 0.461\,\text{kg}$$

12.10 Four-cylinder in-line engine balance

Figure 12.12 shows a layout that is widely used in the construction of four-cylinder in-line engines. The reciprocating parts are identical for each cylinder and the cranks are arranged to provide uniform firing intervals, and in most cases the distances between the cylinders are equal. With this design of engine only the primary forces are balanced; the secondary forces and other harmonic forces are unbalanced. In engines operating at high speed, the secondary forces can affect the operation of the vehicle and secondary balancers based on the principle described in Section 12.8 may be incorporated into the engine design to counter any ill effect.

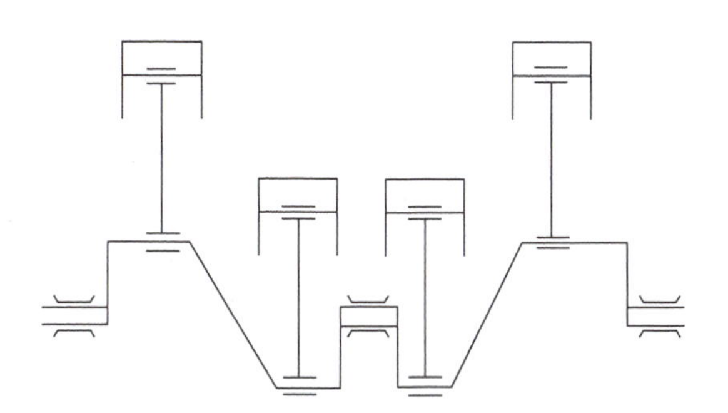

Fig. 12.12 Four-cylinder in-line engine

12.11 Couples and distance between crank throws

Figure 12.13(a) shows a twin-cylinder in-line engine in which the cranks are 180° apart. The primary forces caused by the reciprocating parts produce a turning effect (couple) that causes the engine to rock on its mountings. Figure 12.13(b) shows a four-cylinder engine wherein the cranks are arranged as shown. The cranks are equal distances apart and the couples on each half of the crankshaft oppose each other. The couples balance out inside the engine and no effect is felt at the engine mountings.

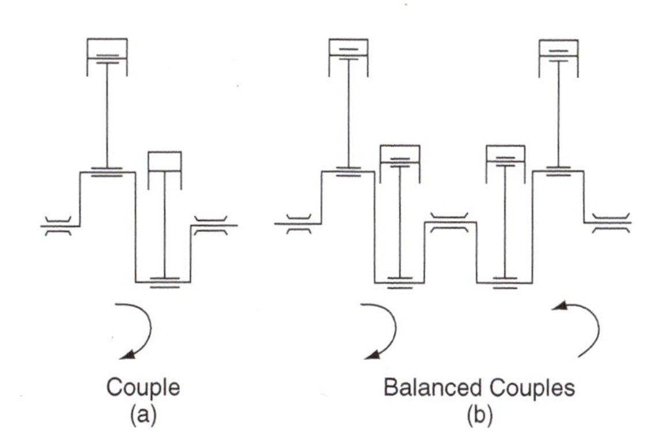

Fig. 12.13 Couples

12.12 Simple harmonic motion (SHM)

In order to show that a motion is simple harmonic the following must occur: (1) its acceleration must be directly proportional to its displacement from the fixed point in its path; and (2) the acceleration must always be directed towards the fixed point.

The traditional approach to a study of simple harmonic motion is to consider a point P that is constrained to move in a circular path about a point O.

With reference to Figure 12.14, let point P move along a circular path with constant velocity v and let N be the projection of P on to the diameter AB.

The velocity of N is equal to the horizontal component of v.

$$\text{velocity of N} = v\sin\theta, \text{ since } v = \omega R$$
$$= \omega R \sin\theta$$
$$= \omega NP$$
$$= \omega\sqrt{(OP^2 - ON^2)}$$
$$= \varpi\sqrt{(R^2 - x^2)}$$

The maximum velocity of $N = v = \omega R$ **(12.7)**

That is when $\theta = 90°$ or $270°$

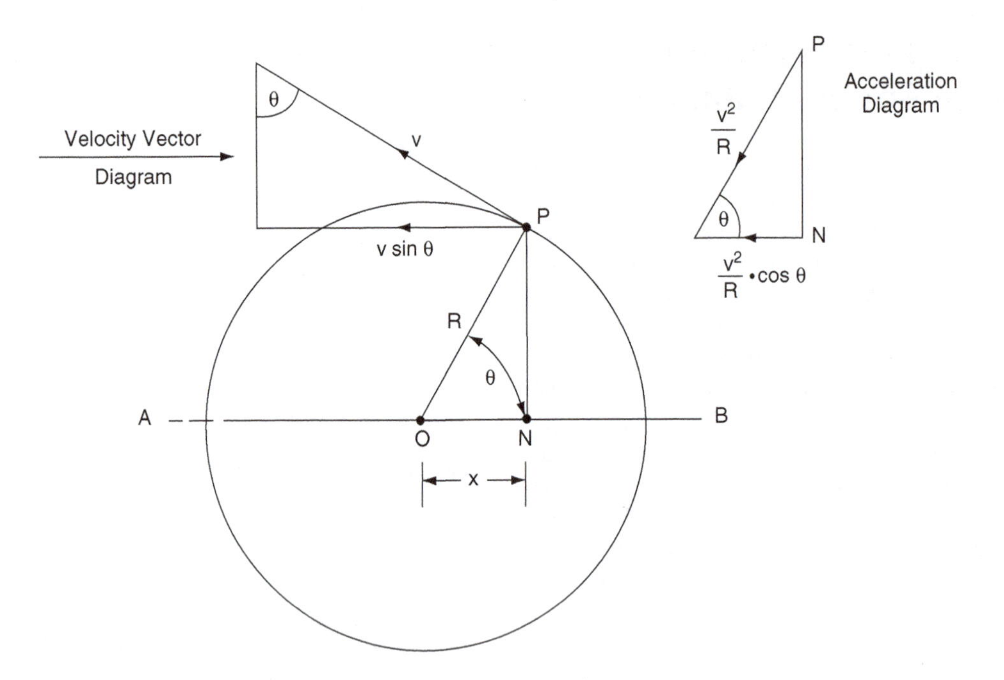

Fig. 12.14 Simple harmonic motion (SHM)

Because the point P moves along a circular path about centre O it has a centripetal acceleration of v^2/R and the only acceleration that N may have is the horizontal component of this centripetal acceleration, and that is acceleration $= v^2/R\cos\theta$.

$$\text{The acceleration of N} = \omega^2 R\cos\theta$$
$$= \omega^2 x \qquad \textbf{(12.8)}$$

$$\text{The maximum acceleration of N} = \omega^2 R\,\text{m/s}^2$$
$$\text{occurs when } \theta = 0° \text{ or } 180° \qquad \textbf{(12.9)}$$

Equation (12.7) shows that the acceleration of N is proportional to the displacement x and is always directed towards O, so the motion of N is simple harmonic.

The **amplitude** of the vibration is the maximum displacement of N from mid-position O and this is equal to the radius of the circle R.

The **periodic time, t**, is the time required for one N to complete two strokes and this is the time that P takes to make one revolution.

$$\text{Thus the periodic time } t = \frac{\text{distance covered}}{\text{velocity}}$$
$$= \frac{\text{circumference}}{\text{velocity}}$$
$$= \frac{2\pi R}{\omega R} \text{(because velocity } v = \omega R)$$

$$\text{Periodic time } t = 2\pi/\omega \qquad \textbf{(12.10)}$$

By using Equation (12.7), acceleration $a = \omega^2 \boldsymbol{x}$, where $\boldsymbol{x}$ = displacement, it can be shown that $\omega = \sqrt{(\text{acceleration / displacement})}$

For simple harmonic motion:

$$\text{Periodic time } t = 2\pi\sqrt{(\text{acceleration/(displacement)}} \qquad \textbf{(12.11)}$$

$$\text{Frequency of vibration} = \frac{1}{2\pi}\sqrt{(\text{displacement/(acceleration)}} \qquad \textbf{(12.12)}$$

The frequency is the number of vibrations made in 1 second = 1/t Hz.

Example 12.3

A body moving with SHM has an amplitude of 1.3 m and a periodic time of 3 seconds. Calculate: (a) the maximum velocity; (b) the maximum acceleration; (c) the frequency.

Solution

(a) $t = 2\pi/\omega$, from Equation (12.10)

$\therefore \omega = t/2\pi$

$= 2\pi/3$ radians/s

Max. velocity $= \omega R = (2\pi/3) \times 1.3 = 2.72$ m/s, from Equation (12.7).

(b) Max. acceleration $= \omega^2 R = (4\pi^2/9) \times 1.3 = 5.7\,m/s^2$, from Equation (12.9).

12.13 Applications of SHM

Vibration of a helical coil spring

The coil spring in Figure 12.15 is fixed at the top end, and at the bottom end is suspended a weight of W kg = W g newtons. This weight is free to move in a vertical plane relative to the datum position O. If the weight is pulled down by a distance x and then released, the weight will vibrate (oscillate) to and fro about the datum line O.

If the stiffness of the spring is S newtons per metre the restoring force exerted by the spring = Sx.

$Sx = Wa$, where W is the mass in kg and a = acceleration in m/s^2

$\therefore a = \dfrac{Sx}{W}$, i.e. acceleration is proportional to displacement.

This is the condition for the motion to be SHM.

The periodic time $t = 2\pi\sqrt{}$ (displacement/acceleration)

$\therefore t = 2\pi\sqrt{(W/S)}$

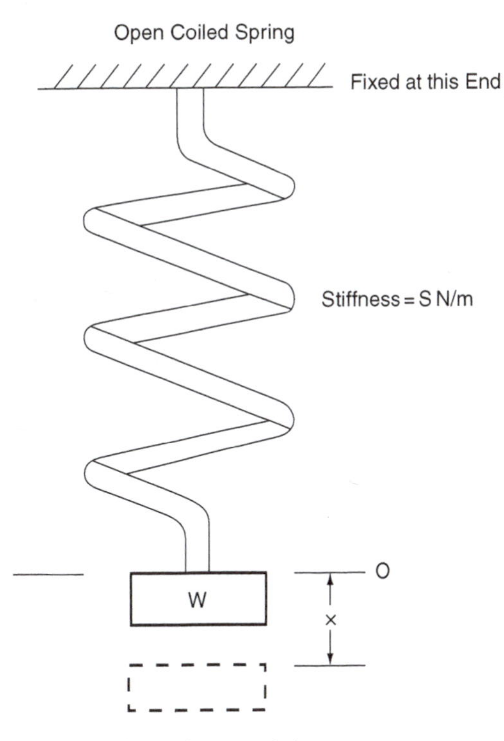

Fig. 12.15 SHM of open coiled spring

Because S = Wg/d, where d = static deflection of the spring and g is the gravitational constant, the periodic time may be written as

$$t = 2\pi\sqrt{(d/g)} \qquad \textbf{(12.13)}$$

Example 12.4

A suspension coil spring has a static deflection of 250 mm. Calculate the frequency of vibration. Take $g = 9.81\,m/s^2$.

Solution

$t = 2\pi\sqrt{(d/g)}$, using Equation (12.13)

$\therefore t = \pi\sqrt{(0.25/9.81)}$ s, working in metres

$= 2\pi.0.16$

$= 1$ s

Frequency $= 1/t = 1/1 = 1$ Hz, or 60 vibrations per minute.

Motor car suspension systems are designed to have the following approximate values for frequencies:

- Front suspension – 60/min to 80/min
- Rear suspension – 70/min to 90/min.

12.14 Torsional vibration

Figure 12.16 shows a flywheel attached to a shaft. If the flywheel is rotated so that the shaft is twisted and then released so that the shaft untwists, the flywheel will vibrate either side of a datum position. The movement to and fro can be shown to be simple harmonic motion.

As the motion is SHM, the periodic time $t = \frac{1\sqrt{(\text{displacement}/\text{acceleration})}}{2\pi}$.

This can be shown to be $t = 2\pi\sqrt{(I/\mu)}$, where I is the moment of inertia of the flywheel and μ is the torsional stiffness of the shaft.

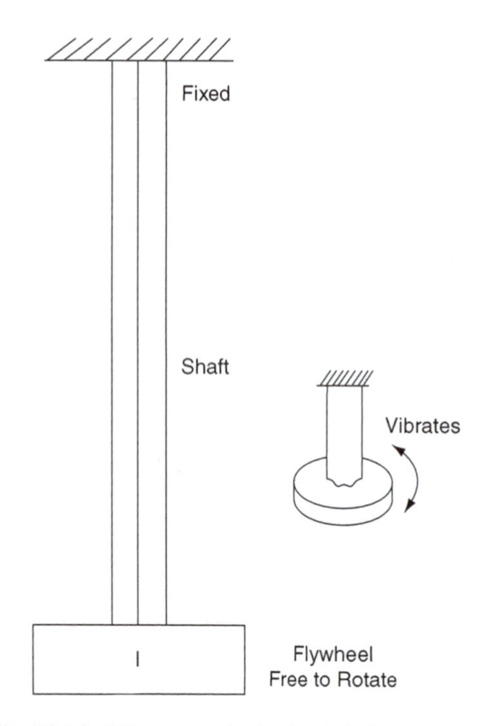

Fig. 12.16 Vibration of flywheel and shaft

Example 12.5

A flywheel for a large engine has a mass of 305 kg and a radius of gyration of 0.38 m. The flywheel is mounted on a shaft that has torsional stiffness of 300 kN m. Determine the periodic time and the frequency of vibration of the flywheel.

Solution

The periodic time $t = 2\pi\sqrt{(I/\mu)}$

$$\mu = 300\,\text{kN m}$$

$$I = m k^2 = 305 \times 0.38 \times 0.38 = 44\,\text{kg m}^2$$

$$\therefore t = 2\pi\sqrt{(44/300\,000)}$$

$$= 0.077\text{s}$$

Frequency of vibration $f = 1/t = 1/0.077 = 13\,\text{Hz}$.

12.15 Free vibrations

In the cases of the spring and the shaft, the vibrations will continue for an indefinite period of time until they die away due to frictional and other effects. Vibrations such as these occur in many mechanical systems and they are known as free vibrations; the frequency of such vibrations is known as the natural frequency.

Example of free vibrations

Figure 12.17 shows a motor vehicle travelling along a smooth surface that suddenly encounters a bump in the surface. The forward motion of the

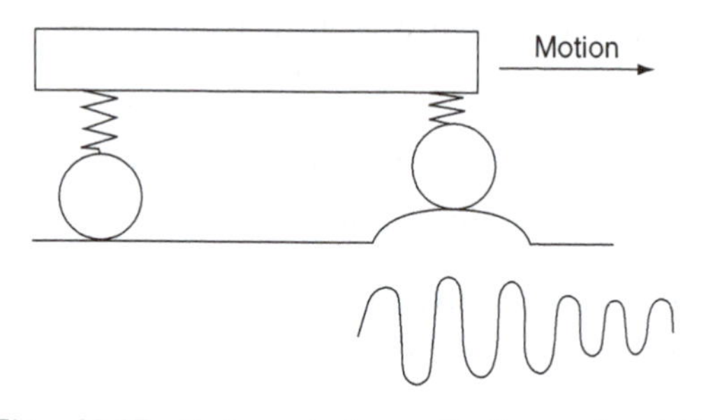

Fig. 12.17 Undamped free vibrations of vehicle suspension

vehicle causes it to rise up over the bump and this imparts a vertical force that is passed to the body of the vehicle through the suspension springs. This vertical force compresses the spring and sets up a free vibration. If the spring is not damped, the vibrating motion will continue to cause the vehicle body to move up and down until friction and other forces cause the vibration energy to be dissipated.

12.16 Forced vibrations

Resonance

When systems such as the loaded spring, and the flywheel and shaft, are subject to periodic disturbing forces such as those that occur in crankshaft rotation and the opening and closing of engine valves, a condition known as resonance arises. Resonance occurs when the frequency of a periodic disturbing force is the same as the natural frequency of the spring, or flywheel, or any other system. At resonance, the amplitude of vibration can become very large and can result in severe damage to a mechanical system, such as an engine and transmission components.

Driveline vibrations

Each firing impulse in a multi-cylinder engine produces a small amount of twist in the crankshaft and when the transmission is engaged this twisting effect is *passed into* the transmission system. Vibrations in the driveline cause noise and wear and various damping methods are employed to reduce these effects.

Damping

Damping is used to control the vibrations that arise from the operation of mechanisms.

Vibration dampers

Figure 12.18 shows a simple form of linear vibration damper that makes use of a piston in a

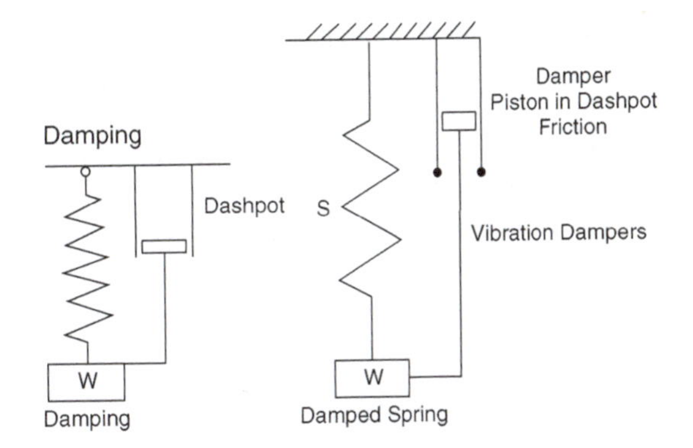

Fig. 12.18 Simple vibration damper

dashpot that contains a viscous fluid. The dashpot and piston removes energy from the spring and thus reduces vibration. This is the basic principle of the dampers used in vehicle suspension systems.

The crankshaft damper shown in Figure 12.19 is an integral part of the belt pulley that is fitted to the front end of a vehicle engine crankshaft. At the outer circumference of the pulley is an annular space that extends around the pulley.

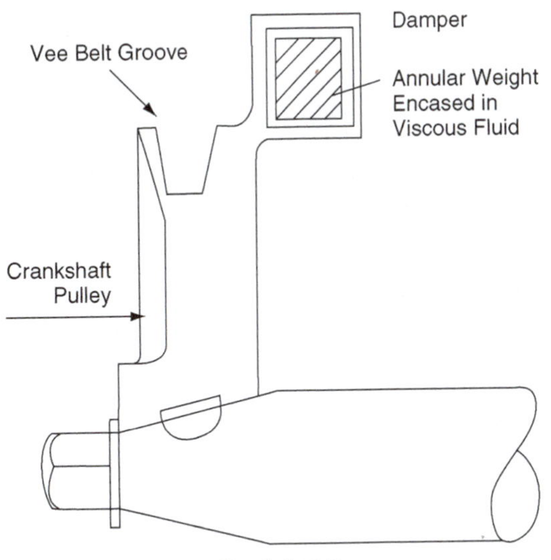

Fig. 12.19 Crankshaft vibration damper

Inside this space is mounted an annular weight; this weight is surrounded by a viscous fluid. The inertia of the annular weight will cause it to remain stationary relative to any vibrational twisting of the shaft and this process will absorb vibrational energy in the friction. The energy thus absorbed is converted into heat and passed into the atmosphere.

Dual mass flywheel

A flywheel that incorporates a torsional vibration damper is known as a dual mass flywheel. In these flywheels there are two main masses that have a small amount of freedom to move relative to each other. The relative movement is controlled by springs, and friction between the two masses provides the damping effect.

Cams

In order to reduce the risk of valve bounce and noise, valve operating cams should possess the following characteristics:

- They should open and close the valves as quickly as possible in order to admit the maximum amount of gas into, or out of, the cylinder.
- They should require as small an external spring force as possible to maintain contact between the cam and the follower during the latter part of the opening stroke and the early part of the return stroke.

To achieve the first of these aims, the acceleration of the follower and valve should be as high as possible as the valve is lifted away from its seat, against the spring force. For the second aim of keeping the valve in contact with the follower, and the follower in contact with the cam, the retardation should be as low as possible.

On the return stroke, when the valve is going back to its seat, the acceleration should be low and the retardation high. These characteristics are shown graphically in Figure 12.20, where the figures are based on an engine speed of 2500 rev/min.

In Figure 12.20 the amount of valve lift and acceleration is plotted against the number of degrees of cam rotation; it should be noted that the cam rotates at half engine speed.

The initial acceleration for the first 15° while the valve is being lifted away from the seat is high. From 15° onwards the acceleration drops, the valve stopping at the fully open position. At the fully open position, on the nose of the cam, the cam is designed to provide a short period of cam rotation while the valve is held open. On the closing movement of the valve, the pattern of acceleration is reversed: in the first part of the movement the initial acceleration is low and the final part is high, the valve being fully stopped by the time it reaches the valve seat. The motion of the valve is determined by the cam profile. The cam, the valve and the follower are designed to work together so that the movement of the valve remains under the control of the cam at all engine speeds. On most production vehicles, the opening and closing profiles of the cam are identical. For some specialist applications, specially designed cams may be used to provide extra valve lift and opening and closing characteristics.

Example 12.6

In a certain engine, a valve and its follower weigh 0.2 kg. The maximum acceleration produced by the cam at a camshaft speed of 2000 rev/min is 3000 m/s^2. Calculate the minimum valve spring force that is required to keep the valve and its follower in contact with the cam.

Solution

The force F that is imparted to the valve and its follower by the acceleration from the cam may be calculated from F = mass × acceleration.

$$F = 0.2\,kg \times 3000\,m/s^2 = 600\,N$$

This is the force provided by the valve spring.

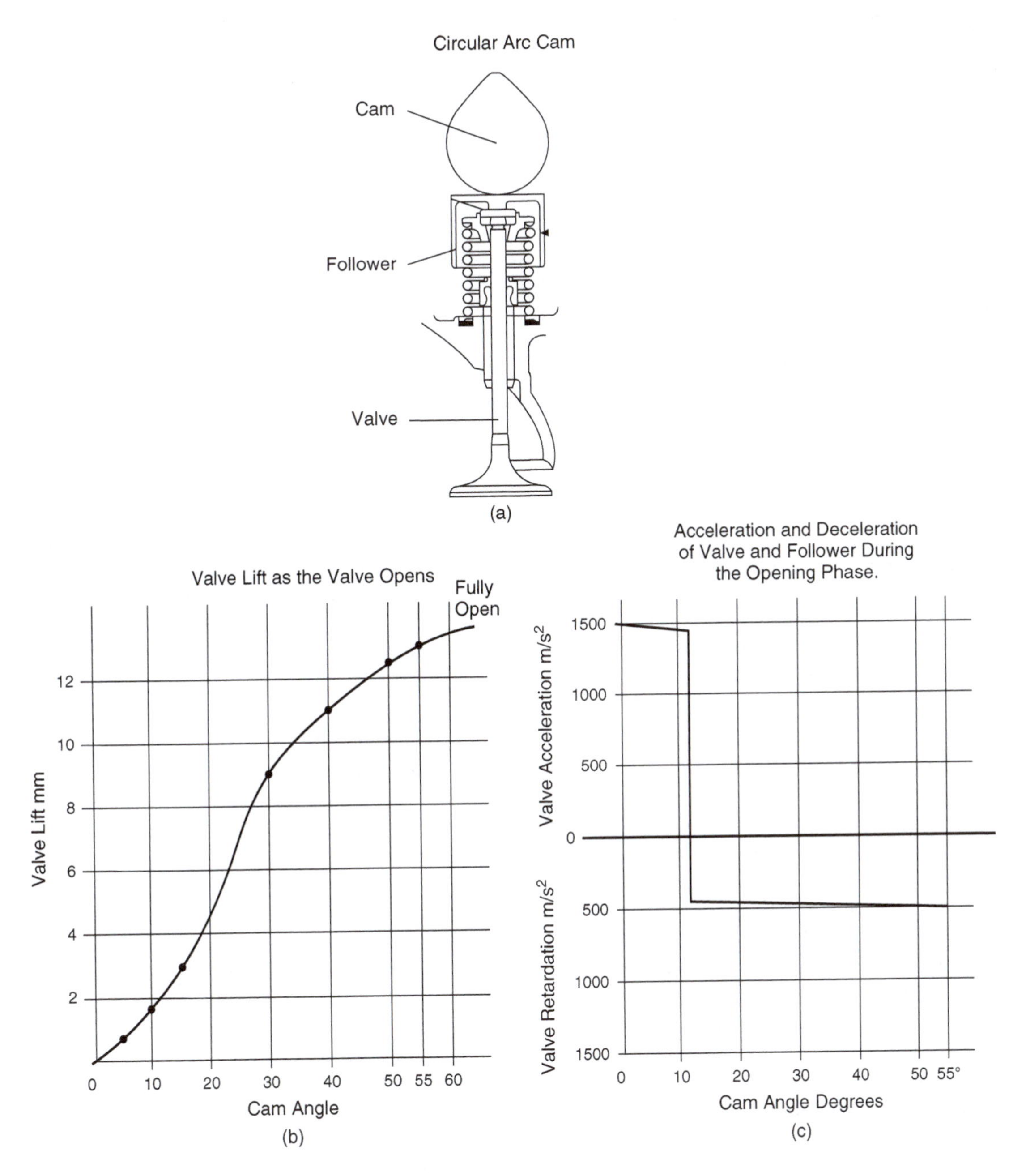

Fig. 12.20 Valve lift

12.17 Summary of formulae

For simple harmonic motion:

The maximum velocity of N = v = ωR. That is, when θ = 90° or 270°.

The maximum acceleration of N = $\omega^2 R\,m/s^2$ occurs when θ = 0° or 180°

Periodic time t = 2π√(acceleration/displacement)

Frequency of vision

$$= \frac{1}{2\pi}\sqrt{(\text{displacement/acceleration})}$$

The frequency is the number of vibrations made in 1 second = 1/t Hz

Periodic time – coiled spring, t = 2π √ (d/g)

Periodic time – flywheel and shaft

As the motion is SHM, the periodic time t = $\frac{1}{2\pi}\sqrt{}$(displacement/acceleration)

This can be shown to be t = 2π√(I/μ), where I is the moment of inertia of the flywheel and μ is the torsional stiffness of the shaft.

12.18 Exercises

12.1 A single-cylinder engine has a stroke of 80 mm and a connecting-rod length of 160 mm. The piston has a mass of 600 g and the connecting-rod mass is 900 g. Assuming that one third of the connecting rod mass is added to the piston mass for calculation purposes, calculate the magnitude of the secondary force at 30 degrees after top dead centre when the engine is running at 3000 rev/min.

12.2 A wheel and tyre assembly has an out-of-balance mass of 0.1 kg acting at a radius of 200 mm from the centre of rotation. Calculate the out-of-balance force created by this out-of-balance mass when the assembly rotates at 600 rev/min.

12.3 Two masses, A and B, are fixed to a disc that rotates at 300 rev/min. A has a mass of 2 kg acting at a radius of 0.2 m, and B has a mass of 1.5 kg acting at a radius of 0.3 m. The angle between the two masses is 120°.
(a) Determine the magnitude of the out-of-balance force.
(b) Determine the magnitude of the balance mass that must be placed at a radius of 250 mm.

12.4 Sketch and describe a crankshaft torsional vibration damper. Explain why such a damper is more likely to be found on 6-cylinder engine than on a 4-cylinder engine.

12.5 Sketch and describe the torsional vibration damper that is to be found at the hub of some clutch plates.

12.6 In a certain engine a valve and its follower weigh 0.2 kg. The maximum acceleration produced by the cam at a camshaft speed of 2000 rev/min is 3600 m/s^2. Calculate the minimum valve spring force that is required to keep the valve and its follower in contact with the cam.

12.7 A single-cylinder engine is to be balanced by placing masses on the crank, as shown in Figure 12.11. The piston and gudgeon pin have a mass of 0.50 kg, the connecting rod mass is 0.60 kg, and the crank-pin mass is 0.14 kg. The crank pin and part of the connecting rod act at an effective radius of 80 mm from the centre of the crank and the balance masses are to be placed at an equal radius on the opposite side of the crank. To achieve a satisfactory degree of balance, 50% of the mass of the reciprocating parts plus the mass of the rotating parts is added in the form of balance masses. For balancing purposes, 70% of the connecting-rod mass is taken to be a rotating mass. Determine the mass of the two balance masses that act at the same radius as the crank pin.

12.8 A vehicle suspension spring has a static deflection under full load of 300 mm. Determine the periodic time and natural frequency of this spring. Take g = 9.81 m/s^2.

12.9 A reciprocating engine has a stroke of 100 mm. The piston weighs 0.7 kg. Assuming that the piston moves with SHM, calculate the force on the piston at each end of the stroke when the engine is running at 3000 rev/min.

12.10 Plot graphs of valve lift and velocity on a base of cam angle, using the data given in Table 12.2.

Table 12.2 Plots of valve lift and velocity on a base of cam angle (Exercise 12.10)

Cam angle	0°	5°	10°	12.6°	20°	30°	40°	50°	55°
Lift (mm)	0	0.43	1.73	2.9	5.81	9.13	11.4	12.5	12.7
Vel. (m/s)	0	1.25	2.5	3.21	2.75	2.02	1.24	0.42	0

12.11 The full lift of a poppet valve that moves with simple harmonic motion is 10 mm. The time taken to open and close the valve is 0.01 s. If the mass of the valve is 0.18 kg, calculate:

(a) the maximum acceleration of the valve;

(b) the force on the valve.

13
Heat and temperature

13.1 Temperature

Temperature is a measure of condition, i.e. how hot or how cold something is. For example, at normal atmospheric pressure, water boils at 100°C and freezes at 0°C and these are conditions that are commonly described as hot and cold. Temperatures are normally measured in degrees Celsius, although it is still quite common to encounter temperatures given in degrees Fahrenheit. One degree Celsius is equal to 1.8 degrees Fahrenheit.

Thermodynamic temperature scale (Kelvin)

In the Kelvin temperature scale, temperature measurement starts at absolute zero. Absolute zero occurs at minus 273°C. To convert Celsius temperatures to Kelvin temperatures it is necessary to add 273 degrees. One degree on the Kelvin scale is the same size as one degree on the Celsius scale. Kelvin temperatures are used in gas law calculations.

Table 13.1 Examples of temperatures

Operating temperature of an engine cooling system (liquid cooled)	80°C to 100°C
Exhaust catalyst operating temperature	400°C to 800°C
Combustion temperature	2000°C
Melting temperature of aluminium	660°C
Melting temperature of copper	1083°C
Melting temperature of titanium	1675°C

Cooling system temperature

Cooling systems on liquid-cooled engines are pressurised and the operating temperature is often greater than 100°C. That is why one should never remove the coolant filler cap without taking the necessary precautions.

13.2 Standard temperature and pressure (STP)

For some scientific purposes, certain values are determined at STP; i.e. at a temperature of **0°C** and a pressure of **1.013 bar**.

13.3 Thermal expansion

When materials are heated they expand. The amount of expansion that is produced by a one degree Celsius rise of temperature in a 1 metre length of material is known as the coefficient of

Table 13.2 Coefficients of linear expansion

Metal	Expansion per 1°C
Aluminium	0.000022
Iron and steel	0.000011
Copper	0.000017
Brass	0.000019
Platinum	0.000008
Invar (iron alloy with 36% nickel)	0.0000001

linear expansion. The coefficient of linear expansion is denoted by the symbol alpha (α). Water is an exception to this general rule because it expands at low temperature, between 4°C and 0°C, which accounts for cracked cylinder blocks and water pipes when anti-freeze precautions are not taken. Some coefficient of linear expansion values are given in Table 13.2.

$$\text{Expansion} = l \times \alpha \times (T_2 - T_1) \qquad \textbf{(13.1)}$$

where l = length of material, α = coefficient of linear expansion, and $(T_2 - T_1)$ is the change in temperature.

Example 13.2
An aluminium alloy rod has a length of 1.2 m at a temperature of 20°C. Determine the increase in length that occurs when the rod is heated to a temperature of 120°C. The coefficient of linear expansion for Al alloy = 0.000022/°C.

Solution
$\text{Exp} = l\alpha(T_2 - T_1) = 1.2 \times 0.000022 \times 100 = 0.00264\,\text{m} = 2.64\,\text{mm}$.

13.4 Heat

Heat is a form of energy that is transferred by virtue of a temperature difference between the heat source and the object that heat is being transferred to. Heat can only be transferred from a hot object to a cold one, unless a device such as a heat pump is used.

Sensible heat

When heat is added to a body, the temperature of the body may rise or the substance may change its state from a solid to a liquid, or from a liquid to a vapour or gas. The heat that is added to a body that causes the temperature to rise is known as sensible heat. The term *sensible* derives from the fact that the presence of heat can be detected (sensed) by a thermometer.

Latent heat

When heat is added to water and its state is changed from ice to water, or from water to steam, the temperature remains constant. The heat involved in these changes of state is known as latent heat, and this applies to other substances undergoing changes of state (e.g. fusion, sublimation or vaporisation). The word *latent* means that the temperature does not change and the presence of additional heat is 'hidden' from the thermometer.

Specific latent heat

Specific means the amount of heat required to produce a change of state in 1 kg of substance. Latent heat is involved whenever a change of state occurs. The specific **latent heat of vaporisation** is the quantity of heat required to convert one kilogram of substance from liquid into vapour. The specific **latent heat of fusion** is the quantity of heat required to convert one kilogram of a solid substance into liquid.

Specific latent heat of water

The specific latent heat of vaporisation of water at atmospheric pressure is **2257 kJ/kg**.

The specific latent heat of fusion of water at atmospheric pressure is **334 kJ/kg**.

Cooling down

When steam condenses back to water, the latent heat of vaporisation is taken out of the steam. When water changes into ice the latent heat of fusion is taken out of the water.

Example 13.3

Calculate the quantity of heat required to melt 5 kg of ice at 0°C, given that the specific latent heat of water at atmospheric pressure is 334 kJ/kg.

Solution

Quantity of heat Q

$$= \text{mass} \times \text{specific latent heat}$$

$$Q = 5\,\text{kg} \times 334\,\text{kJ/kg}$$

$$= 1670\,\text{kJ}$$

$$= 1.67\,\text{MJ}$$

Specific heat capacity

The specific heat capacity of a substance is the quantity of heat required to change the temperature of the substance by one degree Celsius, or Kelvin. The specific heat capacity of water is **4.187 kJ/kg** C. Some specific heat capacity values are given in Table 13.3.

Table 13.3 Approximate values of specific heat capacity

Substance	Specific heat capacity (kJ/kg C)
Water	4.187
Steel	0.494
Copper	0.385
Lead	0.130
Aluminium	0.921
Cast iron	0.544
Air – at constant pressure	1
Ammonia	2.2

Quantity of heat

Each kilogram of substance absorbs its specific heat capacity when it is heated through a one degree Celsius temperature increase. A mass of m kg of the substance will require m × specific heat capacity to raise its temperature by 1°C.

If Q is the quantity of heat required to raise the temperature of m kg of substance whose specific heat capacity is Cp, from T_1 to T_2, the quantity of heat energy required is given by

$$Q = m\,Cp\,(T_2 - T_1) \qquad \textbf{(13.2)}$$

Example 13.4

Calculate the heat absorbed by an aluminium alloy cylinder block of mass 60 kg when its temperature increases from 15°C to 85°C. Take the specific heat capacity of aluminium alloy to be 0.92 kJ/kg C.

Solution

$$Q = mCp(T_2 - T_1)$$

$$m = 60\,\text{kg},\ Cp = 0.92\,\text{kJ/kg C},\ T_2 - T_1 = 70°\text{C}$$

$$\therefore Q = 60 \times 0.92 \times 70$$

$$= 3864\,\text{kJ}$$

13.5 Heat transfer

Conduction

Heat transmission by conduction requires a material medium such as a solid, liquid, or gaseous substance.

Convection

Heat transfer by convection arises from circulating currents that develop when substances such as air and water are heated. Air becomes less dense when heated, and this allows air to rise, so causing the convection current.

Radiation

Heat transfer by radiation does not require a medium. All substances – solids, liquids and gases – emit heat energy by radiation. Hot bodies give off heat in the form of rays in all directions. These rays have characteristics similar to those of light rays. Radiant heat falling on an object may be reflected, absorbed, or transmitted through the object.

13.6 Heating, expansion and compression of gases

Absolute pressure

Absolute pressure = gauge pressure + barometric pressure. Absolute pressures are used in all calculations relating to gases.

Absolute temperature

Absolute temperatures are used in gas law calculations. The absolute temperature scale is known as the Kelvin scale, temperatures being given in degrees K.

Temperature in degrees K

= temperature in degrees C + 273

Example 13.5

The pressure and temperature of air in an engine cylinder on the induction stroke are 25°C and 0.6 bar, respectively. If the barometric pressure is 1.01 bar, determine the absolute temperature and pressure of the air in the cylinder.

Solution

Absolute temperature = 25 + 273 = 298 K

Absolute pressure = gauge pressure + barometric pressure

= 0.6 bar + 1.01 bar = 1.61 bar

13.7 Laws relating to the compression and expansion of gases

A perfect gas is one which obeys the following laws:

Boyle's law: $p_1V_1 = p_2V_2$

Charles' law: $\frac{V_1}{T_1} = \frac{V_2}{T_2}$

Joule's law: the internal energy is a function of temperature only

The specific heat capacities C_p and C_v remain constant at all temperatures

These laws apply to imaginary, ideal gases. Real gases do not obey the laws completely but permanent gases such as oxygen and nitrogen do obey them sufficiently well as to enable the laws to be used for the consideration of the working of practical engine cycles.

Heating a gas at constant volume

Imagine a mass m of gas contained in a cylinder with a piston that is fixed in position, as shown in Figure 13.1. If this gas is heated from a temperature T_1 to a temperature of T_2, the pressure will also rise, but no external work will be done.

$$\text{The amount of heat added} = mCv\,(T_2 - T_1) \qquad \textbf{(13.3)}$$

where m = mass of gas in kg and Cv = specific heat capacity of the gas in kJ/kg K.

Cv for air is approximately **0.71 kJ/kg K**.

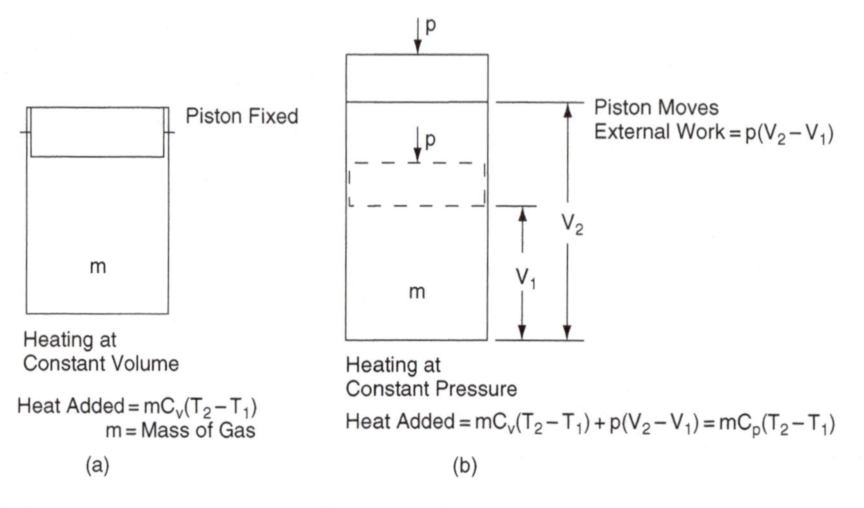

Fig. 13.1 Heating a gas at constant pressure and constant volume

Heating a gas at constant pressure

Imagine the same mass, m, of gas in a cylinder fitted with a movable piston, as shown in Figure 13.1(b) to which heat is applied to raise the temperature from T_1 to T_2. As the heat is applied the temperature of the gas in the cylinder rises but, because the piston is able to move, the pressure inside the cylinder remains constant. In addition to raising the temperature, work is done, by the gas, on the piston. The amount of heat added in this case is

$$Q = m\,Cv\,(T_2 - T_1) + \text{work done by the piston} \quad \textbf{(13.4)}$$

The heat energy required to heat the gas at constant pressure is greater than it is when the volume is constant because of the work done in moving the piston to maintain the constant pressure. This leads to the fact that a greater amount of heat is required to raise the temperature of a given mass of gas from T_1 to T_2 at constant pressure than is required for constant-volume heating because of the external work done by the moving piston. The amount of heat required for constant-pressure heating from T_1 to T_2 is

$$Q = m\,Cp\,(T_2 - T_1) \quad \textbf{(13.5)}$$

where Cp = specific heat capacity at constant pressure. (Cp for air is approximately 1 kJ/kg K.)

Charles' law

When a gas expands while the temperature remains constant, the volume divided by the absolute temperature is a constant, expressed mathematically:

$$\frac{V_1}{T_1} = \frac{V_2}{T_2} \quad \textbf{(13.6)}$$

where the temperatures are absolute K. This relationship is known as Charles' law, after the scientist who first explained it.

Expansion or compression at constant temperature – isothermal

When a fixed mass of gas expands or is compressed in a cylinder such that the temperature remains constant, the pressure multiplied by the volume is a constant. For example:

$$p_1 V_1 = p_2 V_2 = C \quad \textbf{(13.7)}$$

where the pressures are absolute. This relationship was discovered by Robert Boyle in 1661 and it is known as Boyle's law (see Figure 13.3).

The graph produced by this type of expansion is a rectangular hyperbola and such expansions of gases are known as hyperbolic.

General law $pV^n = C$, expansion of gas

This is the type of expansion that occurs in an engine cylinder. During this type of expansion $p_1V_1^n = p_2V_2^n = C$, where the pressures are absolute and the index n lies between 1 and 1.4.

Combined gas law

This law applies when pressure and temperature are changing and it is derived from a combination of Boyle's and Charles' laws. The law is

$$\frac{p_1V_1}{T_1} = \frac{p_2V_2}{T_2}$$

where the temperatures and pressures are absolute. From this equation, it follows that $pV/T = \text{constant}$. This constant is known as R, the characteristic gas constant, and the equation is written as $pV = RT$. When the mass of gas is m, $pV = mRT$.

Example 13.6

The air in an engine cylinder at the beginning of the compression stroke is at atmospheric pressure of 1.02 bar, and a temperature of 23°C. The swept volume of the engine cylinder is 450 cm^3 and the clearance volume is 50 cm^3. If the temperature of the air in the cylinder is 80°C at the end of the compression stroke, calculate the gauge pressure of this air.

Solution

Using the combined gas law $\dfrac{p_1V_1}{T_1} = \dfrac{p_2V_2}{T_2}$

$p_1 = 1.02\,\text{bar}$. $T_1 = 273 + 23 = 296\,\text{K}$, $T_2 = 273 + 80 = 353\,\text{K}$

$$p_2 = \frac{p_1V_1T_2}{T_1V_2}$$

V_1 = swept volume + clearance volume = $450 + 50 = 500\,\text{cm}^3$ V_2 = clearance volume = $50\,\text{cm}^3$

$$\therefore \frac{V_1}{V_2} = \frac{500}{50} = 10$$

$$p_2 = \frac{1.02 \times 10 \times 353}{296}$$

$p_2 = 12.16\,\text{bar} = 12.16 - 1.02 = 11.14\,\text{bar}$, gauge

Adiabatic expansion $pV^\gamma = C$, expansion of gas

When a gas expands or is compressed in such a manner that no heat is added or taken away from the gas the process is said to be adiabatic. In this case the index of expansion is gamma, γ, which is the ratio of the specific heats of the gas. Adiabatic processes form the basis for theoretical engine cycles.

Throttling expansion

This type of expansion occurs when a gas or vapour is forced through a small hole or slightly opened valve. The frictional effect causes the expanded gas or vapour to become drier, with the result that the enthalpy (amount of heat) in the substance is the same after the throttle as it was before it. This fact is made use of in the expansion valve of a vehicle's air-conditioning system. This valve causes the wet vapour to become dryer before entering the evaporator on its way back to the compressor.

Adiabatic index for air

The adiabatic index for air is the ratio of the specific heat capacities. It is used in connection with the study of ideal air cycles such as, Otto, Diesel, Dual combustion, and others. The adiabatic index is denoted by the symbol gamma (γ). Taking the approximate values of C_p and C_v for air, which are given above for air:

$$\gamma = \frac{C_p}{C_v} = \frac{1.00}{0.71} = 1.41$$

General equations relating to expansion of gases

By taking the combined gas equation $p_1V_1/T_1 = p_2V_2/T_2$, together with the general law $pV^n = C$, it can be shown that the following relationship exists:

$$T_1/T_2 = (V_2/V_1)^{n-1} = (p_1/p_2)^{(n-1)/n}$$

Example 13.7

A quantity of gas has a volume of $0.25\,m^3$ at a temperature of 177°C. The gas expands to a volume of $1.25\,m^3$ according to the law $pV^{1.3} = C$. Calculate the final temperature.

Solution

$T_1/T_2 = (V_2/V_1)^{n-1}$; $T_1 = 177 + 273 = 450\,K$, $n = 1.3$, $V2/V1 = 1.25/0.25 = 5$

$$\therefore T_2 = \frac{T_1}{(V_2/V_1)^{n-1}}$$

$T_2 = 450/5^{0.3} = 450/1.62 = 278\,K = 5°C$

13.8 Pressure–volume (pV) diagrams

Work done during expansion or compression of a gas

A pressure–volume diagram is a diagram or graph that represents the pressure and volume at different stages of a process of expansion or compression. The pressure in N/m^2 is plotted vertically and the volume in m^3 is plotted horizontally. Figure 13.2 shows the pV diagram that represents a constant pressure expansion from a volume of V_1 to a volume of V_2.

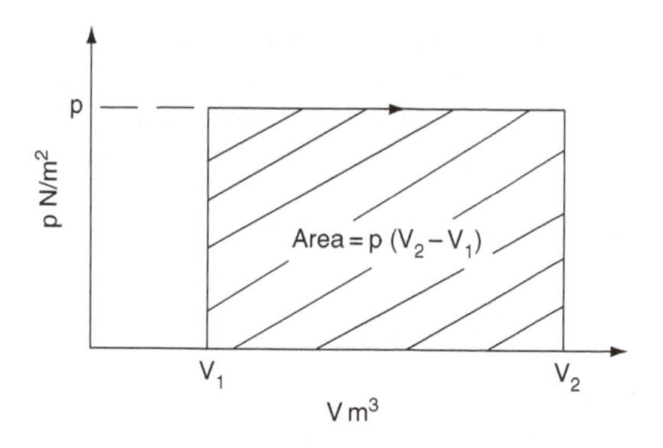

Fig. 13.2 Constant pressure expansion of a gas

Work done during this expansion is represented by the area enclosed by the diagram, which is $p(V_2 - V_1)$ joules, because p is in N/m^2 and $(V_2 - V_1)$ is a volume in m^3.

The product of pressure and volume is

$$\frac{N \times m^3}{m^2} = N\ m \text{ or joules}$$

In general, the area enclosed by a pV diagram represents work done on, or by, the gas.

Work done during hyperbolic expansion

The work done during this process $= p_1V_1 \ln r$, where r is the ratio of expansion $= V_2/V_1$. This expression for work done may be derived by integrating between the limits of V_2 and V_1.

Work done during expansion where $pV^n = C$

Here, by integrating between the limits of V_2 and V_1, the area under the curve proves to be $p_1V_1 - p_2V_2/n - 1$, which gives the formula

$$W = \frac{(p_1V_1 - p_2V_2)}{n - 1} \text{ for work done}$$

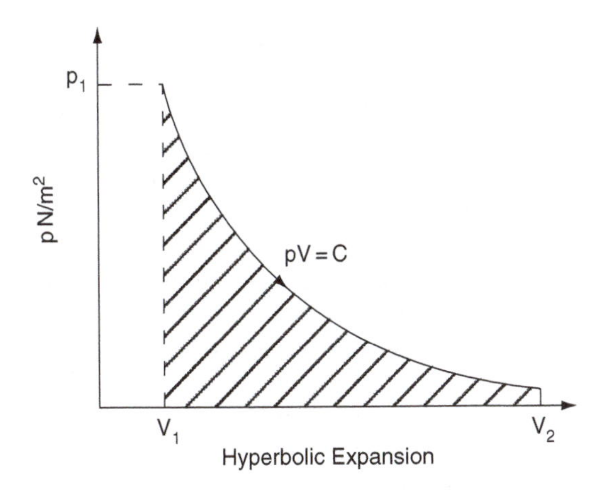

Fig. 13.3 Hyperbolic expansion

Work done during adiabatic expansion where $pV^{\gamma} = C$

This is a special case where $n = \gamma$.

The formula for work done is, thus ,

$$W = \frac{(p_1V_1 - p_2V_2)}{\gamma - 1}$$

Example 13.8

A quantity of gas has a volume of $2\,m^3$ and a pressure of 7 bar. It is expanded in a cylinder until the pressure is 0.5 bar. If the barometric pressure is 1 bar, calculate the work done if the expansion is:

(a) adiabatic; $\gamma = 1.4$
(b) hyperbolic.

Solution

(a) $W = (p_1V_1 - p_2V_2)/\gamma - 1$;
$p_1 = 7 + 1 = 8\,\text{bar}$;
$p_2 = 0.5 + 1 = 1.5\,\text{bar}$.
V_2 is required and is calculated from $p_2V_2{}^{\gamma} = p_1V_1{}^{\gamma}$;
$V_2 = (p_1/p_2)^{1/\gamma}$
$V_2 = (8/1.5)^{1/1.4} \times 2$
$= 5.33^{0.714} \times 2$
$V_2 = 3.8 \times 2 = 7.6\,m^3$.
$W = (8 \times 10^5 \times 2 - 1.5 \times 10^5 \times 7.6)/1.4 - 1$
$W = (11.4 \times 10^5)/0.4 = 28.5 \times 10^5 = 2.85\,\text{MJ}$

(b) $W = p_1V_1 \ln r$, where $r = V_2/V_1$.
To obtain V_2, use Boyle's law, $p_1V_1 = p_2V_2$, which gives $V_2 = p_1/p_2V_1$
$= (8/1.5) \times 2 = 10.67\,m^3$.
$W = 8 \times 10^5 \times \ln(10.67/2)$
$= 8 \times 10^5 \times \ln 5.33$
$= 8 \times 10^5 \times 1.67$
$= 13.39 \times 10^5$
$= 1.339\,\text{MJ}$

13.9 Summary of formulae

Linear expansion $= l\,\alpha(t_2 - t_1)$

Charles' law – constant pressure: $\frac{V_1}{T_1} = \frac{V_2}{T_2}$

Boyle's law – constant temperature: $p_1V_1 = p_2V_2$

Combined gas law: $\frac{p_1V_1}{T_1} = \frac{p_2V_2}{T_2}$

$p_1V_1{}^n = p_2V_2{}^n$

$PV = mRT$ etc.

$$W = \frac{(p_1V_1 - p_2V_2)}{(n-1)}$$

$\gamma = Cp/Cv$

$T_1/T_2 = (V_2/V_1)^{n-1} = (p_1/p_2)^{n-1/n}$

Degrees K = Celsius temperature + 273

13.10 Exercises

13.1 A certain petrol engine has a compression ratio of 8.5:1. During a compression test, a gauge pressure of 12 bar was recorded. At the start of compression the pressure was atmospheric at 1 bar and the temperature of the air in the cylinder was 30°C. Calculate the temperature of the air in the cylinder at the end of the compression stroke.

13.2 At the end of the compression stroke, the air in the combustion space of a certain spark ignition engine occupies a space of $50\,cm^3$. The pressure in the combustion chamber is 13 bar and the temperature is

170°C. If the air–fuel ratio is 14.7:1 calculate the mass of fuel to be injected. Take R for air = 0.287 kJ/kg K and assume the barometric pressure to be 1 bar.

13.3 A certain diesel engine has a compression ratio of 20:1. At top dead centre on the power stroke the gauge pressure in the cylinder is 150 bar. If expansion on the power stroke takes place according to the law $pV^{1.3}$ determine the gauge pressure at the end of this stroke. Atmospheric pressure = 1.014 bar.

13.4 A certain diesel engine has a compression ratio of 18:1. In its naturally aspirated form the air pressure in the cylinder at the commencement of compression is 1.01 bar ABS. The engine is adapted to take a turbocharger and this increases the air pressure in the cylinder to 1.7 bar ABS, at the start of the compression stroke. In both cases the index of the compression is 1.3. Determine the percentage increase in the air pressure at the end of compression. Take atmospheric pressure = 1 bar.

13.5 A single-cylinder petrol engine with a swept volume of 500 cm^3 has a compression ratio of 9.5:1. As part of an engine conversion, the clearance volume is decreased to 50 cm^3. Calculate:

(a) the new compression ratio.

(b) in both cases, the gauge pressure of the air in the cylinder at the end of the compression stroke if the initial pressure is 1.016 bar ABS and the index of compression is 1.35.

13.6 A certain turbocharged diesel engine has a compression ratio of 20:1. The air temperature in the cylinder at the commencement of the compression stroke is 110°C. If the compression follows the law $pV^{1.3} = C$, calculate the temperature at the end of the compression stroke. A later version of this engine has an intercooler that reduces the temperature of the air in the cylinder, at the commencement of compression, to 70°C. Calculate the air temperature in the cylinder at the end of compression in this case and comment on the effect that this reduction in temperature may have on engine power.

13.7 An engine has a swept volume of 750 cm^3 and a compression ratio of 10:1.
Calculate:

(a) the clearance volume;

(b) the work done on a single power stroke if the pressures are 180 bar at the commencement of the power stroke and 0.5 bar at the end of the stroke. Take the barometric pressure = 1 bar and the law of expansion as $pV^{1.25} = C$.

13.8 When fitting a new ring gear to a flywheel, the ring gear is heated so that the expansion produced is sufficient to allow the ring gear to slide on to the periphery of the flywheel, and, when the temperature drops, the ring gear will contract and fit tightly on the flywheel. In order to ensure that the hardness of the gear is not impaired, a thermo-chromatic coloured pencil is used to ensure that the ring gear is heated to 270°C. If the coefficient of linear expansion of the ring gear is 0.000011/°C and the inner diameter is 300 mm, calculate the increase in diameter when the ring gear is heated to 270°C from an original temperature of 25°C.

13.9 A certain gudgeon pin has a diameter of 20.622 mm at 21°C. It is to be fitted into the small end-eye of a connecting rod that has a diameter of 20.59 mm at 21°C. The recommended procedure for performing this fitting operation is to heat the connecting rod. If the coefficient of linear expansion of the connecting rod is 0.000011/°C, calculate the temperature to which the rod must be heated in order to achieve the expansion of 0.05 mm of the gudgeon-pin eye that is required for completion of the fitting process.

13.10 The cooling system of a certain engine contains 6 kg of water. Calculate the quantity of heat absorbed by this water when the temperature increases from 5°C to

80°C. Specific heat capacity of water = 4.2 kJ/kg C.

13.11 During a braking operation, the four brake discs of a motor car absorb 800 kJ of energy. If the original operating temperature of the brake discs was 40°C, calculate the brake discs' temperature at the end of the braking operation. The mass of each brake disc is 2.5 kg and the specific heat capacity of the disc material is 0.5 kJ/kg C. Assume no heat is transferred away from the brake discs.

13.12 A certain small petrol engine has a compression ratio of 9:1. At the start of the compression stroke, the air in the cylinder is at atmospheric pressure and its temperature is 20°C. If the pressure in the cylinder at the end of compression is 12 bar (gauge), determine the temperature of the compressed air in the cylinder. Atmospheric pressure = 1.016 bar.

13.13 2 m^3 of gas at 50°C are heated at constant pressure until the volume is doubled. Determine the final temperature of the gas.

14

Internal combustion engines

14.1 Engine power

Brake power

The engine power that actually reaches the output shaft or flywheel of an engine is known as the brake power. It is the power that is measured by a dynamometer and a dynamometer is also known as a brake, hence the term, brake power. The simplest form of dynamometer is shown in Figure 14.1. Here a rope is wound around the circumference of the engine flywheel; one end of the rope is supported by a spring balance and the other end has weights attached to it. The load that the engine is working against is varied by increasing, or decreasing, the amount of weight. The effective load (force) that the engine is working against is $= W - S$.

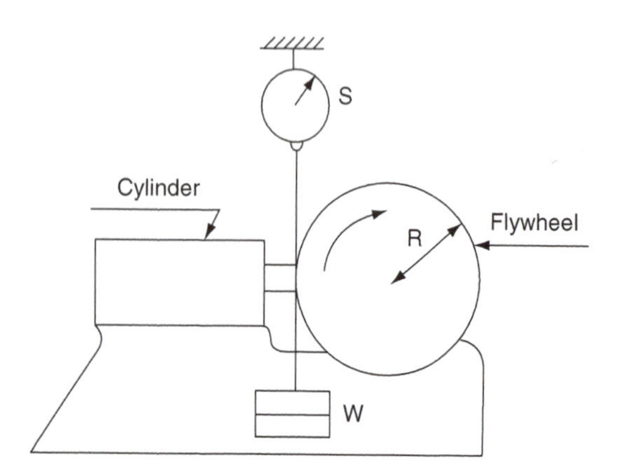

Fig. 14.1 Simple dynamometer for low-speed engines

Rotating the flywheel inside the rope, against the force $(W - S)$ newtons, is equivalent to moving the force $(W - S)$ around the circumference of the flywheel. The distance travelled by the force during one revolution of the flywheel is equal to the circumference of the flywheel. If the radius of the flywheel is R metres, the circumference $= 2\pi R$ metres. From this, it may be seen that:

Work done per revolution of the flywheel
$= \text{force} \times \text{distance}$
$= (W - S) \times 2\pi R$ joules

$(W - S) \times R$ also $= T$, the torque that the engine is exerting. By substituting T for $(W - S) \times R$the work done per revolution becomes $= T \times 2\pi$ joules

Brake power = work done per second
$= (T \times 2\pi) \times$ number of revolutions per second

This is normally written as brake power $b_p = 2\pi TN$, where N = number of revolutions per second.

When the torque is in newton metres (N m), the formula $b_p = 2\pi TN$ gives the power in watts. As engine power is given in thousands of watts, i.e. kilowatts, the formula for brake power normally appears as:

$$b_p = 2\pi TN/1000\,kW \qquad \textbf{(14.1)}$$

14.2 Dynamometers for high-speed engines

Hydraulic dynamometers of the type manufactured by the Heenan Froude company are commonly used for engine testing purposes. The energy from the engine is converted into heat by the action of the dynamometer rotors and the action is somewhat similar to the action of a torque convertor. Figure 14.2 shows a cross-section of the dynamometer; the engine under test is connected to the shaft that drives rotor A. Rotor A acts as a pump that directs pressurised water into the stator F. The stator is connected to a torque arm that registers the torque acting on it by means of weights and a spring balance. The power absorption capacity of the dynamometer is controlled by means of a form of sluice gate E that is interposed between the stator F and rotor A. These sluice gates are controlled manually. The energy absorbed by the dynamometer is converted into heat in the water, which is supplied under pressure through the inlet pipe D.

Electrical dynamometers are also used for high-speed engine testing; these convert the engine power into electricity which is then dissipated through heat or electro-chemical action.

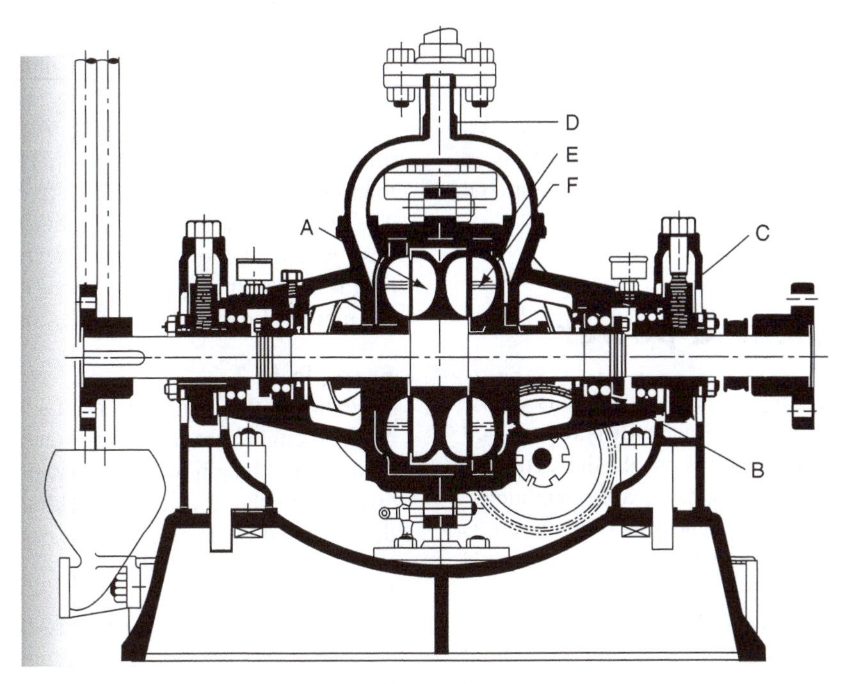

Fig. 14.2 Hydraulic dynamometer

Example 14.1

A certain engine develops a torque of 120 Nm while running at a speed of 3000 rev/min. Calculate the brake power. Take $\pi = 3.142$.

Solution

$b_p = 2\pi TN/1000$.

From the question, $T = 120$ Nm; $N = 3000/60 = 50$ rev/s

Substituting these values in the formula gives $b_p = 2 \times 3.142 \times 120 \times 50/1000 = 37.7$ kW.

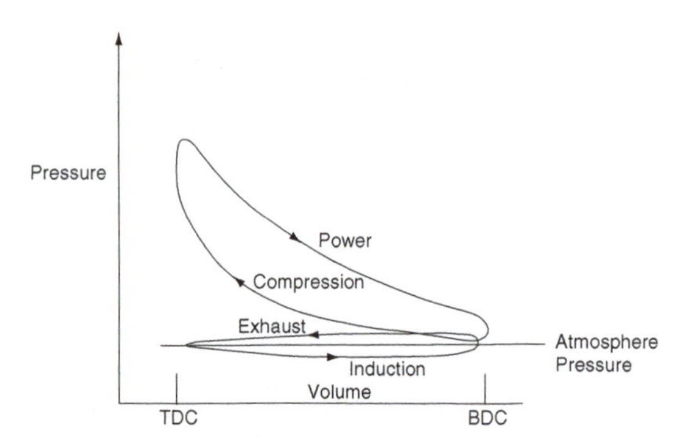

Fig. 14.3 Indicator diagram for a 4-stroke engine

14.3 Horsepower

One horsepower = 746 watts. The conversion from kilowatts to horsepower is:

Power in kilowatts ÷ 0.746. Taking the above example of 37.7 kW, the horsepower equivalent is 37.7 kW ÷ 0.746 = 50.5 bhp.

14.4 PS – the DIN

The abbreviation PS derives from *pferdestrake*, which is German for the pulling power of a horse. 1 PS is slightly less than 1 imperial horsepower.

14.5 Indicated power

Indicated power (ip) is the power that is developed inside the engine cylinders. It is determined by measuring the pressures inside the cylinders while the engine is on test, on a dynamometer. The device that is used to measure the pressure is called an indicator, from which the term indicated power is derived. The indicator diagram for a 4-stroke engine is shown in Figure 14.3.

Because the pressure varies greatly throughout one cycle of operation of the engine, the pressure that is used to calculate indicated power is the mean effective pressure.

14.6 Mean effective pressure

The mean effective pressure is that pressure which, if acting on its own throughout one complete power stroke, would produce the same power as is produced by the various pressures that occur during one operating cycle of the engine.

In Figur 14.4(a), the loop formed by the pressure trace for the exhaust and induction strokes is known as the pumping loop. This represents power taken away and the effective area is determined by subtracting A_2 from A_1. This calculation then gives the indicated mean effective pressure. The resulting area $(A_1 - A_2)$ is then divided by the base length of the diagram. The resulting mean height is then multiplied by a constant that gives the indicated mean effective pressure (imep).

Example 14.2

An indicator diagram taken from a single-cylinder 4-stroke engine has an effective area of 600 mm^2. If the base length of the indicator diagram is 60 mm and the constant is 80 kPa/mm calculate the indicated mean effective pressure.

Solution

Indicated mean effective pressure (imep)

$$= \frac{\text{effective area of indicator diagram}}{\text{base length of the diagram}} \times \text{constant}$$

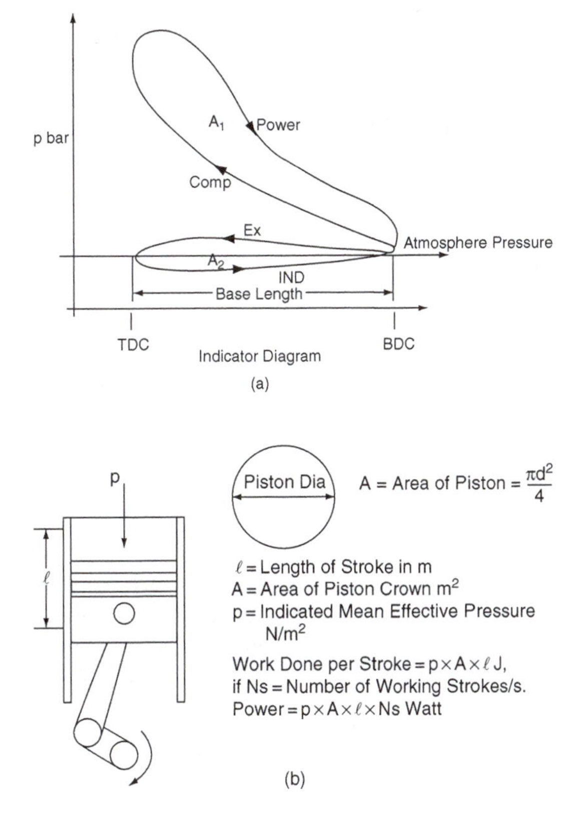

Fig. 14.4 Indicated power calculation

$$\text{imep} = \frac{600}{60} \times 80\,\text{kPa/mm}$$

$$\text{imep} = 800\,\text{kPa} = 8\,\text{bar}$$

14.7 Calculation of indicated power

Indicated power is calculated from the formula

$$\text{Indicated power (ip)} = \frac{\text{PlaN}}{1000}\,\text{kW}$$

where P = mean effective pressure in N/m^2, l = length of engine stroke in m, a = cross-sectional area of cylinder bore in m^2, N = number of working strokes per second.

Example 14.3

In a test, a certain single-cylinder 4-stroke engine develops a mean effective pressure of 5 bar at a speed of 3000 rev/min. The length of the engine stroke is 0.12 m and the cross-sectional area of the cylinder bore is $0.008\,\text{m}^2$. Calculate the indicated power of the engine in kW.

Solution

The engine is single-cylinder 4-stroke, so there is one working stroke for every two revolutions. N, the number of working strokes per second $= 3000 \div 60/2 = 25$; $\text{P} = 5\,\text{bar} = 500\,000\,\text{N/m}^2$, $\text{l} = 0.12\,\text{m}$; $\text{a} = 0.008\,\text{m}^2$.

Substituting these values in the formula gives

$$\text{ip} = 500\,000 \times 0.12 \times 0.008 \times 25/1000\,\text{kW}$$

$$= 12\,\text{kW}$$

Number of working (power) strokes

As mentioned, each cylinder of a 4-stroke engine produces one power stroke for every two revolutions of the crankshaft. In a multi-cylinder engine, each cylinder will produce one power stroke in every two revolutions of the crankshaft. A formula to determine the number of power strokes per minute for a multi-cylinder, 4-stroke engine is:

$$\text{number of power strokes per minute} = \frac{\text{number of cylinders}}{2} \times \text{rev/min}$$

Example 14.4

A 4-cylinder, 4-stroke engine develops an indicated mean effective pressure of 8 bar at 2800 rev/min. The cross-sectional area of the cylinder bore is $0.01\,\text{m}^2$ and the length of the stroke is 150 mm. Calculate the indicated power of the engine in kW.

Solution

The required formula is:

$$\text{ip} = \frac{\text{PlaN}}{1000}\,\text{kW}$$

From the question, N, which is the number of power strokes per second is

$$\frac{\text{number of cylinders}}{2} \times \frac{\text{rev/min}}{60}$$

which gives $N = 4/2 \times 2800/60 = 93.3$ per second.

Length of stroke in metres $= 120/1000 = 0.12\,\text{m}$

$P = 8\,\text{bar} = 800\,000\,\text{N/m}^2$

And the area of the piston $a = 0.01\,\text{m}^2$

Substituting these values in the formula gives:

$$\text{ip} = 800\,000 \times 0.12 \times 0.01 \times 93.3 \div 1000\,\text{kW}$$

$$= 89.6\,\text{kW}$$

Because engine power is still frequently quoted in horsepower, it is useful to know that 1 horsepower $= 746\,\text{watts} = 0.746\,\text{kW}$.

This engine's indicated power is therefore $= 89.6/0.746 = 120.1\,\text{hp}$.

14.8 Cylinder pressure vs. crank angle

The behaviour of the gas in an engine cylinder is an area of study that enables engineers to make detailed assessments of the effects of changes in engine-design and types of fuel, etc. Devices such as pressure transducers and oscilloscopes permit cylinder gas behaviour to be examined under a range of engine operating conditions. The basic elements of an indicator that permit this type of study are shown in Figure 14.5(a). The sparking plug is drilled as shown, and the hole thus made is connected to a small cylindrical container in which is housed a piezoelectric transducer. Gas pressure in the cylinder is brought to bear on this piezo transducer through the drilling in the sparking plug and the small pipe. Gas pressure on the transducer produces a small electrical charge that is conducted to an amplifier, which is calibrated to convert an electrical reading into an input to the oscilloscope that represents gas pressure. An additional transducer is attached to the engine crankshaft. The electrical output from the crank transducer is amplified as necessary and fed to the X input of the oscilloscope.

Figure 14.5(b) shows the type of oscilloscope display that is produced by a high-speed indicator. The display shows part of the compression and power strokes of a diesel engine. Injection commences at point A and combustion starts at point B; point C represents the end of effective combustion. The periods between each of these points are the three phases of combustion.

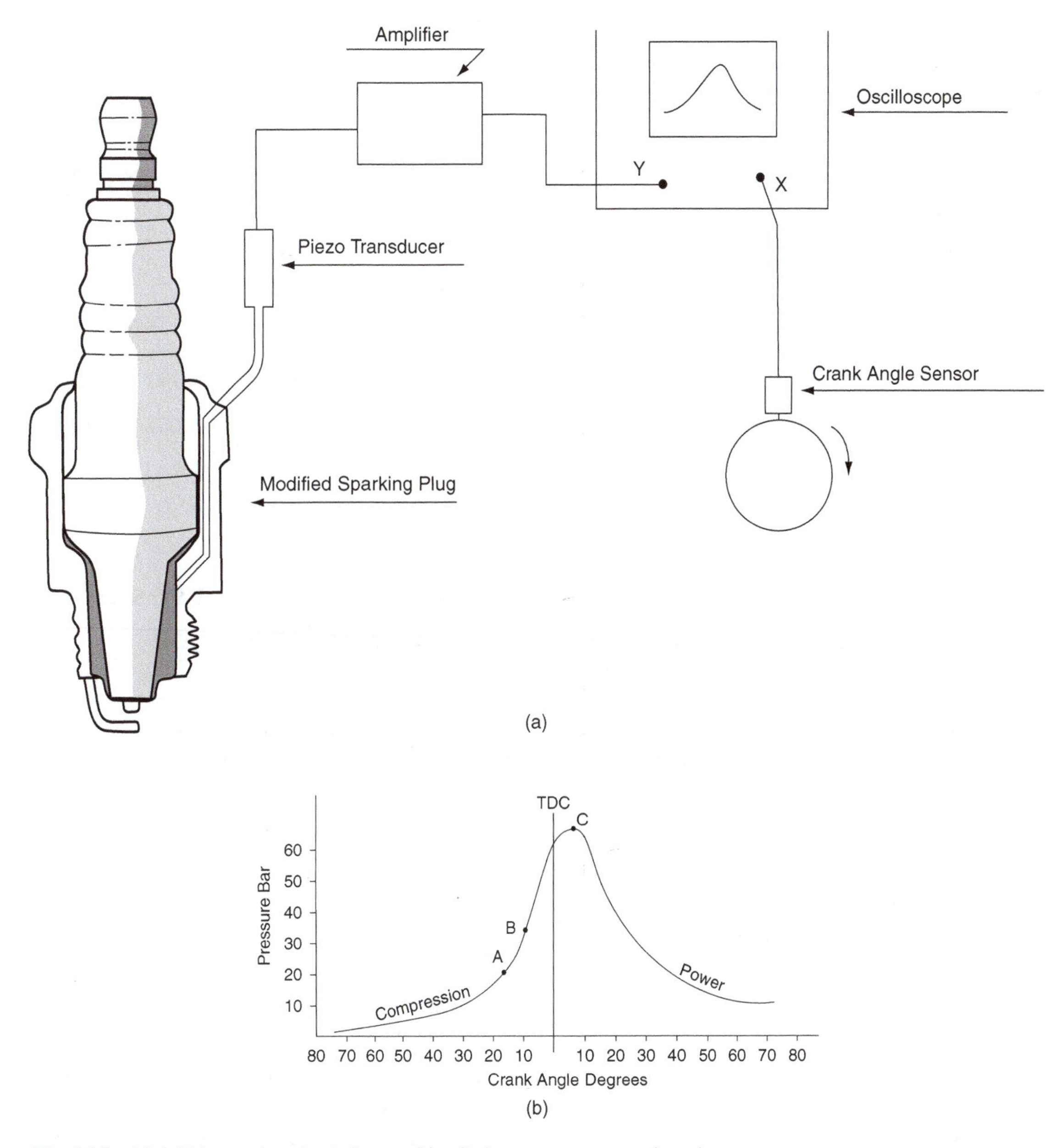

Fig. 14.5 (a) A high-speed engine indicator; (b) cylinder pressure vs. crank angle

14.9 Mechanical efficiency of an engine

The mechanical efficiency of an engine is defined as brake power/indicated power.

Example 14.5

A certain engine develops a brake power of 120 kW at a speed of 3000 rev/min. At this speed, the indicated power is 140 kW. Calculate the mechanical efficiency of the engine at this speed.

Solution

Mechanical efficiency

$$= \frac{\text{brake power}}{\text{indicated power}} \times 100\%$$

$$= 120/140 \times 100 = 85.7\%.$$

14.10 Morse test

Frictional losses in the engine bearings, the valve train and the piston and piston rings are the main causes of the power loss that makes the brake power of an engine smaller than the indicated power. The Morse test is an engine test that gives an approximate value for the frictional losses and which also provides an approximate value for the indicated power of a multi-cylinder engine.

The Morse test is conducted at constant engine speed on a dynamometer. The first phase of the test records the brake power of the engine when all cylinders are firing. Subsequently, one cylinder is prevented from firing and the dynamometer load is adjusted to bring the engine up to the same speed as it was when all cylinders were firing, the brake power then being recorded. The difference between brake power with all cylinders working and that obtained when one cylinder is cut out is the indicated power of the cylinder that is not working. This procedure is repeated for each of the cylinders and the indicated power for the whole engine is the sum of the power of the individual cylinders. A typical Morse test calculation is shown in the following example.

Example 14.6

The results shown in Table 14.1 were obtained when a Morse test was conducted on a 4-cylinder petrol engine.

Table 14.1 Morse test data for a 4-cylinder petrol engine (Example 14.6)

Cylinder cut-out	None	No. 1	No. 2	No. 3	No. 4
Brake power (kW)	60	41	40	43	42
Indicated power (kW)		19	20	17	18

Solution

The total indicated power = sum of each ip of the individual cylinders

$$\text{Total ip} = 19 + 20 + 17 + 18 = 74\,\text{kW}$$

$$\text{Mechanical efficiency} = \frac{\text{bp}}{\text{ip}} \times 100\% = (60/74) \times 100 = 81\%$$

14.11 Characteristic curves of engine performance

Figure 14.6 shows a graph of brake power against engine speed for a petrol engine. The points to note are:

- the graph does not start at zero because the engine needs to run at a minimum idle speed in order to keep running;
- the power increases almost as a straight line up to approximately 2500 rev/min; after that point the power increases more slowly and the curve begins to rise less sharply;
- maximum power is reached at a point that is near the maximum engine speed.

The main reason why the power does not increase in direct proportion to the engine speed is because the amount of air that can be drawn into the cylinder is limited by factors such as valve lift and port design. The amount of air actually drawn into the cylinders compared with the amount that could theoretically be drawn into them is known as the

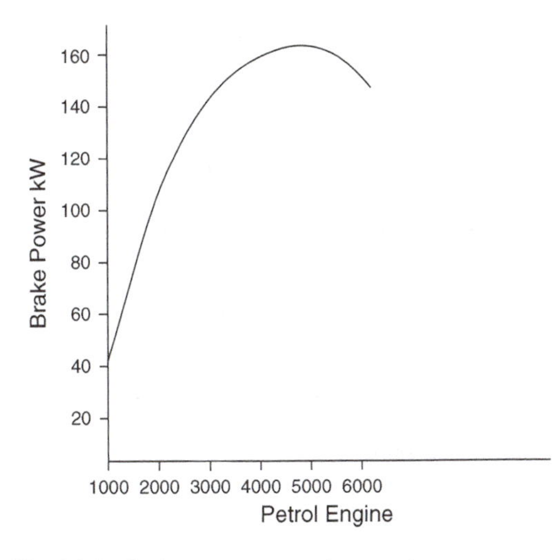

Fig. 14.6 Brake power vs. engine speed

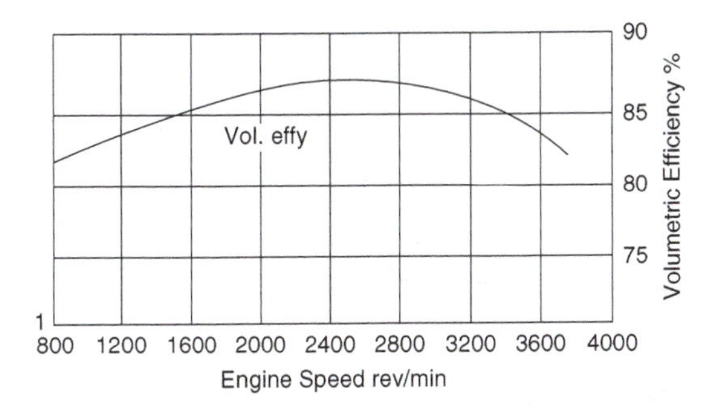

Fig. 14.7 Volumetric efficiency – diesel engine

volumetric efficiency. Above a certain speed, volumetric efficiency drops and this, coupled with frictional and pumping losses, affects the amount of power produced. Figure 14.7 shows a graph of volumetric efficiency plotted against engine speed.

14.12 Volumetric efficiency

Volumetric efficiency

$$= \frac{\text{volume of charge admitted at STP}}{\text{swept volume}}$$

$$= \frac{\text{actual air flow}}{\text{theoretical air flow}}$$

Example 14.7

A 4-cylinder 4-stroke petrol engine with a bore diameter of 100 mm and a stroke of 110 mm has a volumetric efficiency of 74% at an engine speed of 4000 rev/min. Determine the actual volume of air at STP that flows into the engine in 1 minute.

Solution

$$\text{Volumetric efficiency} = \frac{\text{actual air flow}}{\text{theoretical air flow}}$$

Theoretical air flow = swept volume of one cylinder × number of cylinders × $\frac{\text{rev/min}}{2}$

$$\text{Swept vol. of one cylinder} = \frac{\pi d^2 l}{4}$$

$$= 0.7854 \times 0.1 \times 0.1 \times 0.11\,\text{m}^3$$

$$= 8.64 \times 10^{-4}\,\text{m}^3$$

∴ Theoretical volume of air

$$= 8.64 \times 10^{-4}\,\text{m}^3 \times 4 \times 4000/2$$

$$= 6.91\,\text{m}^3/\text{min}$$

Actual air flow

= theoretical air flow × volumetric efficiency

Actual air flow $= 0.74 \times 6.91 = 5.11\,\text{m}^3/\text{min}$

14.13 Torque vs. engine speed

Torque is directly related to brake mean effective pressure (bmep). The graph in Figure 14.8 shows

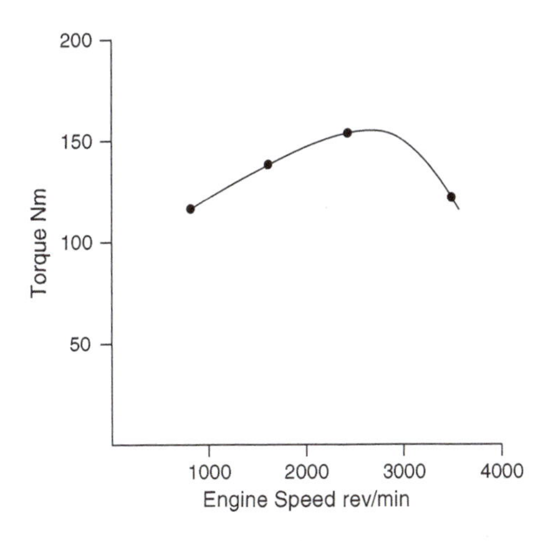

Fig. 14.8 Torque vs. engine speed

that torque varies across the speed range of an engine. Torque may be determined by direct measurement at a dynamometer, or by calculation; for example

$$T = \frac{\text{brake power}}{2\pi n}$$

where the brake power is determined by dynamometer tests.

14.14 Specific fuel consumption vs. engine speed

Brake specific fuel consumption (bsfc) is a measure of the effectiveness of an engine's ability to convert the chemical energy in the fuel into useful work. Brake specific fuel consumption is calculated as shown, for example

$$\text{bsfc} = \frac{\text{mass of fuel consumed per hour}}{\text{brake power}}$$

Figure 14.9 shows how bsfc varies with engine speed.

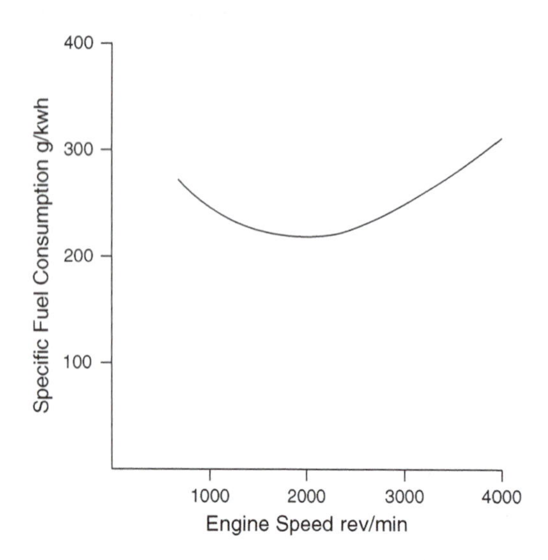

Fig. 14.9 Specific fuel consumption vs. engine speed

Example 14.8

The results shown in Table 14.2 were obtained during a fuel consumption test on an engine.

Table 14.2 Results of fuel consumption test on an engine (Example 14.8)

Brake power (kW)	10	15	20	25	30	35
Total fuel consumed per hour (kg)	4	5.21	6.74	8.4	10.59	12.6

Determine the specific fuel consumption (sfc) in kg/kWh in each case and plot a graph of specific fuel consumption on a base of brake power.

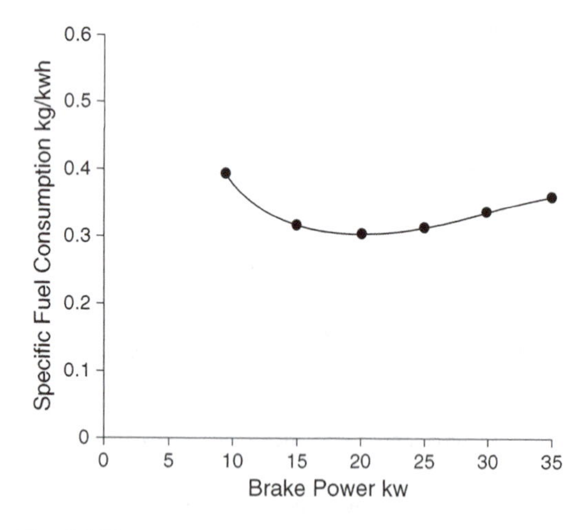

Fig. 14.10

Solution

Specific fuel consumption

$$= \frac{\text{mass of fuel used per hour}}{\text{brake power}}$$

Table 14.3 Specific fuel consumption calculations based on Table 14.2 (Example 14.8)

Brake power (kW)	10	15	20	25	30	35
Specific fuel consumption (kg/kWh)	0.4	0.347	0.337	0.336	0.353	0.36

14.15 Brake power, torque and sfc compared

In order to assess the importance of these curves, it is normal practice to plot brake power, torque and specific fuel consumption on a common base of engine speed. The vertical scales are made appropriate to each variable. Figure 14.11 shows such a graph.

The points to note are:

- the maximum power of 52 kW occurs at 4000 rev/min;
- the minimum specific fuel consumption of 237 g/kWh occurs at 2000 rev/min;
- the maximum torque of 147 N m occurs at 2100 rev/min.

This data highlights details that affect vehicle behaviour; e.g., when the specific fuel consumption is at its lowest value the engine is consuming the least possible amount of fuel for the given conditions – this is considered to be the most economical operating speed of the engine. The maximum torque also occurs at approximately

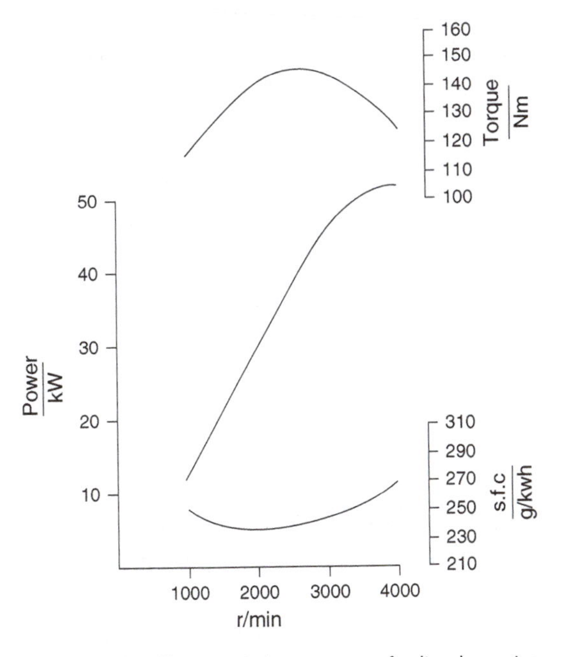

Fig. 14.11 Characteristic curves of diesel engine variables

the same speed, because torque converts to tractive effort; the maximum tractive effort is obtained at the same engine speed as is found for the maximum torque speed.

The maximum power is reached at a high engine speed.

The ratio of maximum power speed to maximum torque speed is known as the engine speed ratio.

The most useful part of the power curve of the engine lies between the maximum torque speed and the maximum torque speed. The difference between maximum power speed and maximum torque speed is known as the engine operating range.

In this case the engine operating speed range = 4000 rev/min − 2100 rev/min = 1900 rev/min.

The actual shapes of the graphs of these engine characteristics are determined by engine design. The graphs for a heavy transport vehicle engine may be expected to be quite different from those for a saloon car.

14.16 Brake mean effective pressure

The brake mean effective pressure (bmep) may be obtained from the brake power curve of the engine as follows: bmep = brake power in kW × 1000 ÷ lan Nm. In this equation, l = length of engine stroke in metres, a = cross-sectional area of the cylinder bore in square metres, and n = the number of working strokes per second.

When bmep is plotted against engine speed, the curve produced is the same shape as the torque curve because torque is related to bmep. Engine performance data such as specific fuel consumption, and its relationship to bmep, at a given engine speed, may be shown in graphical form as in Figure 14.12. Here the engine is run at constant speed, on a dynamometer, and the air–fuel ratio is varied.

The main point to note here is that maximum bmep is developed when the mixture is rich. The minimum fuel consumption occurs when the

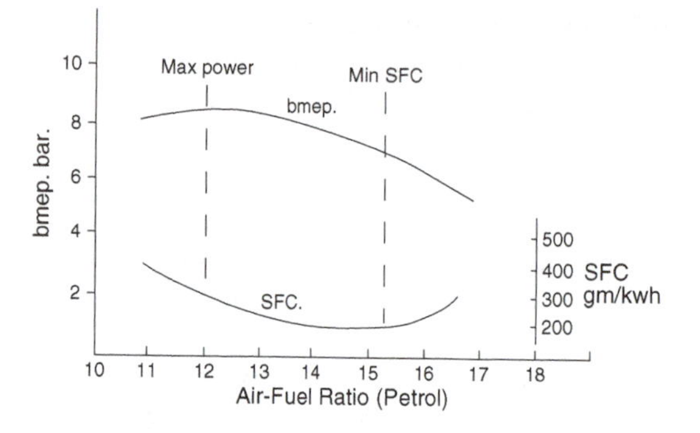

Fig. 14.12 Brake mean effective pressure vs. sfc at constant engine speed

air–fuel ratio is slightly weaker than the chemically correct air–fuel ratio of 14.7:1 for petrol.

14.17 Thermal efficiency

The thermal efficiency of an engine is a term that is used to express the effectiveness of an engine's ability to convert heat energy into useful work. Thermal efficiency is the ratio of energy output of the engine to energy supplied to the engine in the fuel.

Brake thermal efficiency

$$= \frac{\text{brake energy/s}}{\text{mass of fuel/s} \times \text{calorific value of the fuel}}$$

Example 14.9

During a 10-minute dynamometer test on a petrol engine, the engine develops a brake power of 45 kW and uses 3 kg of petrol. The petrol has a calorific value of 43 MJ/kg. Calculate the brake thermal efficiency.

Solution

Fuel used per second $3/600 = 0.005\,\text{kg/s}$

$1\,\text{kW} = 1\,\text{kJ/s}$. Brake energy per second $= 45\,\text{kJ/s}$

Brake thermal efficiency $= 45 \times 10^3 \div 0.005 \times 43 \times 10^6 = 45\,000 \div 215\,000 = 0.209$ or 20.9%.

14.18 Indicated thermal efficiency

Indicated thermal efficiency

$$= \frac{\text{indicated energy/s}}{\text{mass of fuel/s} \times \text{calorific value}}$$

Example 14.10

During a dynamometer test, a certain 4-cylinder, 4-stroke diesel engine develops an indicated mean effective pressure of 8.5 bar at a speed 2000 rev/min. The engine has a bore of 93 mm and a stroke of 91 mm. The test runs for 5 minutes during which time 0.8 kg of fuel are used. The calorific value of the fuel is 43 MJ/kg. Calculate the indicated thermal efficiency.

Solution

Indicated power = P l a N

N = number of working strokes/second

$= \text{number of cylinders}/2 \times \text{revs/s}$

$= 4/2 \times 2000/60 = 66.7$

Length of stroke (l) $= 0.091\,\text{m}$

Area of bore (a) $= \pi/4 \times 0.093 \times 0.093 = 0.0068\,\text{m}^2$

imep, $p = 8.5 \times 10^5\,\text{N/m}^2$

Indicated power

$$= \frac{8.5 \times 10^5 \times 0.091 \times 0.0068 \times 66.7}{1000\,\text{kW}}$$

$= 35.1\,\text{kW}$

Mass of fuel/s $= 0.8/300 = 0.0027\,\text{kg}$

Indicated thermal efficiency

$$= \frac{35.1 \times 1000\,\text{J/s}}{0.0027 \times 43 \times 10^6}$$

$= 0.302$ or 30.2%.

14.19 Brake thermal efficiency petrol vs. diesel

For reasons that are explained in Chapter 15, which deals with engine cycles, the brake thermal efficiency of a diesel engine is normally higher than it is for a comparably sized petrol engine. The higher thermal efficiency of the diesel engine results in good fuel economy. The figures shown in Table 14.4 refer to two vehicles of the same model, one fitted with a diesel engine and the other with a petrol engine of the same size.

Table 14.4 Fuel economy in petrol and diesel engines

Fuel economy	Petrol engine	Diesel engine
Urban cycle miles/US gallon	15	25
Highway cycle	21	34
Composite	18	29

14.20 Heat energy balance

In Example 14.9 the engine has a brake thermal efficiency of 20.9%. This means that 20.9% of the energy supplied to the engine in the fuel is converted into useful work. In order to determine what happens to the remainder of the energy supplied in the fuel, a heat energy balance test is conducted. The energy balance is conducted under controlled conditions in a laboratory. The test apparatus is equipped with a dynamometer and with meters for recording air and fuel supply, cooling water flow and exhaust gas temperature.

Energy supplied to the engine
 = mass of fuel used × calorific value
Energy to the cooling water = mass of water
 × specific heat capacity × temperature increase
Energy to useful work
 = brake power as measured by dynamometer
Specific heat capacity of water = 4.18 kJ/kg C
Energy to the exhaust = mass of exhaust gas
 × specific heat capacity × temperature increase
Specific heat capacity of the exhaust gas
 = 1.012 kJ/kg C approximately
Mass of fuel used as recorded on a fuel flow meter = m_f kg
Calorific value is obtained from the fuel specification
Mass of water flowing through the engine cooling system = m_w kg
Temperature rise of cooling water
 = $Tw_{out} - Tw_{in}$ °C
Temperature rise of exhaust gas is recorded by a thermocouple sensor that is placed in the exhaust stream, close to the engine = $Tex_{out} - Tex_{in}$ °C
Mass of exhaust gas: this is determined by adding the mass of air supplied to the mass of fuel supplied. The mass of air supplied is determined by the airflow meter through which all air for combustion passes.
Mass of exhaust gas = mass of air
 + mass of fuel = m_{ex}

Example 14.11
The data shown in Table 14.5 was obtained during a 1-hour laboratory test on an engine.

Construct a heat balance for this engine. Take the specific heat capacity of the exhaust = 1.01 kJ/kg C; specific heat capacity of the water = 4.2 kJ/kg C.

Table 14.5 Data from engine test (Example 14.11)

Brake power	60 kW
Mass of fuel used	18 kg/h
Air used (air fuel ratio = 16:1 by mass)	288 kg/h
Temperature of air at inlet	20°C
Temperature of exhaust gas	820°C
Mass of water through cooling system	750 kg/h
Cooling water temp in	15°C
Cooling water temp out	85°C
Calorific value of fuel	43 MJ/kg

Solution

Heat energy supplied in the fuel $= 18 \times 43 = 774$ MJ/h

Heat energy to the exhaust $= (18 + 288) \times 1.01 \times 10^3 \times (820 - 20) = 247.25$ MJ/h

Heat energy to cooling water $= 750 \times 4.2 \times 10^3 \times (85 - 15) = 252$ MJ/h

Heat energy equivalent of brake power $= 60 \times 1000 \times 3600 = 216$ MJ/h.

Table 14.6 Tabulated results of heat balance calculations (Example 14.11)

Heat supplied in the fuel/h	774 MJ	100%
Heat to brake power	216 MJ	28%
Heat to exhaust	247.25 MJ	32%
Heat to cooling water	252 MJ	32.6%
Heat to radiation etc. (by difference)	58.75 MJ	7.4%

Comment

This balance sheet provides a clear picture of the way in which the energy supplied to the engine is used. It is clear that approximately two thirds of the energy is taken up by the exhaust and cooling system. In turbo-charged engines, some of the exhaust energy is used to drive the turbine and compressor and this permits the brake power output of the engine to rise.

14.21 Effect of altitude on engine performance

The density of air decreases as the air pressure decreases and this causes a loss of power in naturally aspirated engines when they are operating at high altitude. It is estimated that this power loss is approximately 3% for every 300 m (1000 ft) of altitude. Turbocharged engines are affected to a lesser extent by this effect.

14.22 Summary of main formulae

Volumetric efficiency

$$= \frac{\text{volume of charge admitted at STP}}{\text{swept volume}}$$

$$\text{Brake power (bp)} = \frac{2\pi T n}{60 \times 1000}$$

where T is torque in Nm, n = rev/min

$$\text{Indicated power (ip)} = \frac{P\,l\,a\,n}{60 \times 1000}$$

where P = imep in N/m^2, l = length of stroke in m, a = area of bore in m^2, n = number of working strokes per min; for 4-stroke engine

$$n = \frac{\text{number of cylinders} \times \text{rev/min}}{2}$$

$$\text{Mechanical efficiency} = \frac{bp}{ip} \times 100$$

Brake thermal efficiency

$$= \frac{\text{energy to brake power}}{\text{energy supplied in the fuel}}$$

$$= \frac{bp \times 1000 \times 3600}{\text{mass of fuel/h} \times \text{cal. val}}.$$

Brake specific fuel consumption

$$= \frac{\text{mass of fuel used per hour}}{bp}$$

14.23 Exercises

14.1 A 4-cylinder, 4-stroke engine with a bore diameter of 80 mm and a stroke length of 75 mm develops an indicated mean effective pressure of 6.9 bar at an engine speed of 3600 rev/min. Calculate the indicated power at this engine speed.

14.2 The results shown in Table 14.7 were obtained when a Morse test was conducted

Table 14.7 Results of a Morse test conducted on a 6-cylinder engine (Exercise 14.3)

Cylinder cut-out	None	No. 1	No. 2	No. 3	No. 4
Brake power (kW)	120	85	87	88	86

on a 6-cylinder engine operating at a constant speed of 3000 rev/min. Calculate the indicated power of the engine and the mechanical efficiency at this speed.

14.3 (a) Calculate the brake power of a 6-cylinder engine that produces a bmep of 7 bar at a speed of 2000 rev/min. The bore and strokes are equal at 120 mm. (b) If the mechanical efficiency at this speed is 80%, calculate the indicated power.

14.4 A diesel engine develops a brake power of 120 kW at a speed of 3000 rev/min. At this speed the specific fuel consumption is 0.32 kg/kWh. Calculate the mass of fuel that the engine will use in 1 hour under these conditions.

14.5 As a result of a change in design, the bmep of a 4-cylinder, 4-stroke engine rises from 7.5 bar to 8 bar at an engine speed of 2800 rev/min. The bore and stroke are 100 mm and 120 mm respectively. Calculate the percentage increase in brake power at this speed that arises from the design changes.

14.6 A 6-cylinder, 4-stroke petrol engine develops an indicated power of 110 kW at a speed of 4200 rev/min, the imep at this speed being 9 bar. Calculate the bore and stroke given that the engine is 'square', i.e. the bore = stroke.

14.7 An engine develops a torque of 220 Nm during an engine trial at a steady speed of 3600 rev/min. The trial lasts for 20 minutes during which time 8.3 kg of fuel are consumed. The calorific value of the fuel is 43 MJ/kg. Calculate the brake thermal efficiency.

14.8 The data in Table 14.8 shows the results of a Morse test conducted on a 4-cylinder engine. Calculate the indicated power and the mechanical efficiency of the engine.

Table 14.8 Results of a Morse test conducted on a 4-cylinder engine (Exercise 14.8)

Cylinder cut-out	None	No. 1	No. 2	No. 3	No. 4
bp (kW)	25	17	16	18	16

14.9 A 4-cylinder, 4-stroke petrol engine with a bore diameter of 100 mm and a stroke of 110 mm has a volumetric efficiency of 70% at an engine speed of 4200 rev/min. Determine the actual volume of air at STP that flows into the engine in 1 minute.

14.10 The data in Table 14.9 shows the results obtained during a dynamometer test on a 4-cylinder diesel engine. Calculate the brake thermal efficiency for each value of brake power and plot a graph of (a) specific fuel consumption on a base of bp and (b) brake thermal efficiency on the same base. Calorific value of the fuel = 44 MJ/kg.

Table 14.9 Results obtained during a dynamometer test on a 4-cylinder diesel engine (Exercise 14.10)

bp (kW)	10	20	40	60	80
Specific fuel cons. (kg/kWh)	0.32	0.29	0.27	0.28	0.30

15 Theoretical engine cycles

Ideal engine cycles such as the Otto cycle, the diesel cycle and others are known as air cycles because the working fluid that drives the engine is assumed to be air. In order to study engine theory, the processes that make up the engine operating cycle are represented on a graph of pressure and volume. The graph is referred to as a pV diagram. The pV diagram is, in effect, an indicator diagram for an ideal engine. Throughout the cycle of operation, the working fluid is assumed to be air, and the air is assumed to be a perfect gas. The heat is assumed to be introduced by means of a heat source that is applied to the end of the cylinder, and the residual heat is removed by means of a cold source applied to the cylinder.

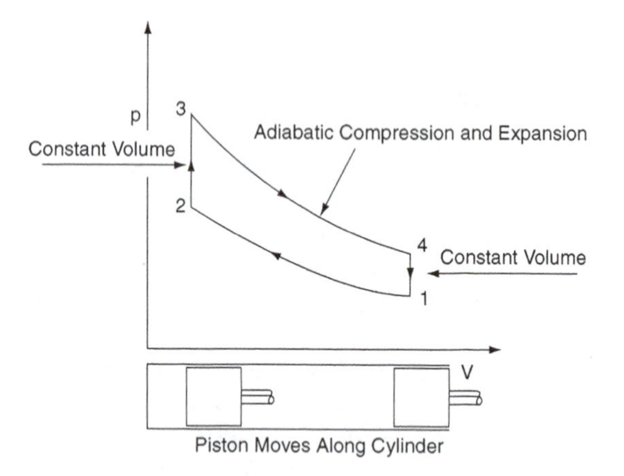

Fig. 15.1 The Otto (constant volume) cycle

15.1 The constant volume cycle (Otto cycle)

The theoretical constant volume cycle was first proposed by Beau de Rochas in 1862. However, the first working engine that used the Beau de Rochas principles was designed by Otto and the constant volume cycle has been known ever since as the Otto cycle. Figure 15.1 shows the pressure volume indicator diagram for the Otto cycle.

The cylinder is filled with air and the piston is used to compress this air and to transmit the work that arises from expansion. Heat is added to the air by means of a hot source that is attached to the cylinder; at the low-pressure end of the cycle the remaining heat in the air is removed by attaching a cold source to the cylinder.

Working round the diagram we start at point 1, which corresponds to bottom dead centre in a petrol engine. As the piston is pushed along the cylinder the volume decreases from V_1 to V_2. The air in the cylinder is compressed adiabatically up to point 2, which is top dead centre. During this adiabatic compression no heat is added to the air in the cylinder and none is lost through the cylinder wall – in effect there is no heat energy transfer between the air in the cylinder and the surroundings. This compression corresponds to the compression stroke in a petrol engine.

At point 2, the piston has reached the end of its stroke and heat energy is added to the air in the cylinder. This causes the temperature and pressure of the air in the cylinder to rise to the level represented by point 3. The volume of the air in the cylinder remains constant during this heating of the air from T_2 to T_3. The heat energy is added by attaching a hot body to the cylinder. The process corresponds to the combustion phase in a petrol engine.

At point 3, the heat source is removed and the gas expands adiabatically to point 4, doing work on the piston in the process. During adiabatic expansion there is no heat transfer between the air in the cylinder and its surroundings. This expansion corresponds to the power stroke in a petrol engine.

At point 4, the remaining energy in the air in the cylinder is removed, at constant volume, until the pressure returns to point 1. This process is performed by attaching a cold body to the cylinder. This part of the cycle corresponds to the exhaust phase in a petrol engine.

The useful work done during one cycle is represented by the area enclosed by the pV diagram. It can be shown that this useful work is equal to the heat energy supplied to the air in the cylinder between points 2 and 3 minus the heat rejected from this air between points 4 and 1 on the diagram.

$$\text{Useful work} = mCv(T_3 - T_2) - mCv(T_4 - T_1)$$

where m = mass of air in the cylinder and Cv = specific heat capacity of air at constant volume.

Thermal efficiency of the theoretical Otto cycle

$$\text{Thermal efficiency} = \frac{\text{useful work done}}{\text{heat energy supplied}}$$

Thermal efficiency

$$\eta = \frac{mCv(T_3 - T_2) - mCv(T_4 - T_1)}{m\,Cv(T_3 - T_2)}$$

$$\eta = \frac{(T_3 - T_2) - (T_4 - T_1)}{(T_3 - T_2)} \qquad \textbf{(15.1)}$$

This can be further simplified to give

$$\eta = 1 - \frac{(T_4 - T_1)}{(T_3 - T_2)}$$

Thermal efficiency in terms of compression ratio r

By using the relationship $T_1/T_2 = (V_2/V_1)^{n-1} = (p_1/p_2)^{(n-1)/n}$, and the fact that the compression ratio $r = V_1/V_2$, the temperatures T_1, T_2, T_3 and T_4 can be expressed in terms of r. For example $T_1 = T_2(1/r)^{\gamma-1}$. When these substitutions are made, the equation

thermal efficiency

$$= 1 - \frac{(T_4 - T_1)}{(T_3 - T_2)} \text{ can be expressed as}$$

$$\text{thermal efficiency} = 1 - \frac{1}{r^{\gamma-1}}$$

The thermal efficiency that is calculated from this formula is known as the air standard efficiency of an engine operating on the Otto cycle.

$$\text{Air standard efficiency} = 1 - \frac{1}{r^{\gamma-1}} \qquad \textbf{(15.2)}$$

where γ = adiabatic index for air, r = compression ratio.

Example 15.1

A petrol engine operating on the Otto cycle has a compression ratio of 8:1. Given the adiabatic index for air $\gamma = 1.41$, calculate the air standard thermal efficiency.

Solution

Air standard thermal efficiency

$$\eta = 1 - \frac{1}{r^{\gamma-1}}$$

$$r = 8, \ \gamma = 1.41$$

$$\eta = 1 - \frac{1}{8^{0.41}}$$

$$= 1 - 0.43$$

ASE $\eta = 0.57 = 57\%$

The ideal, or air standard, efficiency is 57%. This result shows that the larger r becomes, the smaller the fractional part of the thermal efficiency equation becomes and this results in higher thermal efficiency.

Effect of compression ratio on thermal efficiency

A graph of thermal efficiency plotted against compression ratio for an Otto cycle engine shows how the effect of increased compression ratio affects thermal efficiency (Figure 15.2). For the lower compression ratios in the range 5:1 to 10:1, a small increase in compression ratio leads to a marked increase in ideal efficiency.

The air standard efficiency is the efficiency that an Otto cycle engine could achieve should the ideal conditions be achieved, and it thus serves as a benchmark against which practical engines can be compared.

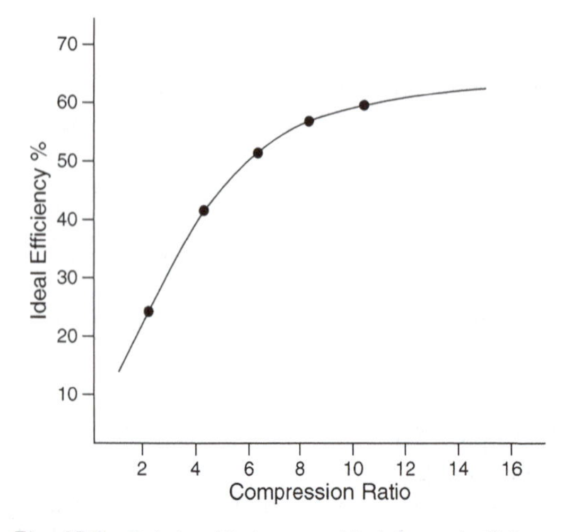

Fig. 15.2 Relationship between ideal thermal efficiency and compression ratio

15.2 Relative efficiency

$$\text{Relative efficiency} = \frac{\text{actual thermal efficiency}}{\text{air standard efficiency}} \quad \textbf{(15.3)}$$

Example 15.2

The following results were obtained during a test to determine the indicated thermal efficiency of an engine: indicated power 50 kW, fuel used per hour = 15 kg. Calorific value of the fuel = 43 MJ/kg. Engine compression ratio = 8:1; $\gamma = 1.41$. Calculate:

(a) the indicated thermal efficiency;
(b) the air standard efficiency;
(c) the relative efficiency.

Solution

(a) Indicated thermal efficiency

$$= \frac{\text{energy of indicated power}}{\text{energy supplied in fuel}}$$

Energy of indicated power = 50 kJ/s

Energy supplied in fuel

$$= \text{mass of fuel/second} \times \text{calorific value}$$

$$= (15/3600) \times 43\,\text{MJ/s}$$

$$= 179.17\,\text{kJ/s}$$

∴ Indicated thermal efficiency

$$= \frac{50\,\text{kJ/s}}{179.17\,\text{kJ/s}} = 0.28 = 28\%$$

(b) Air standard efficiency (ASE)

$$= 1 - \frac{1}{r^{\gamma-1}}$$

$$= 1 - \frac{1}{8^{0.41}}$$

$$= 1 - 0.43$$

$$= 0.57 = 57\%$$

(c) Relative efficiency

$$= \frac{\text{actual thermal efficiency}}{\text{air standard efficiency}}$$

$$= \frac{0.28}{0.57}$$

$$= 0.49 = 49\%$$

15.3 Diesel or constant pressure cycle

In the theoretical diesel cycle the pressure–volume diagram is of the form shown in Figure 15.3.

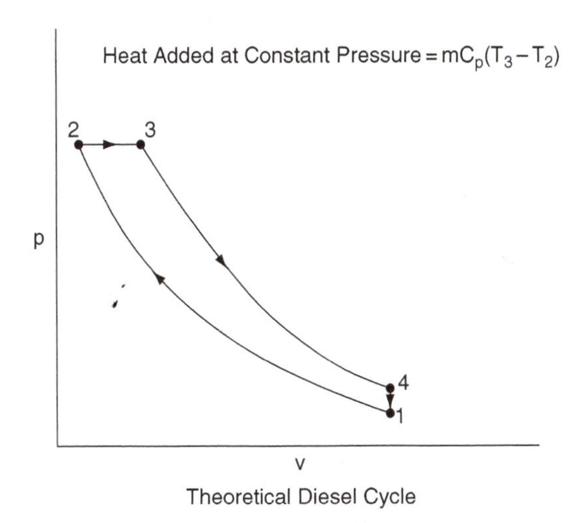

Fig. 15.3 pV diagram for the diesel cycle

Working round the diagram, starting at point 1: air is compressed adiabatically – no heat is added to the air in the cylinder and none is rejected, up to point 2. At point 2, heat energy is added and the air in the cylinder expands, doing work on the piston; the volume increases to point 3 while the pressure remains constant. Work is done on the piston during this phase. No further heat is added after point 3 and the air in the cylinder expands adiabatically doing work on the piston up to point 4. At point 4 the remaining heat energy in the air in the cylinder is removed while the volume remains constant and the air in the cylinder is restored to its original state at point 1.

Because the compression and expansion processes are adiabatic, during which no heat is added or rejected, the work done during the cycle = heat added – heat rejected.

The air standard thermal efficiency of the diesel

$$\text{cycle} = \frac{\text{heat added} - \text{heat rejected}}{\text{heat added}}$$

$$= \frac{mCp(T_3 - T_2) - mCv(T_4 - T_1)}{mCp(T_3 - T_2)}$$

$$= 1 - \frac{1(T_4 - T_1)}{\gamma(T_3 - T_2)}$$

By using the relationship $T_1/T_2 = (V_2/V_1)^{n-1} = (p_1/p_2)^{(n-1)/n}$, and the fact that the compression ratio $r = V_2/V_1$, the thermal efficiency of this cycle, which is also known as the air standard efficiency, can be shown to be:

$$\text{ASE} = 1 - \frac{(\rho^{\gamma} - 1)}{\gamma(\rho - 1)} \cdot \frac{1}{r^{\gamma-1}} \qquad \textbf{(15.4)}$$

where r is the compression ratio, ρ is a factor derived from the fuelling design of the engine, and γ = adiabatic index for air.

Example 15.3

A certain diesel engine has a compression ratio of 14:1. The fuelling factor ρ = 1.78. Calculate the air standard efficiency; take γ = 1.4.

Solution

$$ASE = 1 - \frac{(\rho^{\gamma} - 1)}{\gamma(\rho - 1)} \cdot \frac{1}{r^{\gamma-1}}$$

Taking $r = 14$, $\rho = 1.78$, $\gamma = 1.4$

$$ASE = 1 - \frac{(1.78^{1.4} - 1)}{1.4(1.78 - 1)} \cdot \frac{1}{14^{0.4}}$$

$$= 1 - 0.396$$

$$= 0.604 = 60.4\%$$

15.4 The dual combustion cycle

Large, low-speed (100 rpm) diesel engines such as those used for marine propulsion or stationary power generation operate reasonably closely to the ideal diesel cycle. However, smaller, lighter-weight engines that are necessary for road vehicles are required to operate at higher speeds and it is considered that the modified diesel cycle known as the dual combustion cycle more accurately represents the thermodynamic processes that occur in high-speed compression ignition engines.

Operation of dual combustion cycle

In this cycle, the pV diagram of which is shown in Figure 15.4, the combustion takes place in two stages, one at constant pressure and one at constant volume. Working round the diagram from point 1, which corresponds to bottom dead centre on the compression stroke, the air in the cylinder is compressed adiabatically up to point 2. At point 2 heat is added, increasing the pressure while the volume remains constant until point 3 is reached. At point 3 further heat is added at constant pressure, expansion taking place up to point 4. From point 4 to point 5, the air in the cylinder expands adiabatically. At point 5 the remaining heat is removed at constant volume to restore the air to its original condition ready to repeat the cycle.

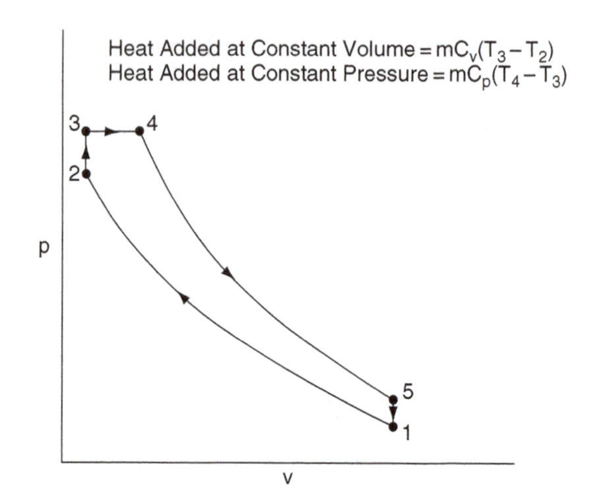

Fig. 15.4 pV diagram for the dual combustion cycle

During the adiabatic compression and expansion processes no heat is added or rejected, and the work done during the cycle = heat added – heat rejected.

The air standard efficiency of the engine operating on this cycle = work done ÷ heat supplied.

Air standard efficiency

$$= \frac{\text{heat added} - \text{heat rejected}}{\text{heat added}}$$

$$= \frac{mCv(T_3 - T_2) + mCp(T_4 - T_3) - mCv(T_5 - T_1)}{mCv(T_3 - T_2) + mCp(T_4 - T_3)}$$

This expression can, by substitution, be shown to be

$$ASE = 1 - \frac{1 \times F}{r^{\gamma-1}} \qquad \textbf{(15.5)}$$

where r is the compression ratio and F is a factor derived from the fuelling arrangements for the engine.

Example 15.4

A compression ignition engine operating on the dual combustion cycle has a compression ratio of 16:1. The fuelling factor F for this engine is 1.2. Use Equation (15.5) to calculate the ideal thermal efficiency of this engine and take $\gamma = 1.4$.

Solution

From Eqn (15.5), ideal thermal efficiency (ASE)

$$= 1 - \frac{1 \times F}{r^{\gamma-1}}$$

$$= 1 - \frac{1 \times 1.2}{15^{0.4}}$$

$$= 1 - (0.338 \times 1.2)$$

$$= 1 - 0.41$$

$$= 0.59 = 59\%$$

15.5 Comparison between theoretical and practical engine cycles

Some of the factors that are taken into account when comparing practical engine cycles with theoretical ones are as follows.

- The air and combustion gases in the cylinder are not perfect gases.
- Adiabatic processes are not realised in practice.
- Instantaneous application and removal of heat energy are practically impossible.

The diagrams in Figure 15.5(a–c) are reasonable representations of actual engine cycles. In both the Otto cycle diagram and the diesel and dual combustion cycle diagrams the principal difference between the theoretical and the actual is the loop at the bottom of the diagram. This loop arises from the exhaust and induction strokes. The exhaust stroke expels heat from the cylinder and it replaces the constant volume heat rejection of the theoretical cycle. The other part of this loop arises from the induction stroke. This loop formed by the exhaust and induction strokes is known as the pumping loop. The area of this loop represents work done by the air in the cylinder and when subtracted from the loop formed by the compression, power and heat addition parts of the diagram, the effective area of the diagram is reduced. This leads to a considerable difference in power produced per stroke of the actual cycle compared with the theoretical one. These, and similar considerations, lead to the main differences between the efficiencies of practical engines and those produced by calculations based on ideal cycles.

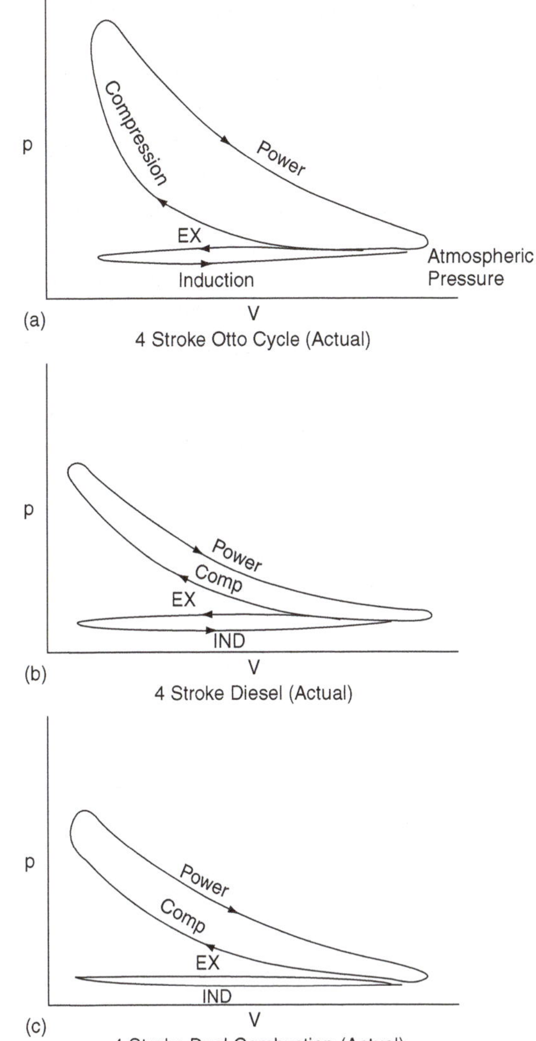

Fig. 15.5 Indicator diagram for (a) petrol engine; (b) low-speed diesel engine; (c) high-speed diesel engine – dual combustion.

15.6 The Stirling engine

The Stirling engine operates by external combustion. The air in the hermetically sealed engine cylinder is heated by means of a heat exchanger known as a regenerator, as opposed to fuel being burned in the cylinder. The original Stirling engine was designed by the Revd Stirling in 1845. Stirling hot-air engines were used in some industrial applications and domestic appliances, but have dropped out of use because steam engines have proven to be more effective. In the twentieth century, several manufacturers worked on the development of Stirling engines for use in motor vehicles because they should ideally produce fewer harmful emissions and have a higher thermal efficiency compared with conventional engines.

The Stirling engine regenerator

The air, or gas, that drives the engine is enclosed in a casing that includes the regenerator. Figure 15.6 shows a simplified form of regenerator that consists of a heater, a cooler, and a wire gauze heat sink. The closed-cycle working air is pumped from end to end of the regenerator by means of the compressor section of the engine. Heat is applied externally from the heater section, this causing the temperature of the engine air to rise to T_1; the air in the engine expansion cylinder is thus heated, causing it to expand and provide engine power. Heat

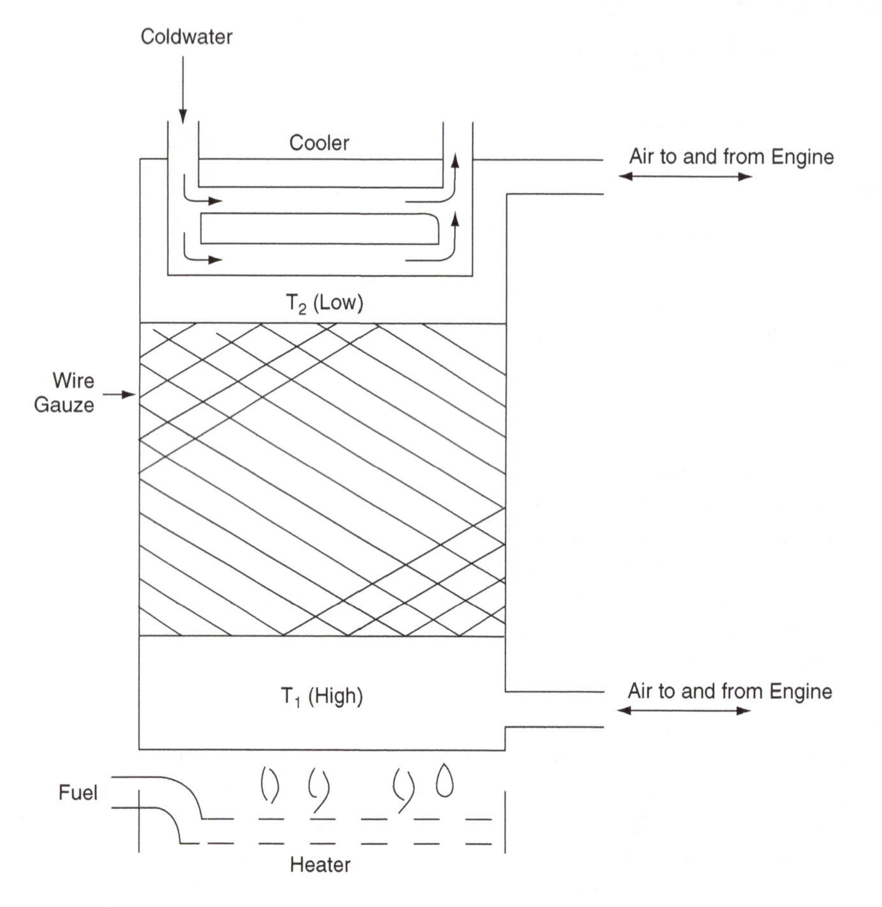

Fig. 15.6 A simplified Stirling engine regenerator

passes through the wire gauze to the cooler end of the regenerator and heat is stored in the wire gauze during the process; the working air at the cool end of the regenerator has a lower temperature of T_2. During another part of the engine operating cycle, the working air is pumped back through the regenerator, from the cool end to the hot end. During this process the working air regains heat from the gauze.

A double-acting Stirling engine

The engine shown in Figure 15.7 has two double-acting cylinders that are interconnected via the regenerators at the ends of each cylinder. The working gas that is sealed inside the engine system may be air, or some other gas such as nitrogen; it is constantly heated and expanded and cooled and compressed as the pistons reciprocate in the

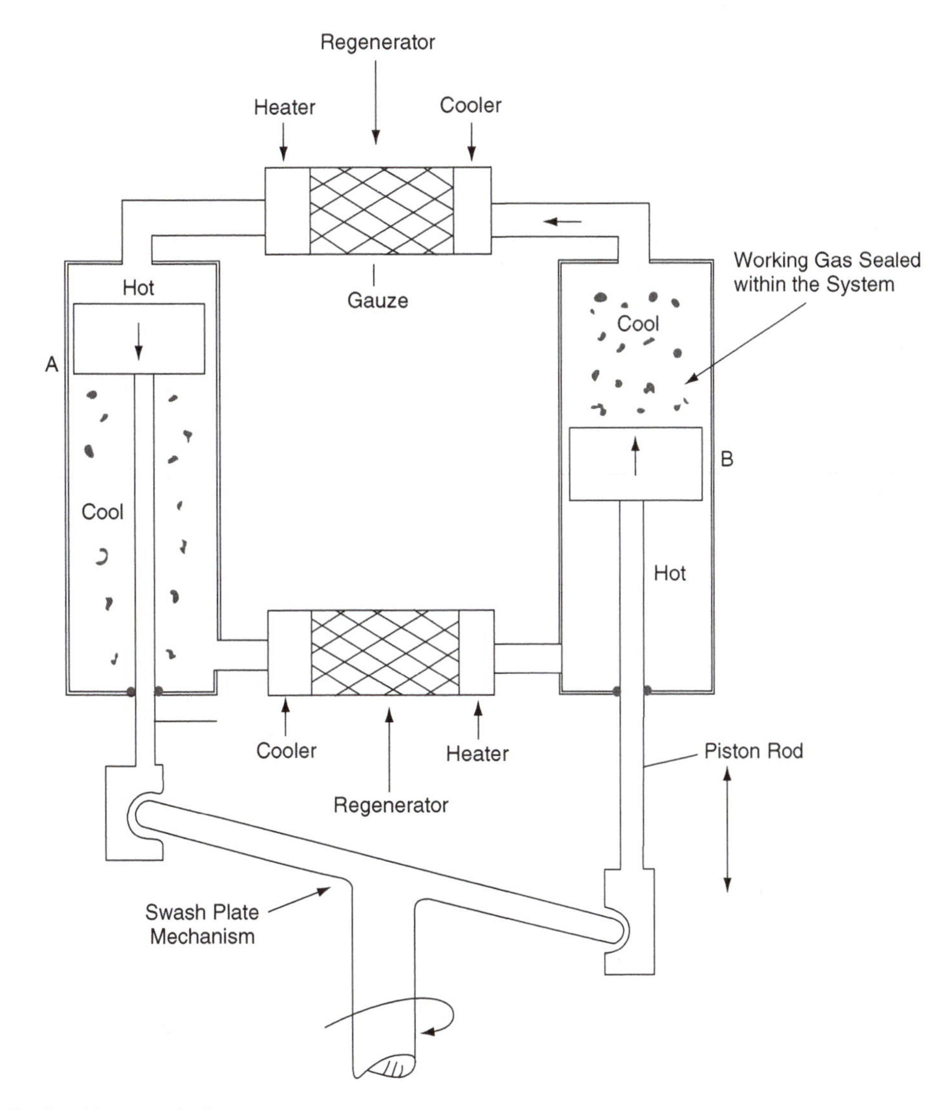

Fig. 15.7 Double-acting Stirling engine

cylinders. The reciprocating linear motion of the piston rods is converted to rotary motion by means of the swash plate. As shown in Figure 15.7, the heated gas is forcing piston A down the cylinder. At the same time the cooler, lower-pressure gas below piston A is being forced through the regenerator where it is reheated as it enters the other cylinder below piston B. At the same time, the cooler gas above piston B is being forced through the top regenerator into the cylinder above piston A. The power output of the engine is determined by the amount of heat energy that is applied at the heater part of the regenerator. The pressure–volume diagram for an engine working on this ideal cycle is shown in Figure 15.8. The heating and cooling of the working gas takes place at constant volume, and the expansion and compression processes are isothermal. The thermal efficiency of this ideal engine is given by the expression

$$\eta = \frac{T_1 - T_2}{T_1} \quad \textbf{(15.6)}$$

This is the same as the Carnot efficiency which is, in theory, the highest efficiency that can be achieved by any heat engine. One attraction of the Stirling engine is the prospect of achieving a very high thermal efficiency; another is the fact that external combustion permits a wide range of fuels to be used.

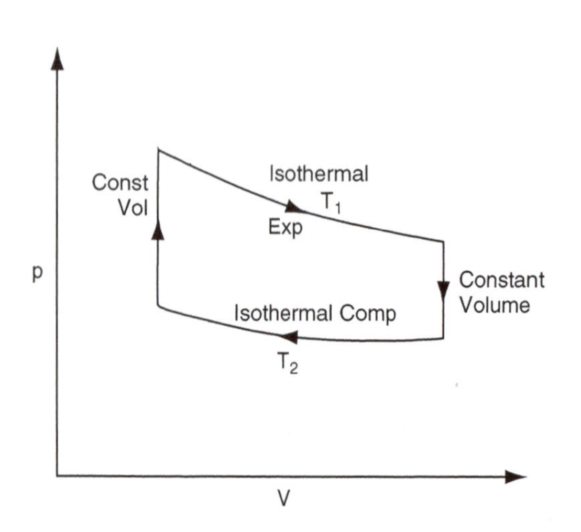

Fig. 15.8 Ideal Stirling cycle

Example 15.5

A test on Stirling engine produced the following results: air at 1 bar pressure and temperature of 25°C is compressed isothermally and then heated at constant volume until the pressure is 9 bar and the temperature is 300°C. Calculate the thermal efficiency.

Solution

The highest temperature $T_1 = 300 + 273 = 573\,\text{K}$

The lowest temperature $T_2 = 25 + 273 = 298\,\text{K}$

$$\text{Thermal efficiency } \eta = \frac{T_1 - T_2}{T_1} = \frac{573 - 298}{573} = 0.48 = 48\%$$

15.7 The gas turbine

While several designs of heavy vehicle have used gas turbine engines of the type used in aircraft, the early promise of these engines has not materialised and the principal use of gas turbines in road transport is in the turbocharger, which has found widespread application. The constant-pressure type of gas turbine engine utilises a compressor to compresses air prior to fuel being injected to increase its temperature. The heated air then passes into the turbine where the energy of the stream of gas is converted to rotational energy that drives the power shaft.

The theoretical cycle for the constant-pressure gas turbine cycle is shown in Figure 15.9b.

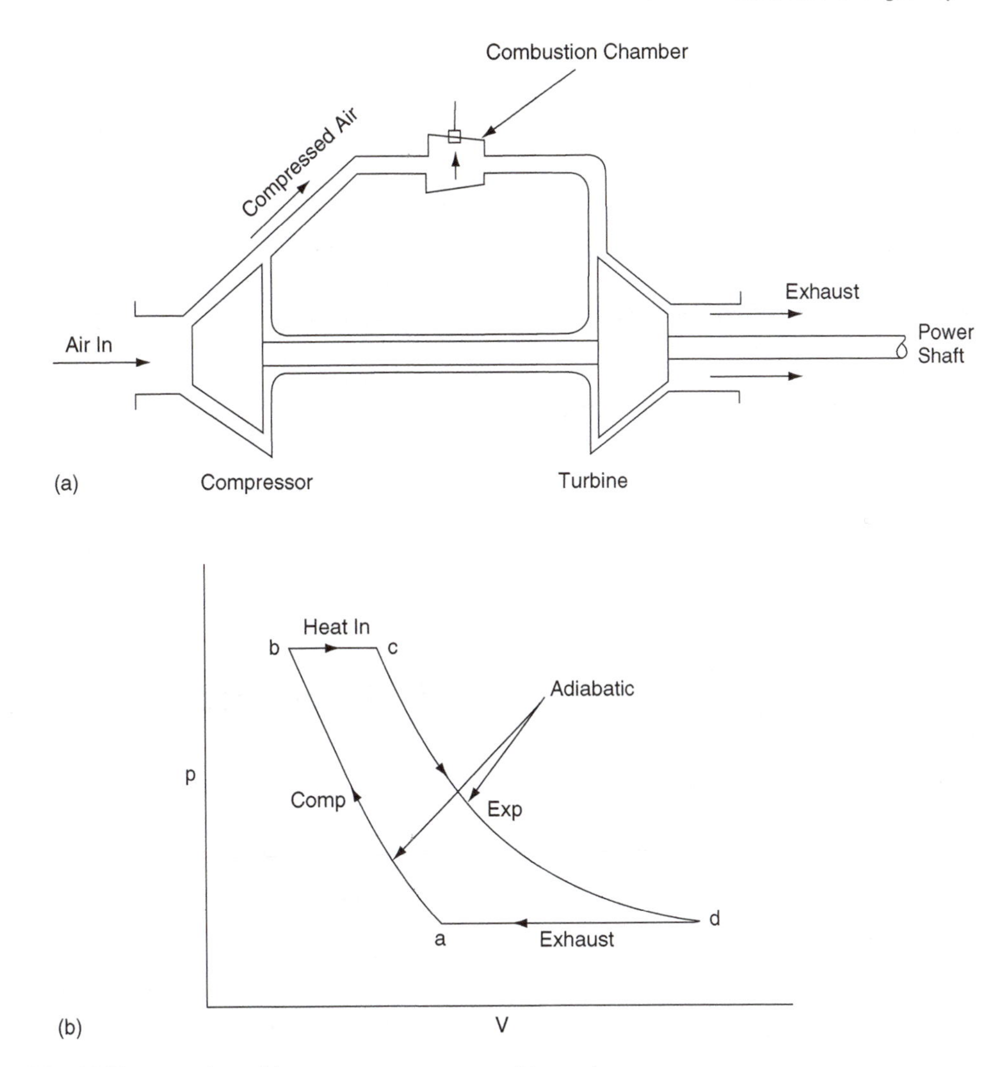

Fig. 15.9 (a) The gas turbine; (b) constant pressure gas turbine cycle

Working round this diagram from point a, air is drawn into the compressor where it is compressed adiabatically up to point b. At point b the compressed air is heated by fuel being burned in the combustion chamber; this heating continues to point c. From point c the hot gas expands adiabatically through the turbine where it is converted into rotational energy at the power shaft. At the end of expansion the remaining heat is removed at constant pressure in the exhaust, shown as d to a on the pV diagram.

The thermal efficiency of this ideal cycle

$$= \frac{\text{heat added} - \text{heat rejected}}{\text{heat added}}$$

$$= \frac{\text{mCp(Tc} - \text{Tb)} - \text{mCp(Td} - \text{Ta)}}{\text{mCp(Tc} - \text{Tb)}}$$

$$= 1 - \frac{(\text{Td} - \text{Ta})}{(\text{Tc} - \text{Tb})}$$

By using the relationship between the pressure, volume and temperature of a gas this equation for thermal efficiency can be expressed in terms of compression ratio and the adiabatic index for air.

Thermal efficiency of constant pressure

$$\text{gas turbine} = 1 - \frac{1}{r^{\gamma-1}} \qquad \textbf{(15.7)}$$

where r = compression ratio of the compressor and γ = adiabatic index for air.

The compressor is attached to the same shaft as the turbine and a considerable amount of turbine power is used in driving the compressor. The power available to drive machinery or to propel a vehicle = turbine power − compressor power. Example 15.6 gives an indication of the performance of gas turbine plant.

Example 15.6

A test on a gas turbine power plant produced the following results:

Work done by turbine = 248.24 kJ/kg of fuel
Work done by compressor = 192.56 kJ/kg of fuel
Energy supplied in fuel = 334 kJ/kg of fuel.

Calculate the thermal efficiency of the plant.

Solution

$$\text{Thermal efficiency} = \frac{\text{useful energy output}}{\text{energy input}}$$

The useful energy output

$$= \text{turbine energy} - \text{compressor energy}$$

$$= (248.24 - 192.56)\,\text{kJ/kg}$$

$$= 55.68\,\text{kJ/kg}$$

Energy input = energy supplied in fuel
= 334 kJ/kg

$$\text{Thermal efficiency} = \frac{55.68}{334} = 16.7\%$$

15.8 Summary of formulae

Thermal efficiency Otto (constant volume) cycle

$$\eta = 1 - \frac{(T_4 - T_1)}{(T_3 - T_2)}$$

$$\text{Air standard efficiency Otto cycle} = 1 - \frac{1}{r^{\gamma-1}}$$

$$\text{Relative efficiency} = \frac{\text{actual thermal efficiency}}{\text{air standard efficiency}}$$

Diesel cycle air standard efficiency ASE

$$= 1 - \frac{(\rho^{\gamma} - 1)}{\gamma(\rho - 1)} \cdot \frac{1}{r^{\gamma-1}}$$

Compression ignition engine – dual combustion

$$\text{cycle efficiency ASE} = 1 - \frac{1 \times F}{r^{\gamma-1}}$$

Carnot and Stirling engine theoretical cycle

$$\text{efficiency } \eta = \frac{T_1 - T_2}{T_1}$$

Thermal efficiency of constant-pressure gas

$$\text{turbine} = 1 - \frac{1}{r^{\gamma-1}}$$

15.9 Exercises

15.1 A single-cylinder 4-stroke petrol engine has a bore of 79 mm and a stroke of 100 mm; the clearance volume is 60 cm^3. Calculate:
(a) the compression ratio;
(b) the air standard thermal efficiency; take γ for air = 1.41.

15.2 A heavy-duty oil-engine operating on the constant-pressure diesel cycle has a compression ratio of 15:1. Given $\rho = 1.2$ and $\gamma = 1.4$, calculate the ideal air standard efficiency of the engine.

15.3 An indirect-injection diesel engine with a compression ratio of 22:1 operates on the dual combustion cycle. Given that the fuelling factor F = 1.2 and $\gamma = 1.4$, calculate the ideal air standard efficiency of the engine.

15.4 The gas turbine engine of a certain motor vehicle has an actual thermal efficiency of 18%. The fuel consumption of the turbine is 0.013 kg/s. If the calorific value of the fuel is 43 MJ/kg, calculate the power available to propel the vehicle.

15.5 Go to the turbokart.com website and study the description that is given for the AGT 1500 gas turbine, which is a power unit for the M1 American battle tank. Also, use your search engine to look at Rover Gas Turbine at allpart.com, which gives a clear description of a Chrysler gas turbine engine.

15.6 The following results were obtained during a test to determine the indicated thermal efficiency of an engine: indicated power 110 kW, fuel used per hour = 30 kg. Calorific value of the fuel = 43 MJ/kg. The engine compression ratio = 8:1; $\gamma = 1.41$. Calculate:
(a) the indicated thermal efficiency;
(b) the air standard efficiency;
(c) the relative efficiency.

15.7 A certain 4-stroke petrol engine has a compression ratio of 9:1. During a test on a dynamometer the engine returns an indicated thermal efficiency of 28%. Calculate:
(a) the air standard efficiency of the engine;
(b) the relative efficiency.
Take $\gamma = 1.41$.

15.8 A 4-cylinder 4-stroke petrol engine has a bore of 120 mm and a stroke of 110 mm. If the compression ratio is 8:1 determine:
(a) the clearance volume;
(b) the air standard efficiency.

15.9 Sketch the indicator diagrams for the ideal Otto cycle and the ideal diesel cycle. Describe the main differences between the two cycles.

16 Fuels and combustion & emissions

16.1 Calorific value

The calorific value of a fuel is the amount of energy that is released by the combustion of 1 kg of the fuel. Two figures for calorific value of hydrocarbon fuels are normally quoted, these being the higher or gross calorific value, and the lower calorific value. These two values derive from the steam that arises from the combustion of hydrogen. The higher calorific value includes the heat of the steam whereas the lower calorific value assumes that the heat of the steam is not available to do useful work. The calorific value that is quoted for motor fuels is that which is used in calculations associated with engine and vehicle performance. For example, the calorific value of petrol is approximately 44 MJ/kg. The approximate values of the properties of other fuels are shown in Table 16.1.

16.2 Combustion

Modern fuels for motor vehicles, such as petrol and diesel fuel, are principally hydrocarbons. That is to say that they consist largely of the two elements, hydrogen and carbon. The proportions of these elements in the fuel vary, but a reasonably accurate average figure is that motor fuels such as petrol and diesel fuel are approximately 85% carbon and 15% hydrogen.

Products of combustion

When considering products of combustion it is useful to take account of some simple chemistry relating to the combustion equations for carbon and hydrogen.

An adequate supply of oxygen is required to ensure complete combustion and this is obtained from atmospheric air.

Table 16.1 Approximate values of fuel properties

Property	LPG	Petrol	Diesel oil	Methanol
Relative density	0.55	0.74	0.84	0.8
Usable calorific value	48 MJ/kg	44 MJ/kg	42.5 MJ/kg	20 MJ/kg
Octane rating	100	90	–	105
Chemical composition	82% C; 18% H_2	85% C ; 15% H_2	87% C; 12.5% H_2; 0. 5% S	38% C;12% H_2 ; 50% O_2
Air–fuel ratio	15.5	14.7	14.9	6.5
Mass of CO_2 per kg of fuel used (kg/kg)	3.04	3.2	3.2	1.4

Relevant combustion equations

For carbon, $C + O_2 = CO_2$. Because of the relative molecular masses of oxygen and carbon this may be interpreted as: 1 kg of carbon requires 2.67 kg of oxygen and produces 3.67 kg of carbon dioxide when combustion is complete.

For hydrogen, $2H_2 + O_2 = 2H_2O$. Again, because of the relative molecular masses of H and O_2 this may be interpreted as: 1 kg of hydrogen requires 8 kg of oxygen and produces 9 kg of H_2O when combustion is complete.

16.3 Air–fuel ratio

Petrol has an approximate composition of 15% hydrogen and 85% carbon. The oxygen for combustion is contained in the air supply and approximately 15 kg of air contains the amount of oxygen that will ensure complete combustion of 1 kg of petrol. This means that the air-to-fuel ratio for complete combustion of petrol is approximately 15:1; a more precise figure is 14.7:1.

Petrol engine combustion

Combustion in spark ignition engines such as the petrol engine is initiated by the spark at the sparking plug and the burning process is aided by factors such as combustion chamber design, temperature in the cylinder, and mixture strength.

Because petrol is volatile each element of the fuel is readily supplied with sufficient oxygen from the induced air to ensure complete combustion when the spark occurs. Petrol engine combustion chambers are designed so that the combustion that is initiated by the spark at the sparking plug is able to spread uniformly throughout the combustion chamber.

For normal operation of a petrol engine, a range of mixture strengths (air/fuel ratios) are required from slightly weak mixtures – say 20 parts air to one part of petrol for economy cruising, to 10 parts of air to one part of petrol for cold starting. During normal motoring, a variety of mixture strengths within this range will occur; for example, acceleration requires approximately 12 parts of air to 1 part of petrol. These varied conditions plus other factors such as atmospheric conditions that affect engine performance lead to variations in combustion efficiency, and undesirable combustion products known as exhaust emissions are produced. Exhaust emissions and engine performance are affected by conditions in the combustion chamber; two effects that are associated with combustion in petrol engines are (1) detonation and (2) pre-ignition.

Detonation

Detonation is characterised by knocking and loss of engine performance. The knocking arises after the spark has occurred and it is caused by regions of high pressure that arise when the flame-spread throughout the charge in the cylinder is uneven. Uneven flame-spread leads to pockets of high pressure and temperature that cause elements of the charge to burn more rapidly than the main body of the charge. Detonation is influenced by engine design factors such as turbulence, heat flow, and combustion chamber shape. The quality of the fuel, including octane rating, also has an effect. Detonation may lead to increased emissions of CO and NO_x and HC. result.

Pre-ignition

Pre-ignition is characterised by a high-pitched 'pinking' sound, which is emitted when combustion occurs prior to the spark, this being caused by regions of high temperature. These high-temperature zones may be the result of sparking plug electrodes overheating, sharp or rough edges in the combustion region, carbon deposits, and other factors. In addition to loss of power and mechanical damage that may be caused by the high pressures generated by pre-ignition, combustion may be affected and this will cause harmful exhaust emissions.

16.4 Octane rating

The octane rating of a fuel is a measure of the fuel's resistance to knock. A high octane number indicates a high knock resistance. Octane ratings are determined by standard tests in a single-cylinder, variable compression ratio engine. The research octane number (RON) of a fuel is determined by running the test engine at a steady 600 rpm while the compression ratio is increased until knock occurs. The motor octane number (MON) is determined by a similar test, but the engine is operated at higher speed. The RON is usually higher than the MON and fuel suppliers often quote the RON on their fuel pumps. An alternative rating that is sometimes used gives a figure which is the average of RON and MON.

16.5 Compression ignition

In order to produce the high temperature of approximately 400°C that is required to ignite the diesel fuel, the air in the cylinder is compressed to a high pressure of approximately 50 bar. Air at this pressure is dense and the fuel that is injected into the cylinder is broken down into minute particles each of which must penetrate the air charge in order to obtain the oxygen for combustion and to provide uniform burning.

Traditionally, compression ignition engine combustion has been considered to take place in the three phases identified by Sir Harry Ricardo. These three phases are as follows:

1. The delay period, phase A to B in Figure 16.1. This is the period between the commencement of injection to the beginning of the pressure rise. During this period the fuel droplets are absorbing the heat from the air in the cylinder that will lead to combustion.

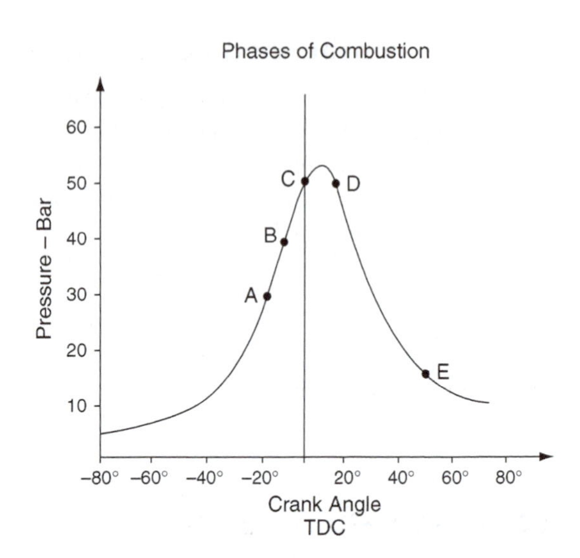

Fig. 16.1 The three phases of combustion in a CI engine

2. B to C represents the amount of luminous combustion before TDC; this luminous combustion carries on after TDC and is shown in Figure 16.1 as C to D, the flame-spread period. As combustion proceeds and the flame spreads through the charge and more of the fuel droplets are ignited, the pressure rises rapidly. The rate of pressure rise during this phase is a major cause of the knock that is a characteristic of compression ignition engines.
3. D to E represents the phase where fuel continues to burn.

Compression ignition engine combustion chambers

Diesel (CIE) engines are classified according to the type of combustion chamber used, namely direct injection and indirect injection (Figure 16.2).

Direct injection

A major part of the combustion space is provided by the specially shaped cavity that is formed in

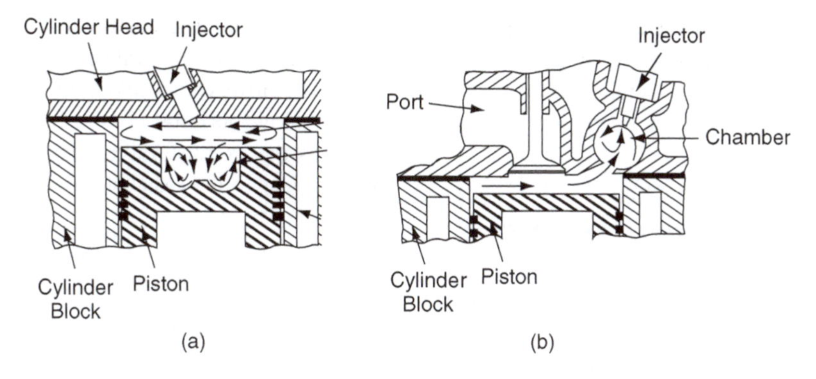

Fig. 16.2 Direct and indirect injection

the piston crown. This cavity is designed to provide swirl, which assists in the combustion. Fuel is sprayed into the combustion space at high pressure in excess of 180 bar. Multi-hole injectors are normally used to ensure maximum spread of the fuel throughout the compressed air.

In the direct injection engine, swirl is normally produced in two planes, vertical and horizontal. The piston is very close to the valves and cylinder head at top dead centre and the small clearance between the top of the piston and the lowest part of the cylinder head is known as the 'bumping clearance'.

Indirect injection

Indirect injection diesel engines are equipped with an additional small chamber into which the fuel is injected. This chamber is known as a pre-chamber and is connected to the main combustion chamber of the engine by means of a small-diameter hole. The pre-chamber is designed to produce swirl so that good combustion is achieved.

16.6 Diesel fuel

Diesel fuel has a calorific value of approximately 45 MJ/kg and it has a specific gravity of approximately 0.8 g/cc. The ignition quality of diesel fuel is denoted by the cetane number; a figure of 50 indicates good ignition properties. Among other properties of diesel fuel that affect normal operation are flash point, pour point and cloud point, or cold filter plugging point.

Flash point

The flash point of a fuel is the lowest temperature at which sufficient vapour is given off to cause temporary burning when a flame is introduced near the surface. A figure of 125°F (52°C) minimum is quoted in some specifications. The flash point is determined by means of apparatus such as the Pensky–Martin flash point apparatus and the test is performed under controlled laboratory conditions.

Pour point

The pour point of a fuel is the temperature at which a fuel begins to thicken and congeal and can no longer be poured. A pour point of 0°F or −18°C is suitable for some conditions.

Cloud point

The cloud point, which is sometimes known as the cold filter plugging point (CFPP), is the temperature at which the fuel begins to have a cloudy appearance and will no longer flow freely through a filtering medium. Cloud point occurs at a temperature of approximately 20°F (11°C) above the cloud point temperature.

Note These figures for diesel fuel are approximate and are presented here as a guide only. Readers who require more detailed information are advised to contact their fuel supplier.

16.7 Exhaust emissions

Exhaust gases are the products of combustion. Under ideal circumstances, the exhaust products would be carbon dioxide, steam (water) and nitrogen. However, owing to the large range of operating conditions that engines experience, exhaust gas contains several other gases and materials such as those given below:

- CO – due to a rich mixture and incomplete combustion;
- NO_x – due to very high temperature;
- HC – due to poor combustion;
- PM – soot and organo-metallic materials;
- SO_2 – arising from combustion of small amount of sulphur in diesel fuel.

Factors affecting exhaust emissions

During normal operation, the engine of a road vehicle is required to operate in a number of quite different modes, as follows:

- modes of road vehicle engine operation;
- idling – slow running;
- coasting;
- deceleration – overrun braking;
- acceleration;
- maximum power.

These various modes of operation give rise to variations in pressure, temperature and mixture strength in the engine cylinder with the result that exhaust pollutants are produced.

Hydrocarbons

Hydrocarbons (HC) appear in the exhaust as a gas and arise from incomplete combustion due to a lack of oxygen. The answer to this might seem to be to increase the amount of oxygen to weaken the mixture. However, weakening the mixture gives rise to slow burning, combustion will be incomplete as the exhaust valve opens and unburnt HC will appear in the exhaust gases.

Idling

When the engine is idling the quantity of fuel is small. Some dilution of the charge occurs because, owing to valve overlap and low engine speed, scavenging is poor. The temperature in the cylinders tends to be lower during idling and this leads to poor vaporisation of the fuel and HC in the exhaust.

Coasting, overrun braking

Under these conditions the throttle valve is normally closed. The result is that no, or very little, air is drawn into the cylinders. Fuel may be drawn in from the idling system. A closed throttle leads to low compression pressure and very little air; the shortage of oxygen arising from these conditions causes incomplete combustion of any fuel that enters the cylinders and this results in hydrocarbon gas in the exhaust.

Acceleration

Examination of torque V specific fuel consumption for spark ignition engine reveals that maximum torque occurs when the specific fuel consumption is high. Because the best acceleration is likely to occur at the maximum engine torque, a richer mixture is required in order to produce satisfactory acceleration; fuelling systems provide this temporary enrichment of appropriate increases in mixture strength to meet demands placed on the engine. This leads to a temporary increase in emissions of HC and CO. As the engine speed rises, combustion speed and temperature increase and this gives rise to increased amounts of NO_x.

High-speed, heavy load running

Here the engine will be operating at or near maximum power. Examination of the power V specific

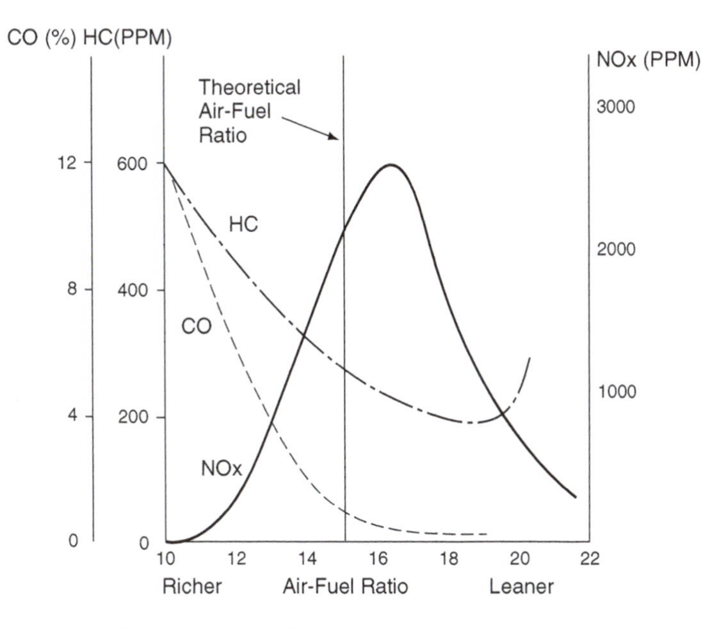

Fig. 16.3 Relationship between air–fuel ratio and exhaust emissions

fuel consumption shows that maximum power is produced at higher specific fuel consumption figures and richer mixtures. Increased emissions of CO and HC are likely to result.

Cruising speed – light engine load

Under these conditions wherein the engine is probably operating at a low specific fuel consumption speed, the mixture strength is likely to be around 16:1 or higher to provide good fuel economy. Other emissions are lower under these conditions (Figure 16.3).

16.8 European emissions standards

In Table 16.2 are shown the European emissions limits standards as per EURO 4.

In addition to these standards, modern vehicles must be equipped with on-board diagnostics (OBD) that notify the driver of the vehicle that a system malfunction is causing the emissions limits to be exceeded.

Table 16.1 EURO 4 emissions limits standards (g/km)

	CO	HC	HC + NO_x	NO_x	PM
Petrol	1.0	0.10	–	0.08	–
Diesel	0.50	–	0.30	0.25	0.025

PM = particulate matter.

Emissions and their causes

Oxides of nitrogen (NO_x)

Oxides of nitrogen are formed when combustion temperatures rise above 1800 K.

Hydrocarbons (HC)

Unburnt hydrocarbons arise from:

- unburnt fuel remaining near the cylinder walls after incomplete combustion being removed during the exhaust stroke;
- incomplete combustion due to incorrect mixture strength.

Carbon monoxide (CO)

CO emissions are caused by incomplete combustion arising from lack of oxygen.

Sulphur dioxide (SO_2)

Some diesel fuels contain small amounts of sulphur, which combines with oxygen during combustion. This leads to the production of sulphur dioxide which can, under certain conditions, combine with steam to produce sulphurous acid (H_2SO_3), which is a corrosive substance.

Particulate matter (PM)

The bulk of particulate matter is soot, which arises from incomplete combustion of carbon. Other particulates arise from lubricating oil on cylinder walls and metallic substances from engine wear. Figure 16.4 gives an indication of the composition of particulate matter.

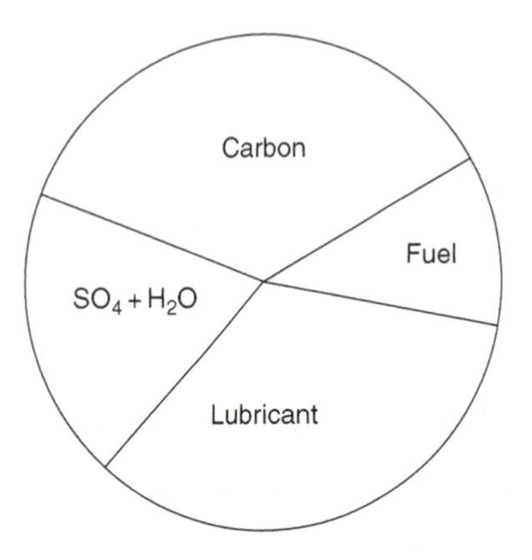

Fig. 16.4 Approximate composition of particulate matter

Carbon dioxide (CO_2)

While CO_2 is not treated as a harmful emission, it is thought to be a major contributor to the greenhouse effect and efforts are constantly being made to reduce the amount of CO_2 that is produced. In the United Kingdom the quantity of CO_2 that a vehicle produces in a standard test appears in the vehicle specification, so it is possible to make comparisons between vehicles on this score. The figure is presented in grams per kilometre; for example a small economy vehicle may have a CO_2 figure of 145 g/km and a large saloon car a figure of 240 g/km. Differential car tax rates are applied to provide incentives to users of vehicles that produce smaller amounts of CO_2.

16.9 Methods of controlling exhaust emissions

Two approaches to dealing with exhaust emissions suggest themselves:

1. **Design measures to prevent harmful emissions being produced**
   In diesel engines, such measures include improved fuel quality, fuel injection in a number of electronically controlled stages, better atomisation of fuel such as occurs at higher injection pressures such as those used in common rail fuel injection systems, and exhaust gas recirculation.
   In petrol engines the approach is somewhat similar. Lean-burn engines employing direct petrol injection is a development that has become reasonably well established. Exhaust gas recirculation is widely used, as is variable valve timing to overcome the combustion problems caused by valve overlap at idling speed.
2. **Post combustion – exhaust emission control**
   In diesel engines this consists of catalytic convertors to oxidise CO and HC, reduction

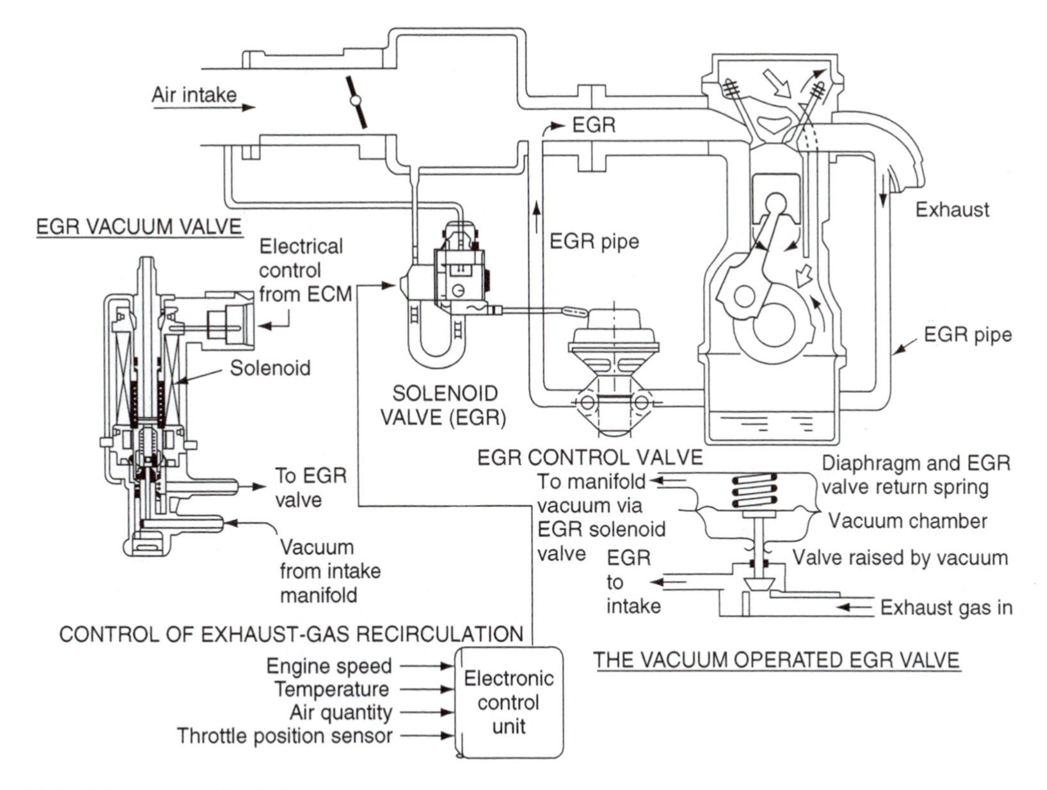

Fig. 16.5 Exhaust gas recirculation system

catalysts to reduce NO_x, and particulate filters to remove particulate matter.

Petrol engines utilise 3-way catalysts and computer-controlled fuelling systems. Other technologies such as evaporative emissions control, to eliminate HC emission from the fuel system, are also used.

Exhaust gas recirculation

The air that is used to provide the oxygen for combustion contains approximately 77% nitrogen by mass. When nitrogen is heated above approximately 1800 K (1527°C), in the presence of oxygen, oxides of nitrogen (NO_x) are formed. These conditions occur in the combustion chamber when excess oxygen is present as happens at an air–fuel ratio of approximately 16:1. If the combustion chamber temperature is kept below 1800 K the conditions for the creation of NO_x no longer exist. Exhaust gas recirculation is a method that is used to keep combustion chamber temperatures below the critical figure. A proportion of exhaust gas is redirected from the exhaust system to the induction system by means of electronically controlled valves. Figure 16.5 shows the layout of an engine management system that incorporates exhaust gas recirculation.

Catalysts

Catalysts are materials that assist chemical reactions but are not themselves changed in the process.

Platinum

At the correct temperature, in excess of 300°C, platinum acts as a catalyst that aids the conversion of CO to CO_2, and HC to H_2O and CO_2. In order for the exhaust system catalytic convertor to function correctly, the air–fuel ratio is maintained close to that which is chemically correct (stoichiometric) by means of exhaust gas sensors and electronic control.

Rhodium

Rhodium is a catalyst that reduces NO_x to N_2.

These metals are added to a ceramic honeycomb structure that exposes the exhaust gases to the maximum area of the catalysing material. Figure 16.6 shows the layout of an emission control system for a petrol engine vehicle.

Selective reduction catalysts

The power output of diesel engines is controlled by the quantity of fuel that is injected and these engines operate with excess air over much of the operating range. Excess air and high combustion temperatures give rise to NO_x. The selective catalyst reduction system (SCR) is a methodology that is used in heavy vehicles to reduce NO_x emissions.

A liquid such as urea is introduced into the exhaust stream where it works in conjunction with a catalytic converter to change NO_x into nitrogen gas and H_2O (water vapour).

Diesel particulate filters

Particulate matter that arises from combustion in diesel engines consists mainly of particles of carbon with some absorbed hydrocarbons. Filtration of the exhaust products is a technique that is widely used for the removal of particulate matter from exhaust gases before they are passed into the atmosphere. Various forms of filtering medium are used to trap the PM and, in common with most filtering processes, cleaning of the filter to avoid blockage is required. The cleaning process in diesel particulate filters (DPFs), which consists primarily of controlled burning to convert the carbon (soot) into carbon dioxide, is known as regeneration. Among the methods of regeneration are:

Passive systems The heat in the exhaust gas, acting with the materials used in the construction of the particulate filter, produces sufficient temperature to remove the filtered deposits. External sources of heat are not required.

Active systems Heat is supplied either by injecting fuel into the exhaust stream, or by secondary injection of fuel into the engine; this

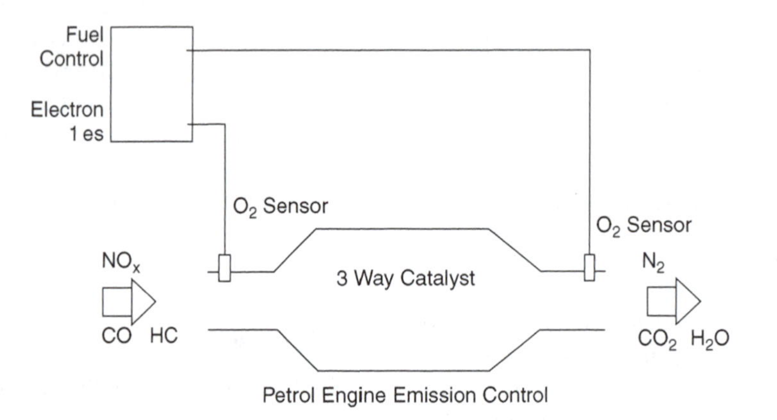

Fig. 16.6 A 3-way catalyst emission control system

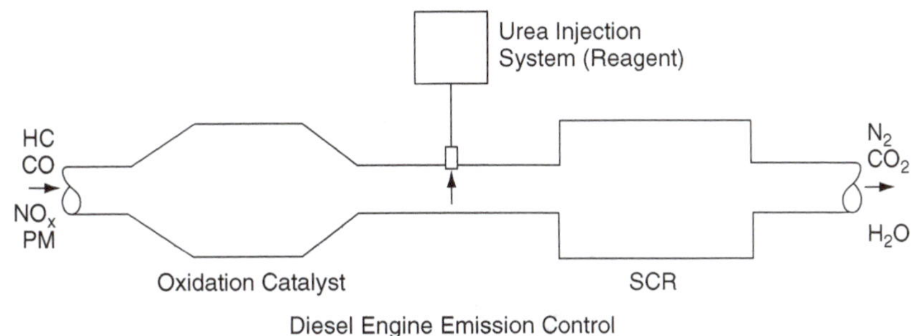

Fig. 16.7 Selective catalyst reduction (SCR) system

creates the temperature that is required to burn off the particulate matter that accumulates in the filter.

16.10 Biofuels

It is currently generally accepted that the world's oil resources are finite and that they are being depleted at a rapid rate. Attention is concentrating on alternative fuels and methods of propulsion for motor vehicles. Alcohols such as methyl alcohol, or methanol as it is commonly known, is produced from vegetable matter. Biofuels are said to be environmentally valuable because their products of combustion are, very roughly, H_2O and CO_2. It is argued that the CO_2 from the combustion of these fuels is consumed by the vegetation that is producing the crop that, in turn, will make the next supply of fuel – this process is referred to as a 'closed carbon cycle'.

Methanol is not strictly a hydrocarbon because it contains some oxygen. The calorific value of methanol is approximately 26 MJ/kg. It has a higher latent heat value than that of petrol and higher resistance to detonation. While the higher latent heat value and relatively high ignition temperature of methanol indicate that higher compression ratios can be used, there is a disadvantage in that vaporisation at low temperatures is poor and this can lead to poor cold starting capability.

16.11 Liquefied petroleum gas (LPG)

Petroleum gases such as butane and propane are produced when oil is refined to produce liquid fuels such as diesel (derv) and petrol. Once liquefied and stored under pressure, LPG will remain liquid until is exposed to atmospheric pressure. The chemical composition of propane is C_3H_8 and that of butane, C_4H_{10}; the relative densities at 15°C are approximately 0.5 and 0.57 respectively while the calorific value is slightly higher than that of petrol at approximately 46 MJ/kg. A considerable industry exists to support the conversion of road fuelling from petrol and diesel (derv) to LPG. In some countries, including the UK, favourable tax systems support the use of LPG.

16.12 Hydrogen

Compressed hydrogen may be stored on a vehicle and used in an internal combustion engine. Among the advantages claimed for it are that there are no carbon dioxide emissions and that the products of combustion are primarily water. The main future use of hydrogen as a propellant for vehicles is thought to be as a source of energy in fuel cells. The electricity produced in these cells is used as a power source for an electric motor replacing the internal combustion engine

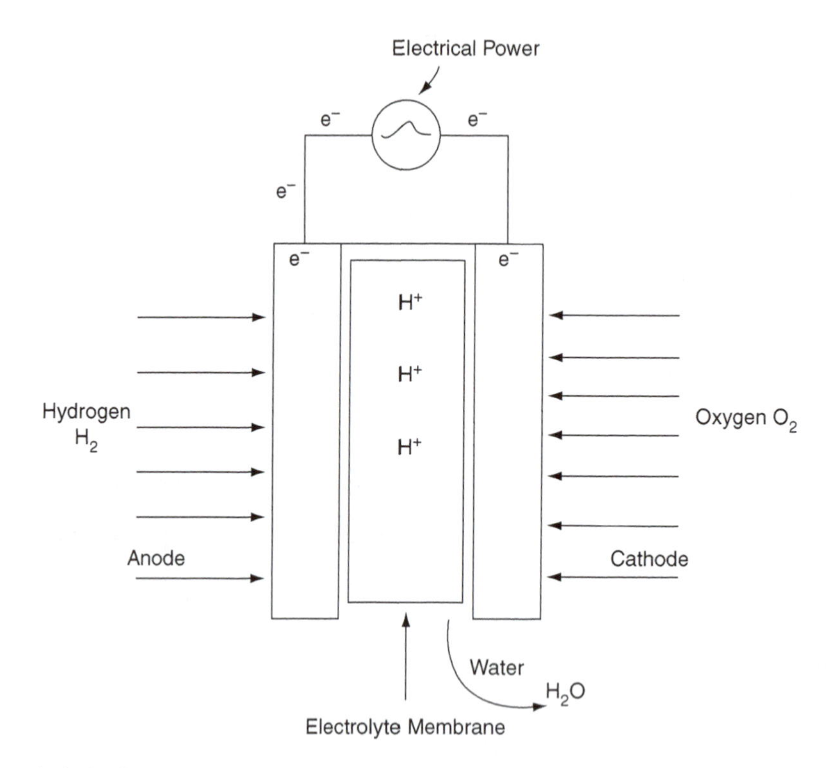

Fig. 16.8 A simple fuel cell

of the vehicle. A simple fuel cell is shown in Figure 16.8. It consists of two electrodes, an anode and a cathode, that are separated by an electrolyte. Hydrogen acts on the anode and oxygen from the atmosphere acts on the cathode. The catalytic action of the anode causes the hydrogen atom to form a proton and an electron; the proton passes through the polymer electrolyte to the cathode and the electron passes through an external circuit to the cathode. This action provides an electric current in external circuit. In the process the hydrogen and oxygen combine to produce water, which is the principal emission.

16.13 Zero emissions vehicles (ZEVs)

The operation of fuel cells for vehicle propulsion does not involve combustion in an engine and the normal products of combustion and associated pollutants are not produced. The main product of the electrochemical processes in the fuel cell is water, and consequently vehicles propelled by fuel cells and electric motors are known as zero emissions vehicles.

Exercises

16.1 Describe with the aid of sketches the procedure and the equipment required to convert a fuel injection petrol engine so that it will operate satisfactorily on LPG.

16.2 Assess the advantages of methanol as a motor fuel and describe the modifications to a vehicle and its fuel system that may be required when converting from petrol to methanol.

16.3 Sketch and describe an exhaust gas recirculation system.

16.4 Sketch and describe an exhaust gas oxygen sensor. Explain how such a sensor is used to help control exhaust emissions when used in conjunction with a 3-way catalyst.

16.5 Describe a selective catalyst reduction system that is used on some diesel vehicles to help reduce NO_x emissions.

16.6 Enter the Johnson–Matthey website and study the various exhaust treatment techniques that are described there.

16.7 Compare exhaust gas recirculation and selective catalyst reduction as two alternative methods of reducing NO_x emissions. Pay particular attention to initial costs, weight of equipment, maintenance factors, effect on engine power, and availability of urea as a reagent, and mention any other factors that you think should be considered.

16.8 With the aid of sketches, describe a low-emissions hybrid vehicle that makes use of a diesel engine and electric motor propulsion.

16.9 Describe a vehicle that uses hydrogen as a gas to provide the fuel to run an adapted 4-stroke engine.

16.10 Conduct some research to satisfy yourself about the effect on fuelling requirements that arise when considering the merits and demerits of converting a spark ignition engine to run on alcohol fuel. Pay particular attention to the calorific value and general effect on engine performance.

17 Electrical principles

17.1 Electric current

An electric current is a flow of electrons through a conductor such as a copper wire.

17.2 Atoms and electrons

All materials are made up of atoms. An atom has a **nucleus** around which one or more **electrons** circle in different orbits. An electron has a **negative** electric charge and a nucleus has a **positive** one.

The electrons orbit around the nucleus in rings that are known as **shells**, as shown in Figure 17.1(a). **The hydrogen atom has a central nucleus and a single orbiting electron.** Figure 17.1(b) shows another type of atom which has a nucleus and six electrons. Two of the electrons orbit around an inner circle and the other four around an outer circle.

The electrons in the outer shells are less strongly attracted to the nucleus than are those closer to it. These outer electrons are can be moved out of their orbit and they are called **free electrons**.

It is the movement of free electrons that causes an electric current. Free electrons occur in most metals.

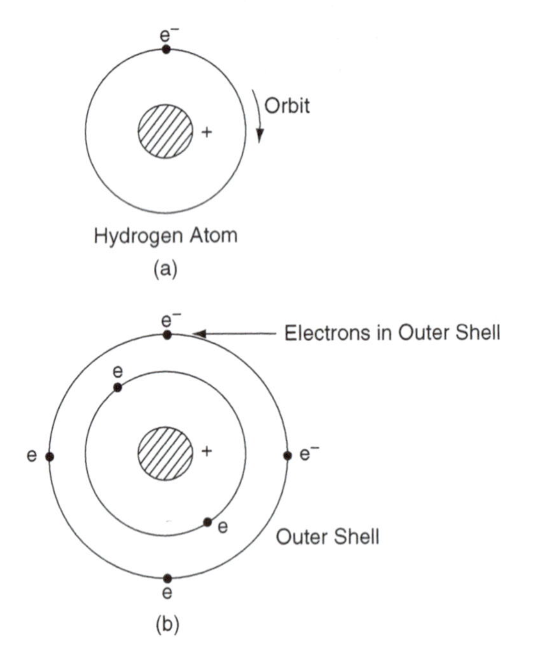

Fig. 17.1 The atom and electron orbits

17.3 Conductors and insulators

Conductors

Materials in which free electrons can be made to move are known as conductors. Most metals are good conductors.

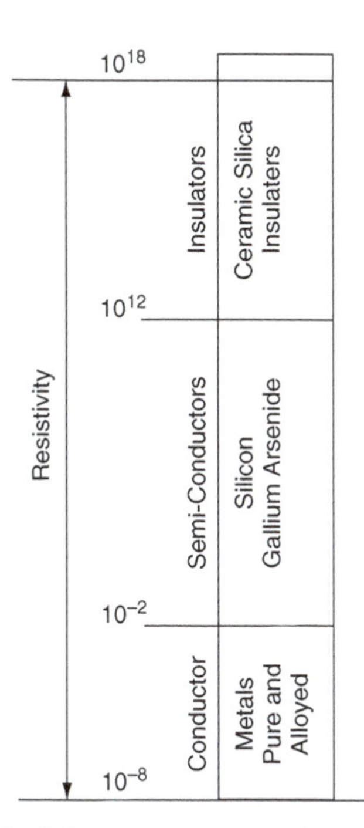

Fig. 17.2 Relative conductivity scale for various materials

Semiconductors

Semiconductors are materials whose free electrons' ability to move falls between that of conductors and insulators.

Insulators

Materials in which free electrons are not readily moved are known as insulators.

The relative conductivities of various materials are shown in Figure 17.2.

17.4 Electromotive force

The force that moves free electrons is called an electromotive force (emf). The symbol E is used to denote emf and it is measured in volts (V).

Electromotive forces are produced by power sources and there are three principal types of power source:

1. chemical – batteries;
2. magnetic – generators (alternators and dynamos);
3. thermal – thermocouples; used in temperature measurement.

17.5 Electrical power sources – producing electricity

Chemical power source

The wet cell shown in Figure 17.3(a) consists of two plates made of different metals to which are attached the terminals (poles). The liquid which surrounds the metal plates is called the electrolyte and it conducts electricity. The electric potential (voltage) at the terminals is caused by a chemical reaction between the metal plates and the electrolyte.

Magnetic power source

Figure 17.3(b) shows the basic elements of a dynamo. When the loop of wire is moved in a circular motion between the poles of a magnet the electric potential difference (volts) exists at the two ends of the loop.

Thermal power source

Figure 17.3(c) shows two wires of different metals, such as copper and iron, that are twisted together at one end. When the twisted end is heated, a small emf exists at the opposite ends of the wires. This type of device is called a thermocouple and it is used to measure high temperatures such as the temperature of exhaust gas emitted from an engine.

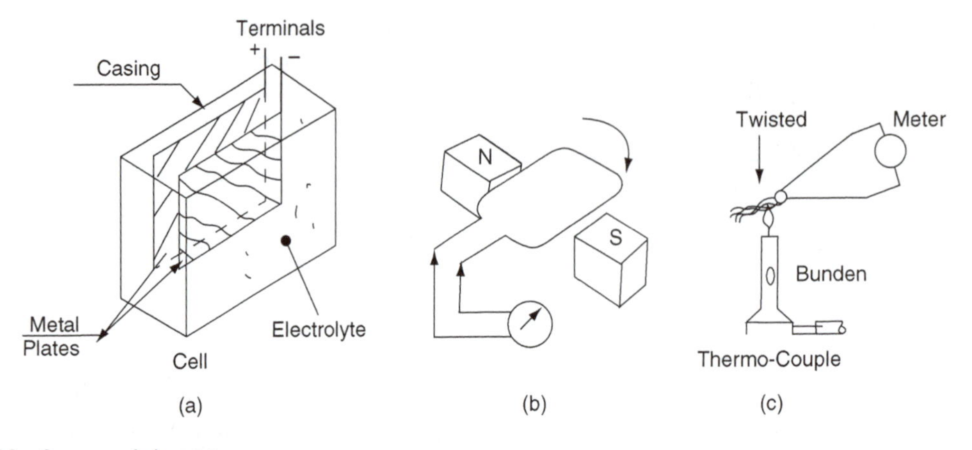

Fig. 17.3 Sources of electricity

17.6 Effects of electric current – using electricity

1. **Heating effect** When electric current flows through a conductor, the conductor heats up. This effect is seen in electric-fire elements, light bulbs (lamps) and similar devices.
2. **Chemical effect** The main battery on a vehicle makes use of the chemical effect of an electric current to maintain its state of charge, and to provide a source of electric current as required.
3. **Magnetic effect** The magnetic effect of an electric current is made use of in components such as starter motors, alternators, solenoids and ignition coils, and in many other devices.

17.7 Electrical circuits

In order to make use of a power source it must be connected to a circuit.

Circuit principles

The symbols shown in Figure 17.4 are used to represent circuit components such as batteries and switches.

A simple circuit

The electrical pressure at any point in an electrical system is called the potential. The potential difference (p.d.) is the difference in electrical pressure between two points in an electrical system. When two points of different p.d. are joined together by an electrical conductor an electric current will flow between them.

The circuit shown in Figure 17.5 consists of a cell (battery) that is connected to a lamp (bulb) and a switch by means of conducting wires. When the switch is closed (on), the emf (voltage) at the battery will cause an electric current to flow through the lamp and the switch, back to the cell, and the lamp will illuminate.

Points to note

A potential difference between two points in a circuit can exist without a current. Current cannot exist in a circuit without a potential difference.

Direction of current flow

By convention, electric current flows from high potential (+, positive) to low potential (−, negative).

Name of Device	Symbol	Name of Device	Symbol
Electric Cell		Lamp (Bulb)	
Battery		Diode	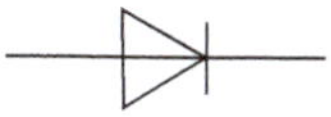
Resistor	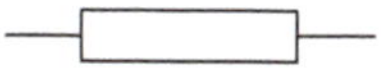	Transistor (npn)	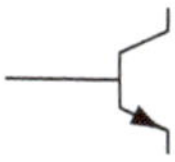
Variable Resistor		Light Emitting Diode (LED)	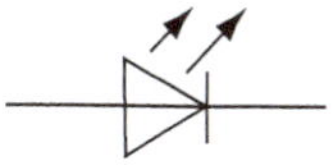
Potentiometer	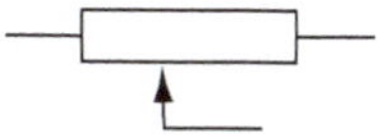	Switch	
Capacitor		Conductors (Wires) Crossing	
Inductor (Coil)		Conductors (Wires) Joining	
Transformer	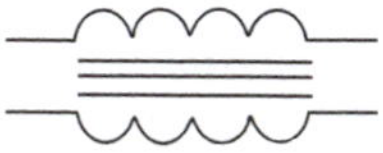	Zener Diode	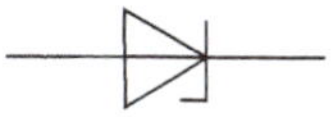
Fuse	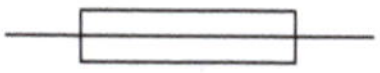	Light Dependent Resistor (LDR)	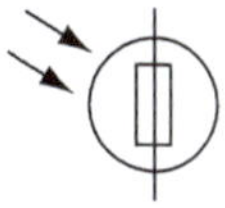

(a) A Selection of Circuit Symbols

Fuse	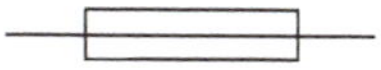	Fuse	
Lamp (Bulb)		Lamp (Bulb)	

Approved Symbol — Symbol Sometimes Used

(b) Non-Standard Circuit Symbols

Fig. 17.4 Circuit symbols

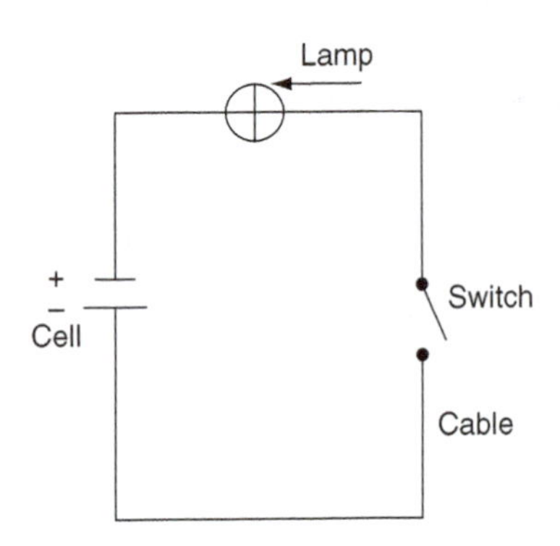

Fig. 17.5 Simple circuit

Power in watts = volts × amperes

$$W = V \times I \tag{17.1}$$

Example 17.1

An electrical component has a current of 5 A flowing through it. If the voltage drop across the component is 10 V, calculate the power used by the component.

$$W = V \times I$$
$$= 10 \times 5$$

Power in watts = 50 W.

17.8 Electrical units

Volt

Potential difference and electromotive force is measured in volts (V).

Ampere

Current – Electric current is measured in amperes (A), and it is the number of electrons that pass through a cross-section of the conductor in 1 second. The symbol used for electric current is I.

Ohm

Resistance – All materials resist the passage of electricity to some degree. The amount of resistance that a circuit, or electrical component, offers to the passage of electricity is known as the **resistance**. Electrical resistance is measured in ohms and it is denoted by the symbol Ω.

Watt

Power in a circuit is measured in watts. It is calculated by multiplying the current in the circuit by the voltage applied to it.

17.9 Ohm's law

Figure 17.6 shows a circuit that can be used to examine the relation between current, voltage and resistance in a circuit when the voltage is varied while the resistance and its temperature are held constant.

The graph in Figure 17.6(b) shows how the current increases as the voltage is raised. This graph shows that the current in this circuit is directly proportional to the voltage producing it, provided that the temperature of the circuit remains constant. This statement is known as Ohm's law and it is normally written in the form

$$V = I \times R \tag{17.2}$$

where V = voltage, I = current in amperes, and R is the resistance in ohms.

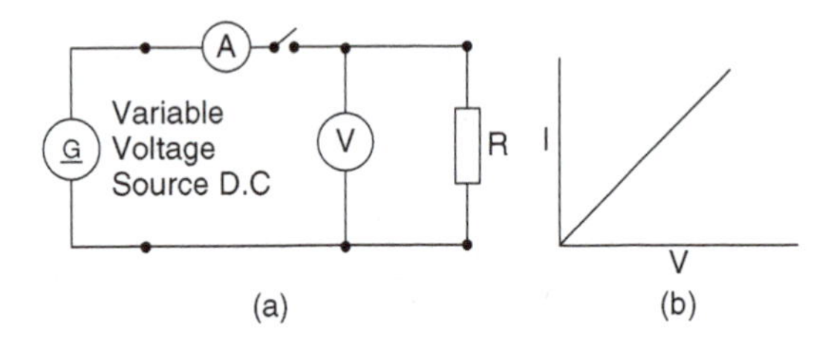

Fig. 17.6 Ohm's law circuit

17.10 Resistors in series

Example 17.2

Figure 17.7(a) shows a circuit that has a 12-volt supply and two resistors that are connected by a conducting cable. When resistors are connected as shown, they are in **series** and the resistance of the circuit is found by adding the two resistance values together.

In this case $R = R_1 + R_2$

$$R = 4\,\Omega + 8\,\Omega$$

$$R = 12\,\Omega$$

Ohm's law can now be applied to find the current in the circuit.

Transposing Equation (17.2) gives $I = \frac{V}{R}$

$$I = \frac{12\,V}{12\,\Omega}$$

$$I = 1\,A$$

In Fig. 17.7(b), the total resistance R

$= R_1 + R_2 + R_3$

$R = 3\,\Omega + 5\,\Omega + 10\,\Omega$

$R = 18\,\Omega$

The current in this case is $I = \frac{V}{R}$

$$= \frac{12}{18}$$

$$I = 0.67\,A$$

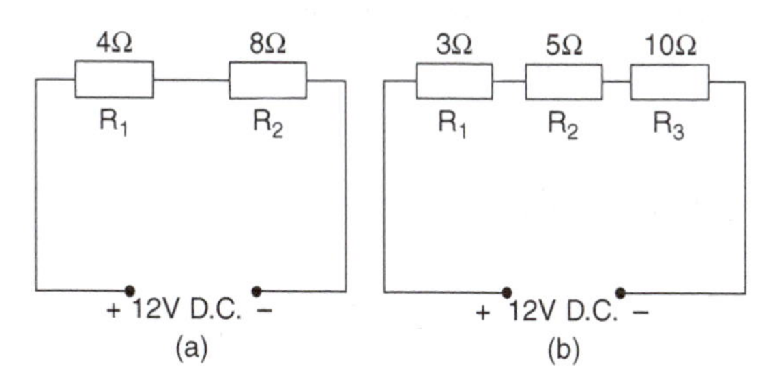

Fig. 17.7 Resistors in series

In a series circuit the current is the same in all parts of the circuit.

17.11 Resistors in parallel

Example 17.3

Figure 17.8 shows a circuit in which two resistors are placed in parallel. This ensures that the same p.d. (voltage) is applied to each resistor.

Ohm's law can be used to find the current in each resistor as follows:

For resistor R_1, $I_1 = \frac{V}{R_1}$

$$= \frac{12}{6}$$

$$I_1 = 2\,A$$

For resistor R_2, $I_2 = \frac{V}{R_2}$

$$= \frac{12}{4}$$

$$I_2 = 3\,A$$

The total current drawn from the supply

$$I = I_1 + I_2 = 2 + 3 = 5\,A$$

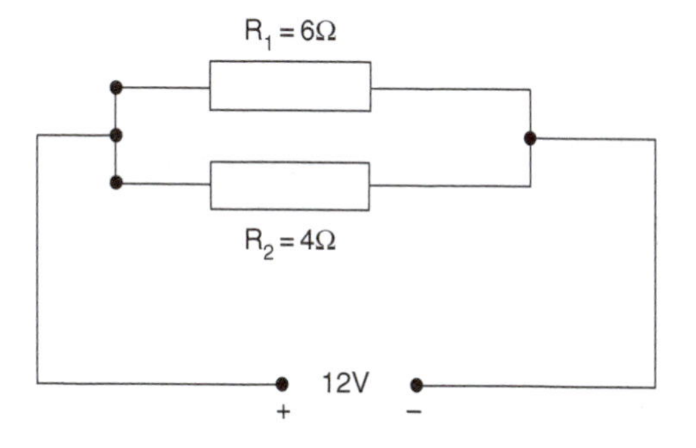

Fig. 17.8 Resistors in parallel

17.12 Alternative method of finding total current in a circuit containing resistors in parallel

The equivalent series resistance of the resistors in parallel can be found as follows:

Let R = the equivalent series resistance; then

$$\frac{1}{R} = \frac{1}{R_1} + \frac{1}{R_2}$$

Using the rules for fractions, the equivalent series resistance

$$R = \frac{R_1 R_2}{R_1 + R_2}$$

Example 17.4

The circuit in Figure 17.8 has the two resistors $R_1 = 6\,\Omega$ and $R_2 = 4\,\Omega$ in parallel.

Using Equation (17.2), the equivalent series resistance R

$$= \frac{6 \times 4}{6 + 4}$$

$$= \frac{24}{10}$$

$$R = 2.4\,\Omega$$

To find the current in the circuit, $I = \frac{V}{R}$

$$= \frac{12}{2.4}$$

$$I = 5\,A$$

17.13 Measuring current and voltage

The ammeter measures current and it is connected in **series**, so all of the current in the circuit flows through it.

The voltmeter is connected in **parallel** and it measures the potential difference between the two points to which it is connected. The p.d. between two points is sometimes referred to as the voltage drop (Figure 17.9).

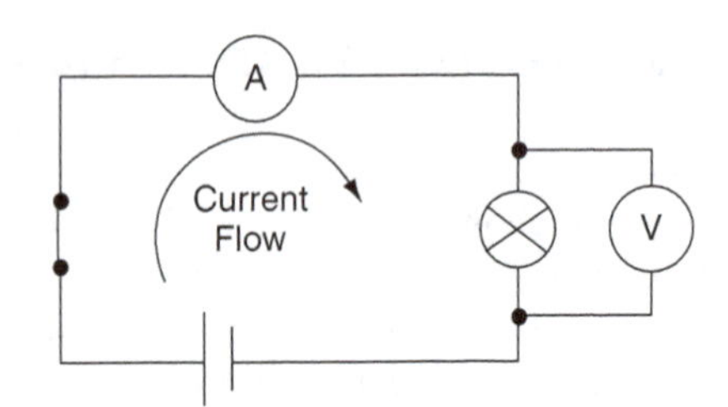

Fig. 17.9 Ammeter and voltmeter connected to a circuit

17.14 Ohmmeter

An ohmmeter is essentially an ammeter that has been adapted to measure resistance (Figure 17.10). The ohmmeter operates by applying a small voltage by means of a battery incorporated into the meter. The current flowing is then recorded on a scale that is calibrated in ohms (Ω).

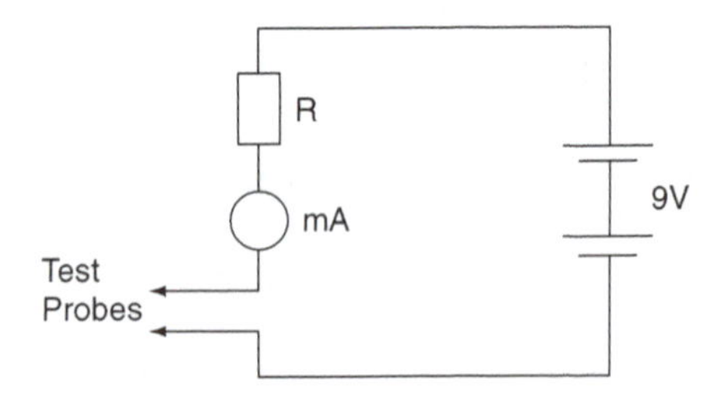

Fig. 17.10 Principle of the ohmmeter

17.15 Open circuit

An open circuit occurs when there is a break in the circuit (Figure 17.11). This may be caused by a broken wire or a connection that has worked loose.

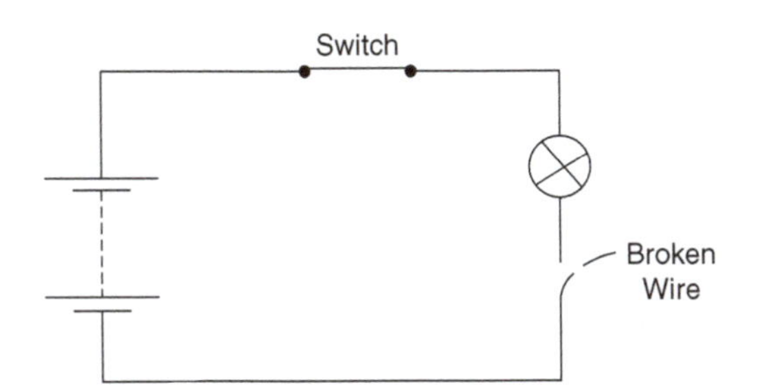

Fig. 17.11 Open circuit

17.16 Short circuit

A short circuit occurs when the conducting part of a wire comes into contact with the return side of the circuit. Because electric current takes the path of least resistance, the short circuit takes the current back to the battery, and therefore the lamp shown in Figure 17.12 will not function.

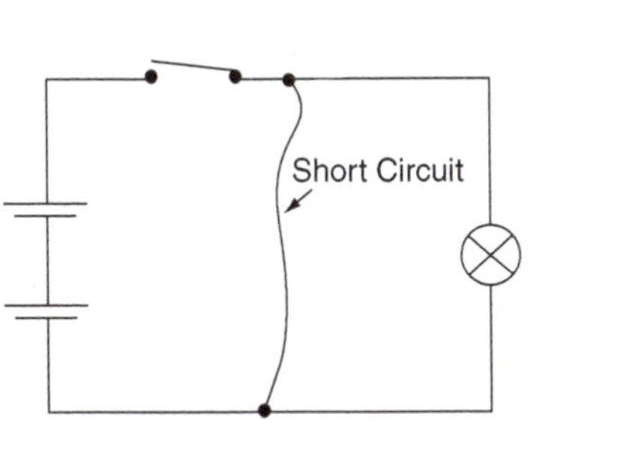

Fig. 17.12 Short circuit

17.17 Temperature coefficient of resistance

When pure metals such as copper, iron and tungsten are heated, the resistance increases. The amount that the resistance rises for each 1°C rise in temperature is known as the temperature coefficient of resistance.

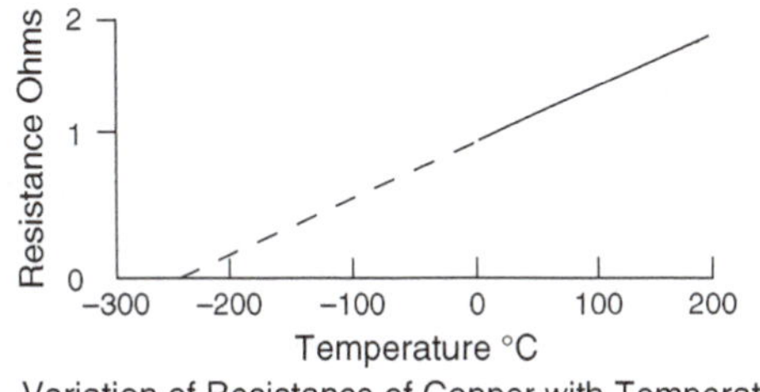

Fig. 17.13 Temperature coefficient of resistance

Negative temperature coefficient

When some semiconductor ceramic materials are heated, their resistance decreases. For each degree rise in temperature there is a decrease in resistance. These materials are said to have a negative temperature coefficient (Figure 17.14) and they are used in temperature sensors in various vehicle systems.

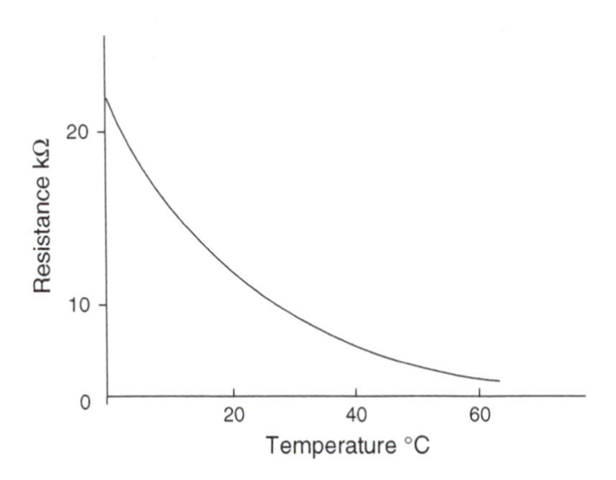

Fig. 17.14 Negative temperature coefficient

17.18 Electricity and magnetism

Permanent magnets

Permanent magnets are found in nature. Magnetite, which is also known as lodestone, is naturally magnetic and ancient mariners used it as an aid to navigation. Modern permanent magnets are made from alloys. One particular permanent (hard) magnetic material named 'alnico' is an alloy of iron, aluminium, nickel, cobalt, and a small amount of copper. The magnetism that it contains arises from the structure of the atoms of the metals that are alloyed together.

Magnetic force is a natural phenomenon. To deal with magnetism, rules about its behaviour have been established. Those that are important to our study are related to the behaviour of magnets and the main rules that concern us are as follows:

- Magnets have north and south poles.
- Magnets have magnetic fields.
- Magnetic fields are made up from lines of magnetic force.
- Magnetic fields flow from north to south.

If two bar magnets are placed close to each other so that the north pole of one is close to the south pole of the other, the magnets will be drawn together. If the north pole of one magnet is placed next to the north pole of the other, the magnets will be pushed apart. This tells us that like poles repel each other, and unlike poles attract each other (Figure 17.15).

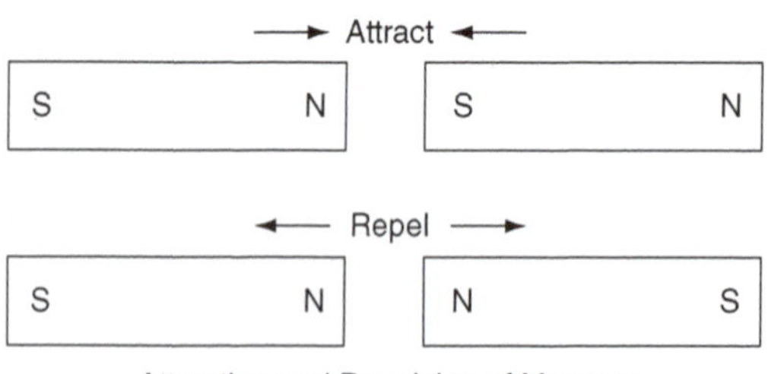

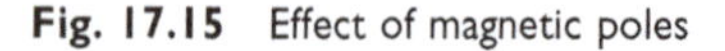

Fig. 17.15 Effect of magnetic poles

The magnetic effect of an electric current

Figure 17.16 shows how a circular magnetic field is set up around a wire (conductor) which is carrying electric current.

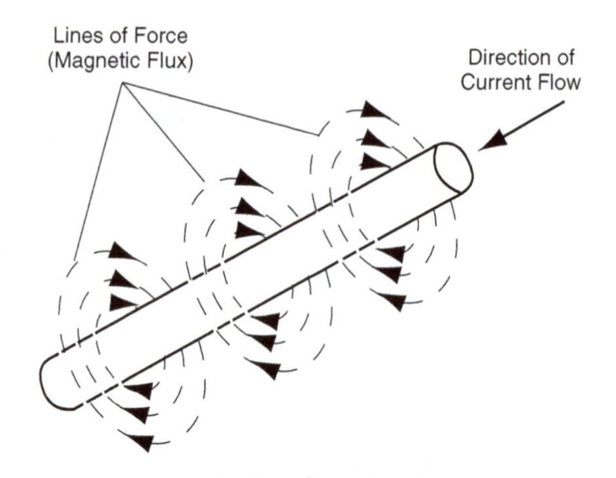

Fig. 17.16 Magnetic field of straight conductor

Direction of the magnetic field due to an electric current in a straight conductor

When a right-hand-threaded bolt is screwed into a nut, or a screw into a piece of wood, the screw is rotated clockwise (this corresponds to the direction, north to south, of the magnetic field); it penetrates the wood and this corresponds to the direction of the electric current in the conductor. The field runs in a clockwise direction and the current flows in towards the wood, as shown in Figure 17.17.

This leads to a convention for representing current flowing into a conductor and current flowing out (Figure 17.18). At the end of the conductor

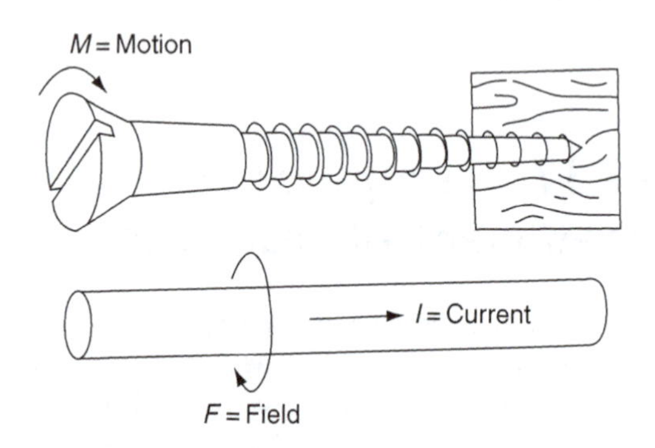

Fig. 17.17 Right-hand-screw rule

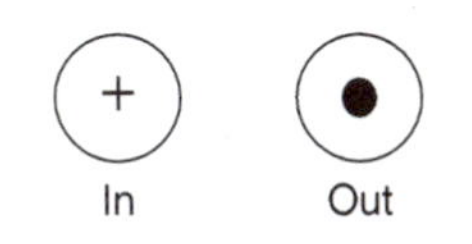

Fig. 17.18 Convention for current direction

where current is flowing into the wire, a '+' is placed, this representing an arrowhead. At the opposite end of the wire, where current is flowing out, a dot is placed, this representing the tip of the arrow.

Magnetic field caused by a coil of wire

When a conductor (wire) is made into a coil, the magnetic field created is of the form shown in Figure 17.19(a). A coil such as this is the basis of a solenoid.

17.19 Solenoid and relay

The solenoid principle is used to operate a relay (Figure 17.19(b)). Relays are used in vehicle systems where a small current is used to control a larger one. When the switch in the solenoid circuit is closed, current flows through the solenoid winding and this causes the magnetic effect to pull the armature towards the soft iron core. This action closes the contacts of the heavy current switch to operate a circuit, for example a vehicle headlamp circuit. When the relay solenoid current is switched off, the magnetic effect dies away and the control spring pulls the armature away from the soft iron core to open the heavy current switch contacts.

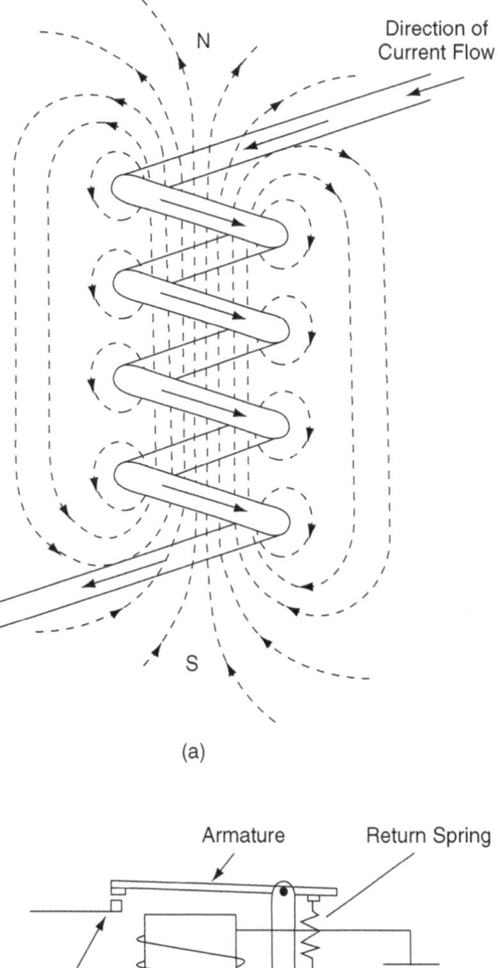

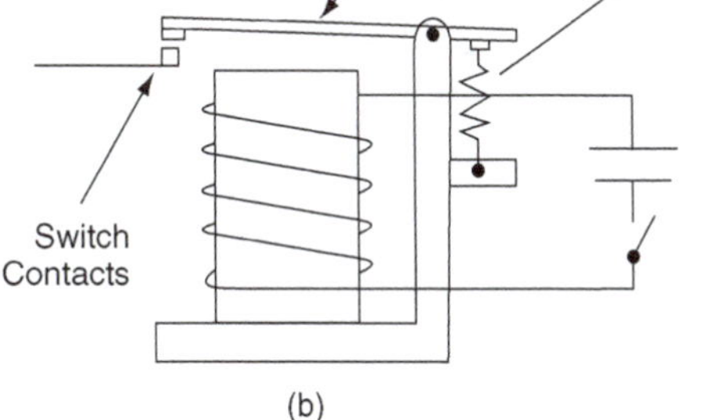

Fig. 17.19 (a) Magnetic field of a coil; (b) relay

17.20 Electromagnetic induction

Figure 17.20 shows a length of wire that has a voltmeter connected to its ends. The small arrows pointing from the north pole to the south pole of the magnet, represent lines of magnetic force (flux) that make up a magnetic field.

The straight arrow on the wire (conductor) shows the direction of current flow and the larger arrow, with the curved shaft, shows the direction in which the wire is being moved.

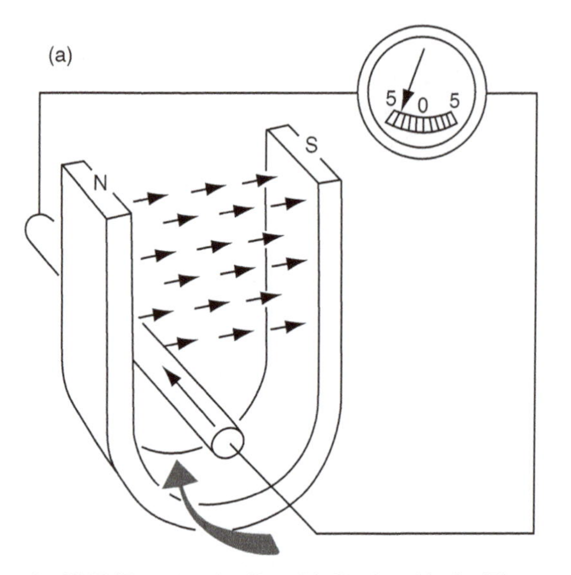

An EMF (Electromotive Force) is Produced in the Wire Whenever it Moves Across the Lines of Magnetic Flux.

Note: The Reversal of Current takes place When Direction of Movement is Reversed.

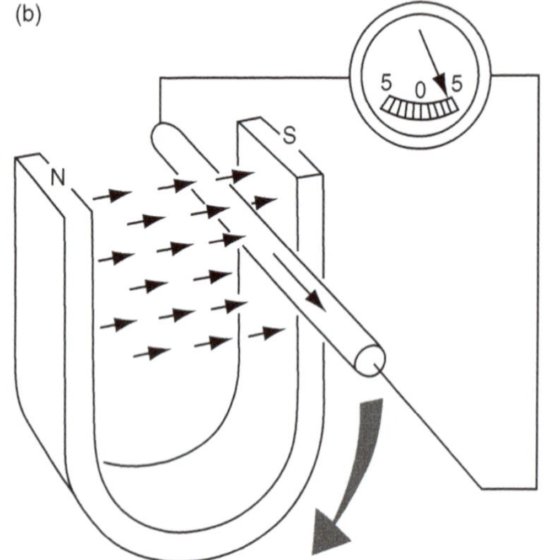

Fig. 17.20 Current flow, magnetic field and motion

Movement of the wire through the magnetic field, so that it cuts across the lines of magnetic force, causes an emf (voltage) to be produced in the wire. In Figure 17.20(a), where the wire is being moved upwards, the electric current flows in towards the page.

In Figure 17.20(b) the direction of motion of the wire is downwards, across the lines of magnetic force, and the current in the wire is in the opposite direction. Current is also produced in the wire if the magnet is moved up and down while the wire is held stationary. Mechanical energy is being converted into electrical energy, and this is the principle that is used in generators such as alternators and dynamos.

17.21 The electric motor effect

If the length of wire shown in Figure 17.20(a) is made into a loop, as shown in Figure 17.21, and electric current is fed into the loop, opposing magnetic fields are set up.

In Figure 17.21 current is fed into the loop via brushes and a split ring. This split ring is a simple commutator. One half of the split ring is connected to one end of the loop, the other half being connected to the other end. The effect of this is that the current in the loop flows in one direction. The coil of wire that is mounted on the pole pieces, marked N and S, creates the magnetic field. These opposing fields create forces. These forces 'push' against each other, as shown in Figure 17.22, and they cause the loop to rotate.

In Figure 17.22 the ends of the loop are marked A and B. B is the end that the current is flowing

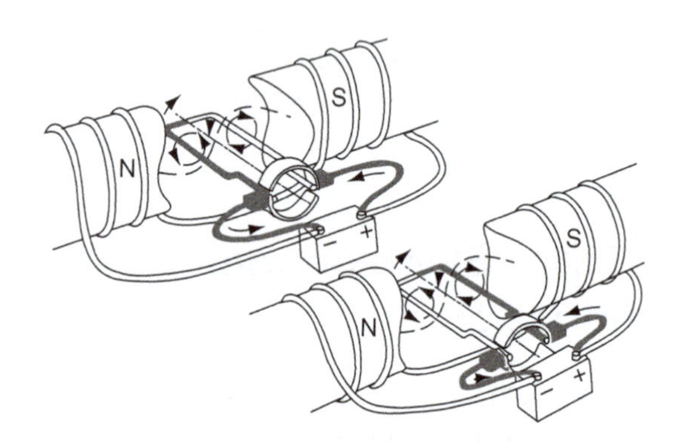

Fig. 17.21 Current flowing through the loop

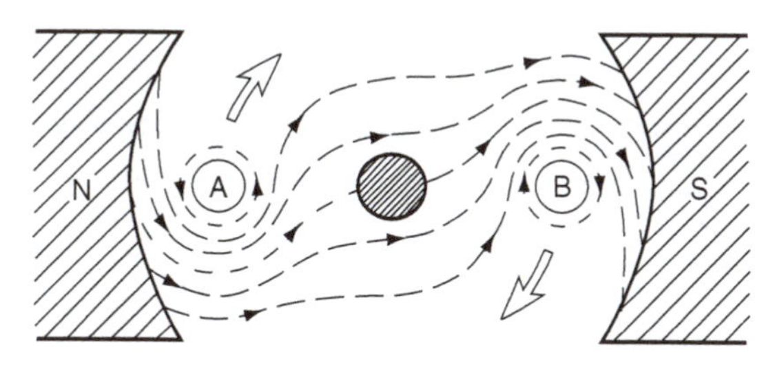

Fig. 17.22 Opposing magnetic fields – the motor effect

into, and A is the end of the loop where the current is leaving.

By these processes electrical energy is converted into mechanical energy, and it is this principle by which electric motors, such as the starter motor, operate. Just to convince yourself, try using Fleming's rule to work out the direction of the current, the field, and the motion.

17.22 Fleming's rule

Consider the thumb and first two fingers on each hand. Hold them in the manner shown in Figure 17.23.

Motion – this is represented by the direction that the thumb (M for motion) is pointing.

Field – the direction (north to south) of the magnetic field is represented by the first finger (F for field).

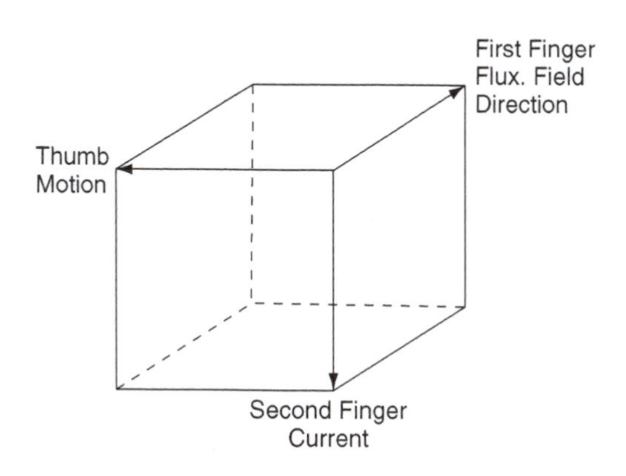

Fig. 17.23 Fleming's rule

Current – think of the hard 'c' in second. The second finger represents the direction of the electric current.

An aid to memory

It may be helpful for vehicle technicians to think of the MG car badge. If you imagine yourself standing in front of an MG, looking towards the front of the vehicle, the badge will present you with an M on your left, and a G on your right. So you can remember M for MOTORS and G for GENERATORS – left-hand rule for motors, right-hand rule for generators.

17.23 Alternating current

Cycle

An alternating voltage rises from zero to a maximum then falling back to zero. Thereafter, it rises to a maximum in the negative direction before returning to zero. This is called a cycle (Figure 17.24).

Period

The time taken to complete one cycle is a period. Period is measured in seconds.

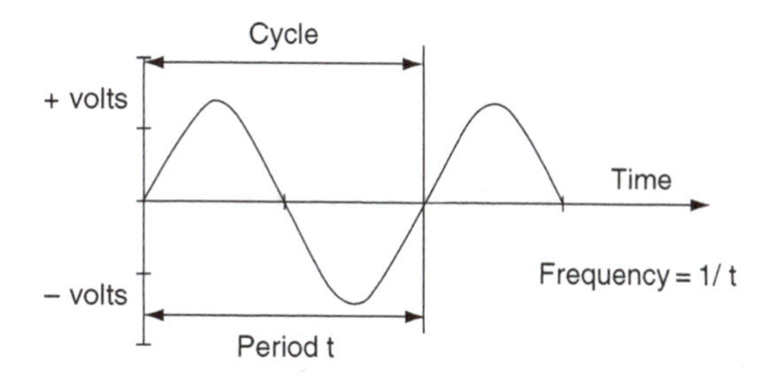

Fig. 17.24 An a.c. waveform

Frequency

Frequency is the number of cycles in one second and is measured in hertz (Hz). 1 Hz = one cycle per second.

17.24 Applications of alternating current

In a vehicle alternator (Figure 17.25) the magnet is the rotating member and the coil of wire is the stationary part in which the electrical energy to recharge the battery and operate electrical/electronic systems is generated. The magnetic rotor is driven from the crankshaft pulley and the stationary part – the stator, which contains the coils – is attached to the alternator case.

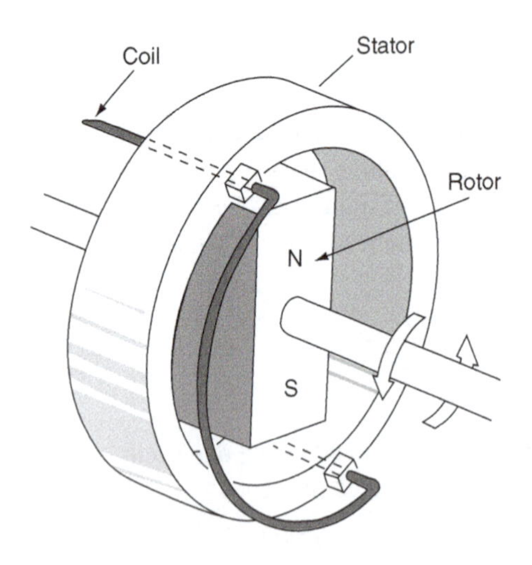

Fig. 17.25 The vehicle alternator

17.25 Transformer

A simple transformer consists of two coils of wire that are wound round the limbs of an iron core, as shown in Figure 17.26. One coil is the primary winding and the other is the secondary winding.

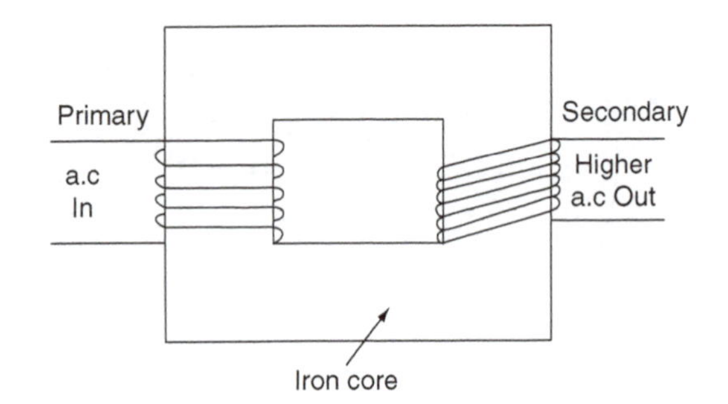

Fig. 17.26 Principle of the transformer

The iron core concentrates the magnetic field that is created by current in the primary winding. The moving lines of magnetic force that are created by the alternating current in the primary winding cut across the coils in the secondary winding and this creates a current in the secondary winding. The change in voltage that is created by the transformer depends on the number of turns in the coils. For example, if there are 10 turns of coil on the primary winding and 50 on the secondary, the secondary voltage will be 5 times the primary voltage. This relationship is known as the turns ratio of the transformer.

17.26 Lenz's law

In an inductive circuit such as that of the solenoid, there is a reaction to the switching off of the current that creates the magnetic field. As the magnetic field dies away, a current is produced that operates in the opposite direction to the magnetising current. This effect is known as Lenz's law, after the physicist who first explained the phenomenon in the 1830s.

Lenz's law states:

> The direction of the induced emf is always such that it tends to set up a current opposing the change responsible for inducing that emf.

17.27 Inductance

Any circuit in which a change of current is accompanied by a change of magnetic flux, and therefore by an induced emf, is said to possess inductance. The unit of inductance is the henry (H), named after the American scientist, Joseph Henry (1797–1878). A circuit has an inductance of L = 1 henry (1 H) if an emf of 1 volt is induced in the circuit when the current changes at a rate of 1 ampere per second. The symbol L is used to denote inductance.

17.28 Back emf

The emf that is created when a device such as a relay is switched off is known as the back emf.

$$\text{emf} = L \times \text{rate of increase, or decrease, of current}$$

$$= L \times \frac{I_2 - I_1}{t}$$

where $I_2 - I_1$ change of current, and t = time in seconds.

The back emf can be quite a high voltage and in circuits where this is likely to be a problem, special measures are incorporated to minimise the effect; an example is the surge protection diode in an alternator.

17.29 Inductive reactance

Figure 17.27(a) shows that when the switch is on, the rising current causes the coil (inductor) to produce a current that opposes the battery current. In Figure 17.27(b), the switch is off and the battery current is falling. The current from the inductor keeps a current flowing in the direction of the original battery current. In an a.c. circuit, this effect gives rise to inductive reactance, which has the same effect as resistance. Inductive reactance is measured in ohms.

$$\text{Inductive reactance } X_l = 2\pi f L$$

where f = frequency, L = inductance in henries.

17.30 Time constant for an inductive circuit

Figure 17.27(c) shows how the current grows from zero to its maximum value when the switch is

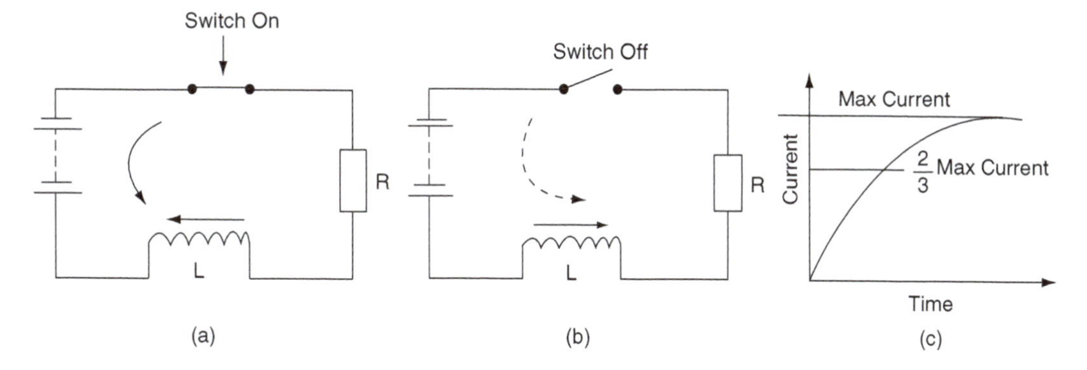

Fig. 17.27 Inductive effects

on. The time taken for the current to reach its maximum value is known as the **time constant**.

$$\text{The time constant } T = \frac{L}{R} \text{ seconds}$$

L = inductance in henries, R = resistance in ohms.

Example 17.5

A circuit contains a coil having an inductance of 2 H is connected in series with a resistance of 4 Ω. Calculate the time constant for the circuit.

Solution

$$\text{Time constant } T = \frac{L}{R} = \frac{2}{4} = 0.5\,\text{s}$$

17.31 Capacitors

A capacitor is a device that acts as a temporary store of electricity. In its simplest form a capacitor consists of two parallel metal plates that are separated by a material called a dielectric. The external leads are connected to the two metal plates as shown in Figure 17.28.

When a voltage is applied to the capacitor, electrons are stored on one plate and removed from the other, in equal quantities. The action sets up a potential difference which remains there when the external voltage is removed. The ability of a capacitor to store electricity is called its capacitance. Capacitance is measured in farads – after Michael Faraday, who worked in London in the nineteenth century.

Capacitance

Capacitance is determined by the properties of the dielectric and the dimensions of the metal plates. It is calculated from the formula

$$C = \frac{\varepsilon_0 \varepsilon_r A}{d} \text{ farads}$$

The terms ε_0 ε_r are properties of dielectric and they are called permittivity. A is the area of the plates and d is the distance between them.

Capacitance in a circuit

Figure 17.29(a) shows a circuit that can be used to examine the charging of a capacitor. When the switch is closed the capacitor is discharged. When the switch is opened, the capacitor will recharge. The voltmeter is observed while the time is recorded. The result is plotted and a graph of the type shown in Figure 17.29(b) is produced. This graph shows how the capacitor voltage changes with time. The time taken for the capacitor to

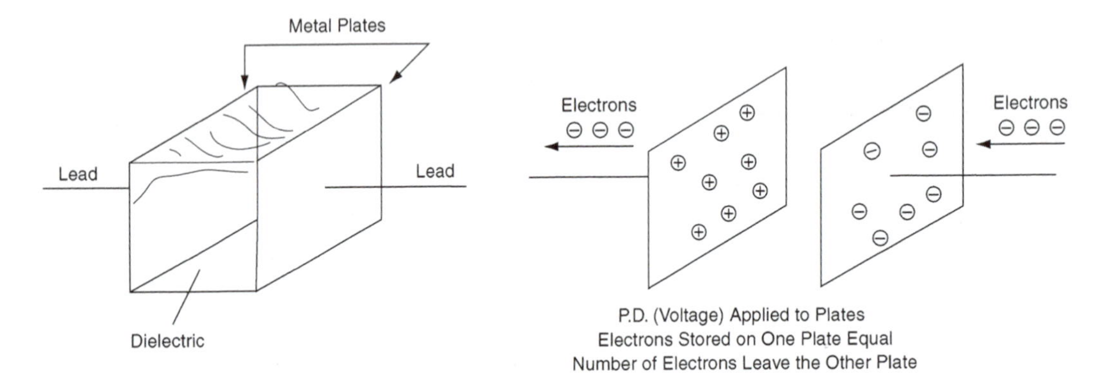

Fig. 17.28 A simple capacitor

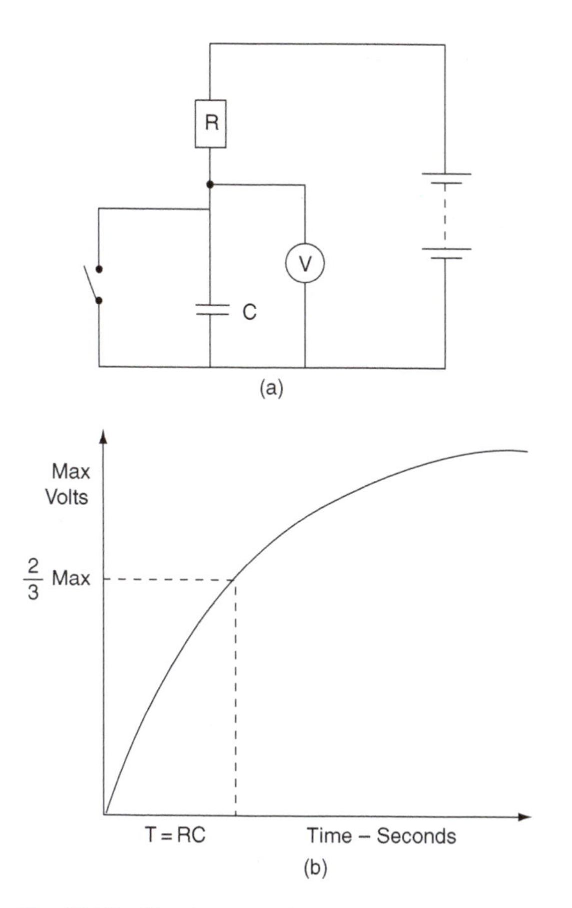

Fig. 17.29 Charging a capacitor

reach two thirds of the supply voltage is known as the time constant for the circuit.

Time constant $T = R \times C$

where R = resistance in ohms, C = capacitance in farads.

Capacitance values are normally given in millionths of a farad, known as microfarads, and denoted by the symbol μF.

Example 17.6

A circuit of the type shown in Figure 17.29(a) has a resistance of 10 000 Ω and a capacitance of 10 000 μF. Calculate the time constant for the circuit.

Solution

Time constant $T = C \times R = 0.01 \times 10000 =$ 100 seconds.

17.32 Capacitors in circuits

Contact breaker ignition circuit

In the contact breaker-type ignition system a capacitor is placed in parallel with the contact breaker (Figure 17.30). When the contact points open, a self-induced high-voltage electric charge attempts to arc across the points. The capacitor overcomes this problem by storing the electric charge until the points gap is large enough to prevent arcing.

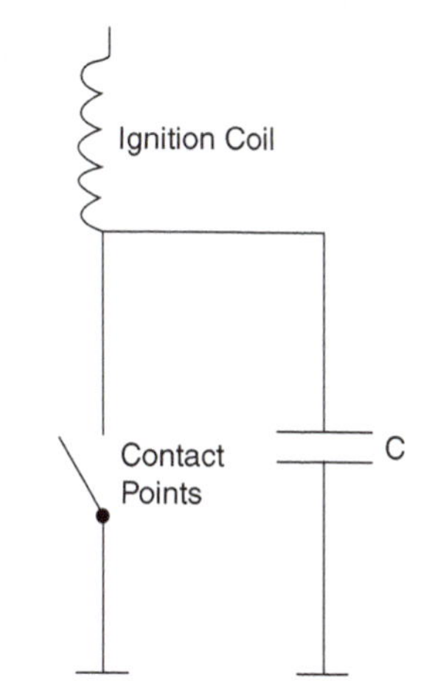

Fig. 17.30 Capacitor in contact breaker ignition circuit

Capacitive discharge ignition system

In the capacitive discharge ignition system the capacitor stores electrical energy until a spark is required (Figure 17.31). To generate a spark, an electronic circuit releases the energy from the

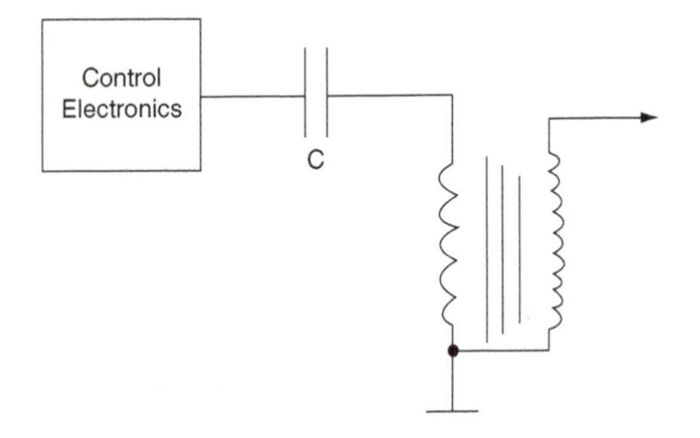

Fig. 17.31 A capacitive discharge ignition system

capacitor into the ignition coil circuit, a high-voltage spark thereby being produced at the sparking plug.

Capacitors in parallel and series

Parallel capacitors

Figure 17.32(a) shows two capacitors connected in parallel. The total capacitance provided by the two capacitors $= C = C_1 + C_2$.

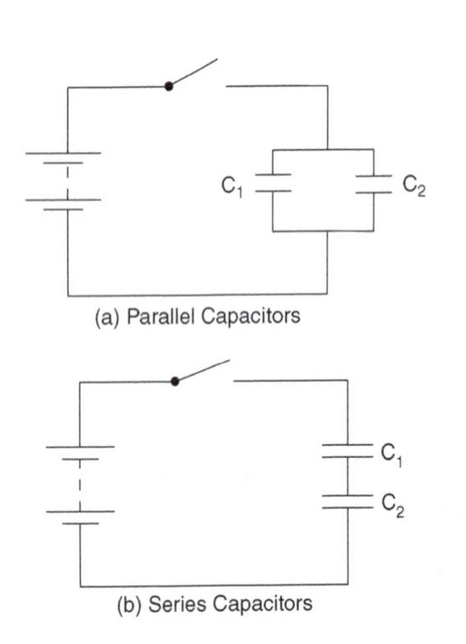

Fig. 17.32 Capacitors in circuits

Series capacitors

Figure 17.32(b) shows the two capacitors connected in series. In this case, the total capacitance C is given by the equation $C = C_1\ C_2/C_1 + C_2$.

Example 17.7

Two capacitors, one with a capacitance of 8 μF and the other with a capacitance of 4 μF are connected (a) in parallel, (b) in series. Determine the total capacitance in each case.

Solution

(a) Total capacitance $C = 8 + 4 = 12\,\mu F$

(b) Total capacitance $C = \dfrac{8 \times 4}{8 + 4} = \dfrac{32}{12} = 2.67\,\mu F$

Impedance

Figure 17.33 shows a circuit that contains a resistor, an inductor (coil) and a capacitor. The combined effect of these three devices in an a.c. circuit produces an effect known as impedance. Impedance acts like a resistance and it is measured in ohms.

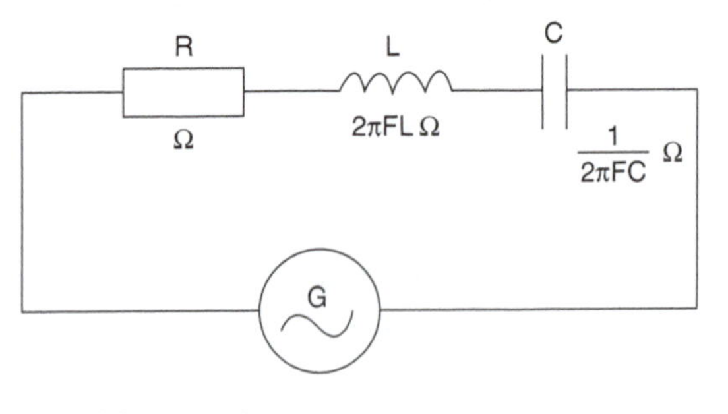

Fig. 17.33 Impedance

17.33 Summary of formulae

Ohm's law: $V = IR$

Power $= VI$ watts

Series resistance, $R = R_1 + R_2 + \cdots$

Parallel resistance, $\frac{1}{R} = \frac{1}{R} + \frac{1}{R_1}$ or $R = \frac{R_1\,R_2}{R_1 + R_2}$

Capacitors in parallel, $C = C_1 + C_2$

Time constant for inductive circuit $t = R \times C$

Capacitance $C = \frac{\varepsilon_0 \varepsilon_r\, A}{d}$ farads

Time constant for inductive circuit t = the time constant $T = \frac{L}{R}$ seconds

Lenz's law: *The direction of the induced emf is always such that it tends to set up a current opposing the change responsible for inducing that emf.*

17.34 Exercises

17.1 In the circuit shown in Figure 17.34, what is the total series resistance?
(a) 1.33 Ω
(b) 6 Ω
(c) 8 Ω
(d) 2 Ω

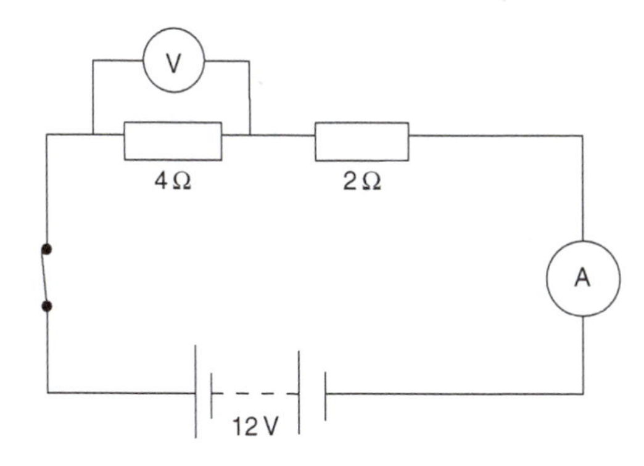

Fig. 17.34

17.2 In the circuit in Figure 17.34, what is the current at the ammeter?
(a) 6 A
(b) 8 A
(c) 2 A
(d) 1 A

17.3 In the circuit of Figure 17.34, what is the voltage drop across the 4 Ω resistor?
(a) 8 V
(b) 12 V
(c) 3 V
(d) 1 V

17.4 In the circuit of Figure 17.35, what is the equivalent series resistance of the two resistors in parallel?
(a) 8 Ω
(b) 1.5 Ω
(c) 3 Ω
(d) 12 Ω

17.5 In Figure 17.35, what is the current at the ammeter?
(a) 1 A
(b) 1.5 A
(c) 8 A
(d) 24 A

17.6 In Figure 17.35, what is the current in the 2 Ω resistor?
(a) 4 A
(b) 6 A
(c) 1.5 A
(d) 1 A

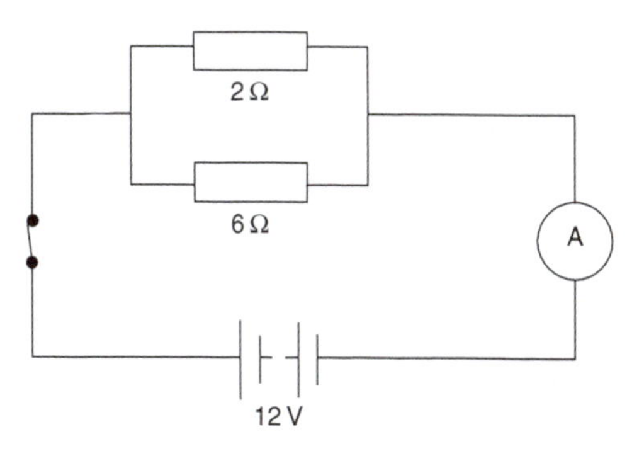

Fig. 17.35

17.7 In the circuit in Figure 17.36, what is the equivalent series resistance of the three resistors?
(a) 16 Ω
(b) 12.4 Ω
(c) 0.55 Ω
(d) 4 Ω

17.8 In the circuit of Figure 17.36, what is the total current flow at the ammeter?
(a) 48 A
(b) 3 A
(c) 0.97 A
(d) 5 A

17.9 In the circuit in Figure 17.36, what is the voltage drop across the 2.4 Ω resistor?
(a) 9.6 V
(b) 3 V
(c) 7.2 V
(d) 4.8 V

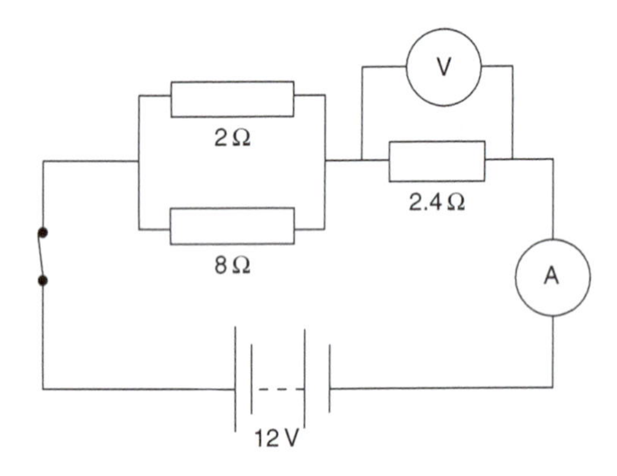

Fig. 17.36

17.10 What is the total capacitance of the two capacitors in parallel as shown in Figure 17.37?
(a) 15 000 μF
(b) 1.5 μF
(c) 250 μF
(d) 0.67 μF

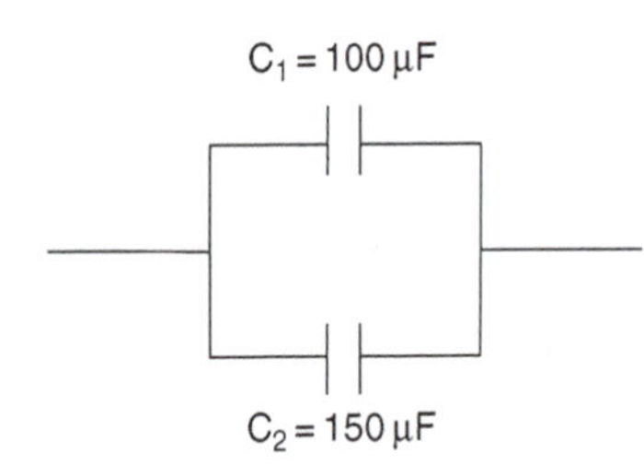

Fig. 17.37

17.11 Does the capacitor in parallel with the ignition points of a coil ignition system, as shown in Figure 17.38:
(a) prevent arcing across the points?
(b) act as a suppressor to reduce radio interference?
(c) reduce erosion at the spark plug contacts?
(d) prevent current overload in the coil primary circuit?

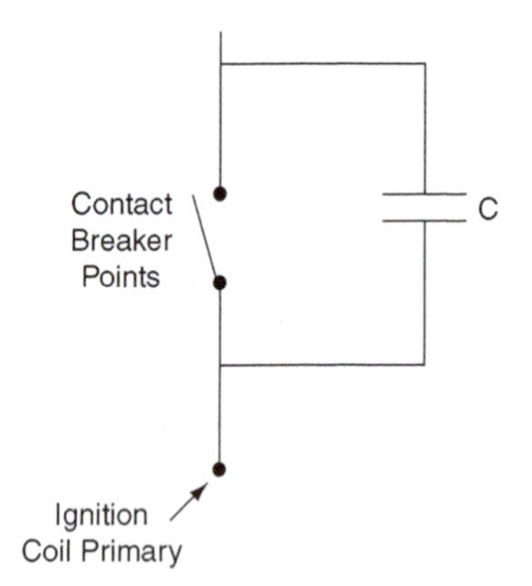

Fig. 17.38

17.12 The time constant for a capacitor in a circuit that contains resistance is the time that it takes to charge the capacitor to
(a) 6% of the charging voltage
(b) 67% of the charging voltage
(c) 33% of the charging voltage
(d) the full charging voltage

17.13 Is Fleming's right hand rule useful for working out the direction of current flow in
(a) the armature winding of a starter motor?
(b) the armature winding of a generator?
(c) the winding of a non-inductive resistor?
(d) a solenoid?

17.14 Impedance in an a.c. circuit is the total opposition to current flow that the circuit causes. Is impedance measured in
(a) watts?
(b) amperes?
(c) ohms?
(d) coulombs?

17.15 In a metal conductor the flow of current is caused by the drift of free electrons. Is the free electrons flow
(a) from the positive to the negative electrode?
(b) from a lower electrical potential to a higher electrical potential?
(c) from the positive to the negative terminal of the battery?
(d) from high electrical potential to low electrical potential?

17.16 A 12-volt electric motor takes a current of 20 A. Is the power of the motor
(a) 1.67 kW
(b) 0.3 kW
(c) 240 W
(d) 32 W

17.17 Does Lenz's law state that
(a) the direction of an induced emf is always such that it tends to set up a current opposing the change responsible for inducing the emf?
(b) V = IR?
(c) t = RC?
(d) the back emf is directly proportional to the capacitance?

17.18 Are electrical conductors materials
(a) in which free electrons can be made to move?
(b) that have a high specific resistance?
(c) that do not heat up when current flows?
(d) that are always physically hard?

17.19 Does a circuit that has a resistance of 10 000 Ω and a capacitance of 10 000 μF have a time constant of
(a) 10 seconds?
(b) 0.1 seconds?
(c) 100 seconds?
(d) 1000 seconds?

18

Electronic principles

18.1 Introduction

Microprocessors are small computers which find use in many vehicle systems such as engine management, anti-lock braking, and traction control (Figure 18.1).

In order to function, the microprocessor requires inputs from sensors. These inputs are used to make decisions which, in turn, are used to make outputs that enable the operation of various actuators. Many of these sensors and actuators operate on electronic principles.

The subject matter in this chapter is designed to provide an insight into basic electronic principles.

18.2 Semiconductors

Materials such as copper have many free electrons in their outer shells. These free electrons can be made to flow in a circuit under the influence of an applied voltage, and they are good conductors. Semiconductor materials such as pure silicon have few electrons in their outer shells, which makes

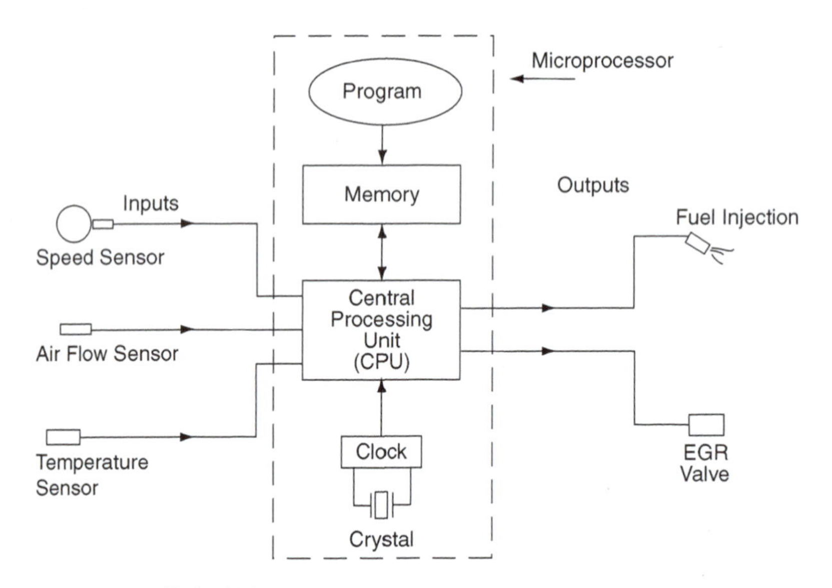

Fig. 18.1 Microprocessor-controlled vehicle systems

them poor conductors of electricity. The conductivity of pure silicon is improved by diffusing impurities into it under high temperature. Impurities that are often used for this purpose are boron and phosphorus, and they are called **dopants**.

Effect of dopants

When boron is diffused into silicon the resulting material is known as a **p-type** semiconductor. When phosphorus is diffused into silicon it is known as an **n-type** semiconductor.

Electrons and holes

Phosphorus doping gives **surplus electrons** to the material. These surplus electrons are mobile. Boron doping leads to atoms with insufficient electrons that can be thought of as **holes**. **P-type** semi-conductors have **holes** whereas**N-type** semi-conductors have surplus **electrons**. When a p.d. (voltage) is applied, electrons drift from − to +, and holes drift from + to −.

18.3 The p–n junction

When a piece of p-type semiconductor is joined to a piece of n-type semiconductor, a junction is formed (Figure 18.2). At the junction an exchange of electrons and holes takes place until **an electrical p.d. which opposes further diffusion** is produced between the two sides of the junction. The region where this barrier exists is known as the **depletion layer and it acts like a capacitor that holds an electric charge**.

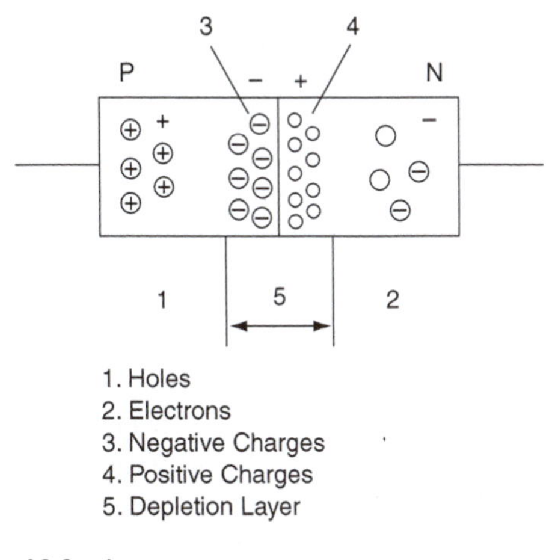

Fig. 18.2 A p–n junction

18.4 Bias

When a p–n junction is placed in a circuit, as shown in Figure 18.3(a), the battery provides a potential difference across the depletion layer. This overcomes the barrier effect and allows electrons to flow through the p–n junction (diode) and the lamp will illuminate. This is known as forward bias.

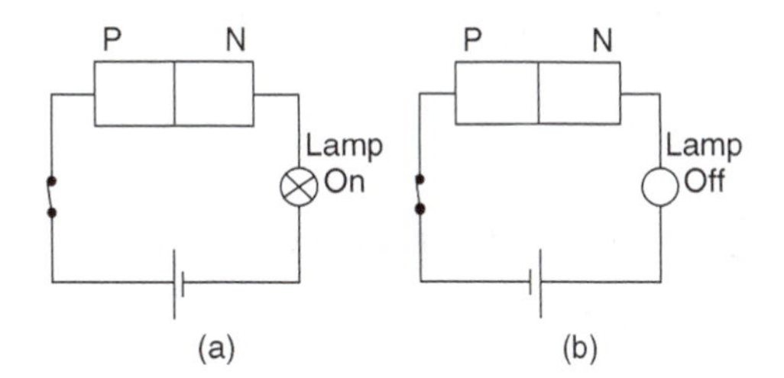

Fig. 18.3 Forward and reverse bias

When the battery connections are reversed, as shown in Figure 18.3(b), a reverse bias is applied at the p–n junction. This reinforces the barrier effect at the depletion layer and prevents a flow of current through the circuit. The result is that the lamp will not illuminate. The voltage required to overcome the barrier potential is known as the knee voltage and it is approximately 0.7 V for a silicon diode.

A junction diode, or just diode, is an effective one-way valve in an electronic circuit.

18.5 Behaviour of a p–n junction diode

Figure 18.4(a) shows a circuit in which the resistor can be changed to vary the current in the circuit and hence the voltage across the diode.

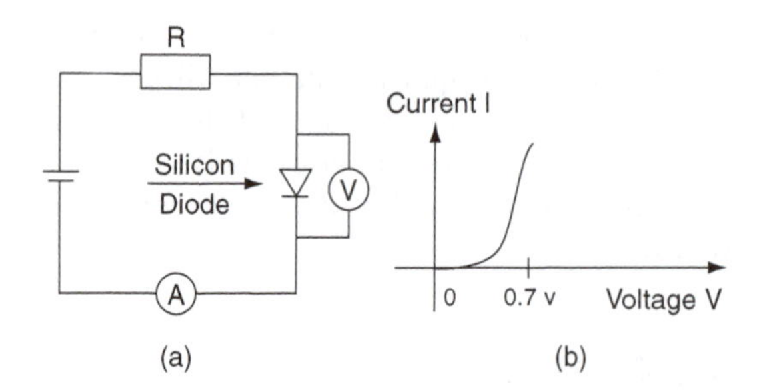

Fig. 18.4 Current–voltage in forward biased diode

A series of readings of current and voltage for various resistance values will produce a graph of the form shown in Figure 18.4(b). This graph shows that there is very little current below 0.7 V but as soon as the voltage reaches 0.7 V the current rises rapidly.

18.6 Diode protection resistor

If the current in a diode is too large, the heating effect will destroy the diode properties. To prevent this happening a resistor is placed in series with the diode and this restricts the current through the diode to a safe level. The diode current may be calculated from

$$I = \frac{Vs - Vd}{R}$$

where R = resistance in ohms, Vs = supply voltage, Vd = diode voltage.

Example 18.1

In the circuit shown in Figure 18.4(a), the resistor value = 1 kΩ, the supply voltage = 12 V, and the silicon diode voltage = 0.7 V. Calculate the current through the diode.

Solution

$$\text{Diode current } I = \frac{12 - 0.7}{1000} = 0.0113\,\text{A} = 11.3\,\text{mA}$$

18.7 Negative temperature coefficient of resistance – semiconductor

When the temperature of a semiconductor is increased, thermal activity causes an increase in electrons and holes. These actions increase the conductivity of the semiconductor. (See Figure 18.5.)

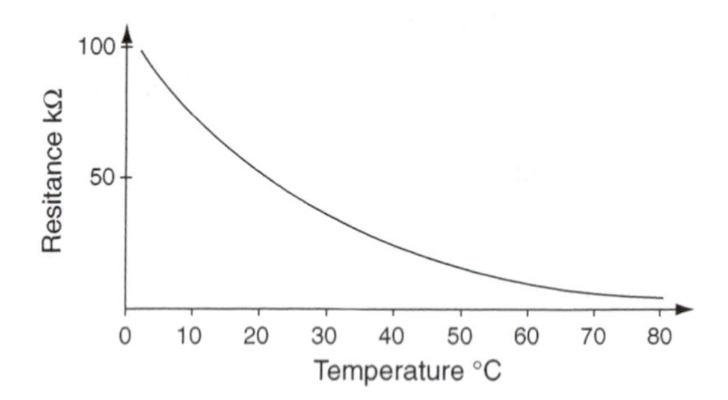

Fig. 18.5 Negative temperature coefficient

18.8 The Zener diode

The Zener diode is a p–n junction diode that is designed to break down at a specified voltage (Figure 18.6). The breakdown mechanism is non-destructive provided that the current is limited to prevent overheating. The voltage at which breakdown occurs and current flows in the reverse direction is known as the Zener voltage. Once current flows in the reverse direction the voltage drop across the Zener diode remains constant at the Zener voltage. Zener voltages normally fall in the range 3 V to 20 V. The actual figure is chosen at the design stage, typical values being 5.6 V and 6.8 V. Two common uses of the Zener diode in automotive practice are voltage surge protection in an alternator circuit and voltage stabilisation in instrument circuits.

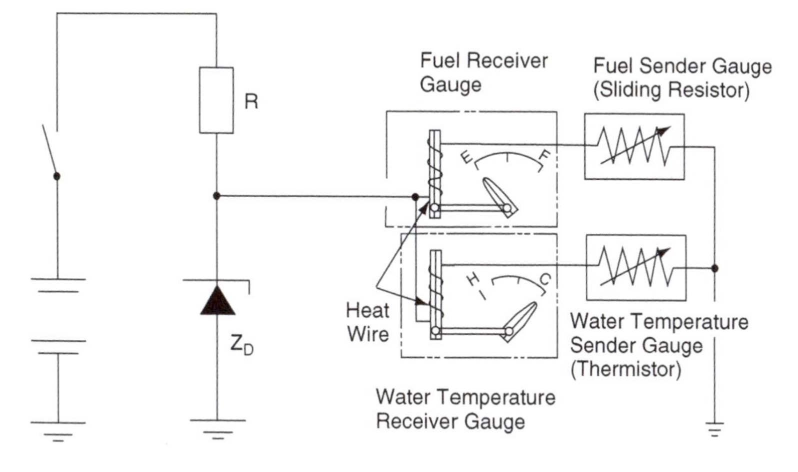

Fig. 18.6 The Zener diode as a voltage regulator

18.9 Light emitting diode (LED)

When electrons have been moved into a larger orbit they may fall back to their original energy level. When this happens, the electrons give up energy in the form of heat, light, and other forms of radiation. The LED utilises this effect, the various types found emitting light of different colours. When elements such as gallium, arsenic and phosphorus are used, colours such as red, green and blue are produced.

Voltage and current in an LED

Most LEDs have a voltage drop of 1.5 to 2.5 volts with currents of 10 to 40 mA. In order to limit the current, a resistor is placed in series with an LED, as shown in Figure 18.7.

Example 18.2

The voltage drop across the LED in Figure 18.7 is 2 V and the maximum permitted current is 20 mA.

The voltage drop across the resistor is $(9-2) = 7\,\text{V}$

Using Ohm's law, $R = \frac{V}{I} = \frac{7}{0.02} = 350\,\Omega$.

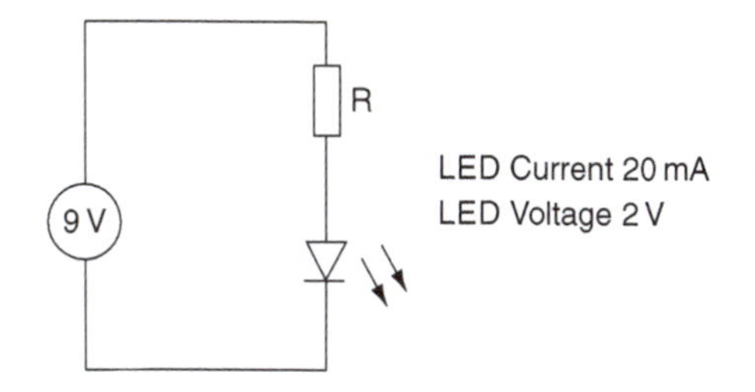

Fig. 18.7 The current-limiting resistor – LED

18.10 Photodiode

When radiation in the form of light falls on a p–n junction diode, some electrons are caused to move thus creating an electric current. The greater the intensity of the light falling on the diode, the greater the current. This makes photodiodes suitable for use in a range of automotive systems such as security alarms and automatic light operation.

18.11 Bipolar transistors

Bipolar transistors use both positively and negatively charged carriers. This form of transistor basically comprises two p–n junction diodes placed back to back to give either a p–n–p type,

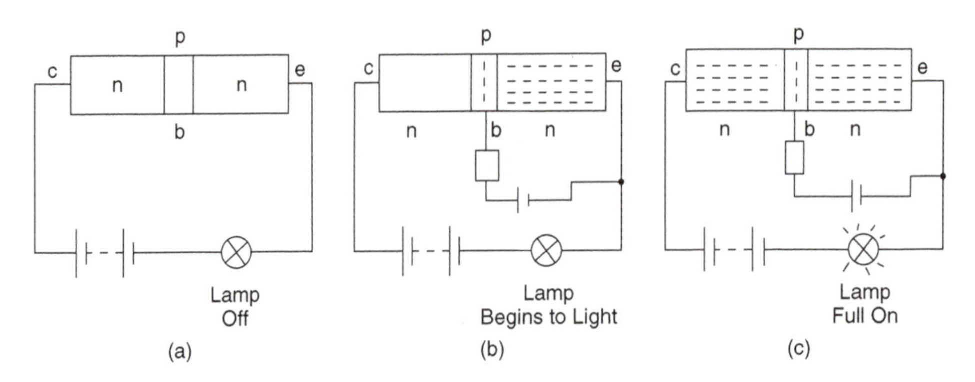

Fig. 18.8 A simple transistor

or an n–p–n type. In both cases the transistor has three main parts, these being the collector, the emitter and the base, as shown in Figure 18.8(a), (b), and (c) respectively.

Basic operation of transistor

Figure 18.8(a) shows an n–p–n transistor in a circuit that includes a battery and a lamp. There is no connection to the base – the transistor is not biased. Virtually no current flows in the circuit and the lamp will not illuminate.

In Figure 18.8(b) a voltage source is applied to the base of the transistor and this causes electrons to enter the base from the emitter. The free electrons can flow in either of two directions; i.e. they can flow either into the base or into the collector, which is where most of them go. In this phase the lamp begins to light.

Once free electrons are in the collector they become influenced by the collector voltage and current flows freely through the main circuit to illuminate the lamp. This phase is shown in Figure 18.8(c).

This switching action takes place at very high speed, which makes transistors suitable for many functions in electronic and computing devices.

Current gain in transistor

Under normal operating conditions the collector current I_c is much larger than the base current I_b. The ratio I_c/I_b is called the gain and is denoted by the symbol β.

$$\beta = \frac{Ic}{I_b}$$

Values of current gain vary according to the type of transistor being used. Low-power transistors have current gains of approximately 100 to 300, high-power transistors less than 100.

Example 18.3

A certain transistor has a current gain of 120. Determine the collector current I_c when the base current $I_b = 0.1\,mA$.

Solution

$$\beta = \frac{Ic}{I_b}$$

$$I_c = \beta \times I_b$$

$$I_c = 120 \times 0.1$$

$$= 12\,mA$$

Current flow in transistors

Figure 18.9 shows the path of current flow in transistors. It is useful to study this when working out how circuits operate.

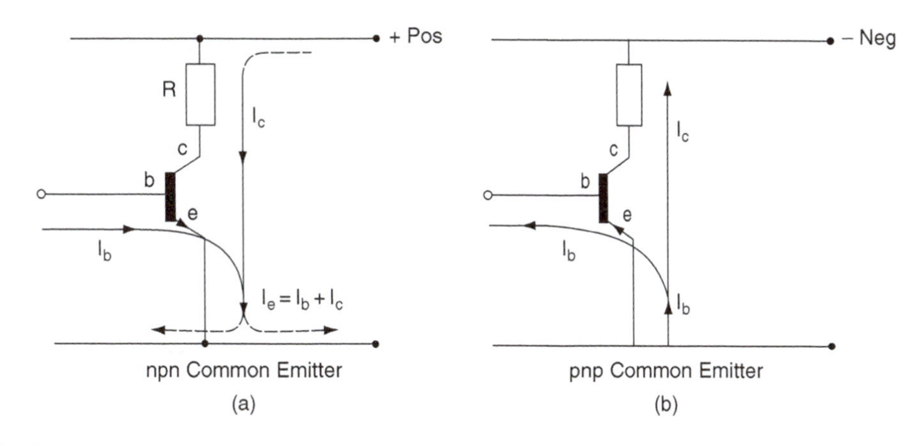

Fig. 18.9 Current flow in transistors

18.12 Transistor circuit used in automotive applications

Voltage amplifier

Some devices such as sensors in vehicle systems produce weak electrical outputs. These weak outputs are amplified (voltage is increased) to make them suitable for use in operating a vehicle system. The circuit shown in Figure 18.10 shows a simple voltage amplifier circuit.

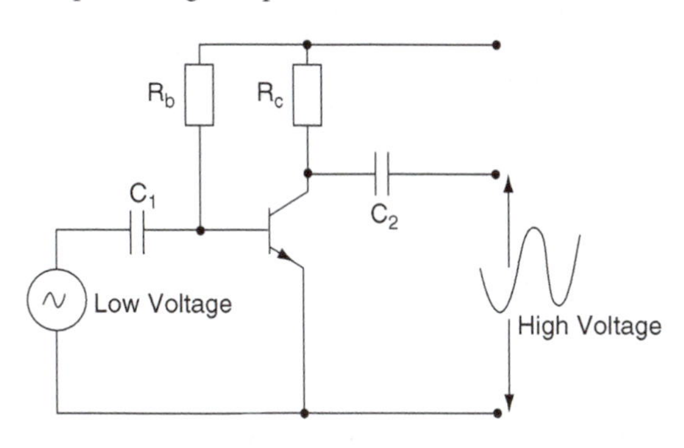

Fig. 18.10 Voltage amplifier

Darlington pair

A Darlington pair is a circuit that contains two transistors, as shown in Figure 18.11. The effect is

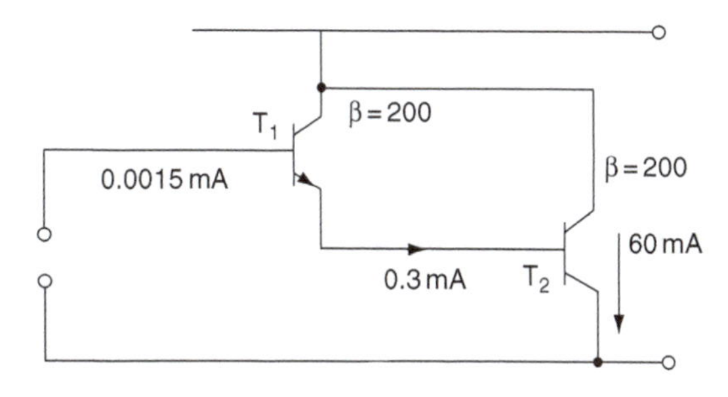

Fig. 18.11 Darlington pair

that the circuit behaves as a single transistor with a very high current gain. The collector current from T_1 is fed to the base of T_2.

In the example shown , the transistors each have a current gain of $\beta = 200$. This gives a collector current at T_1 of 0.3 mA, which becomes the base current of T_2 and the resulting collector current at T_2 is $200 \times 0.3 = 60$ mA.

A Darlington pair is often used to provide the current for the primary winding of an ignition coil.

Heat sink

Semiconductor devices are adversely affected by temperatures outside their designed operating range. In low-power devices the heat is normally dissipated through the structure of the circuit and its surroundings. High-power devices such as the

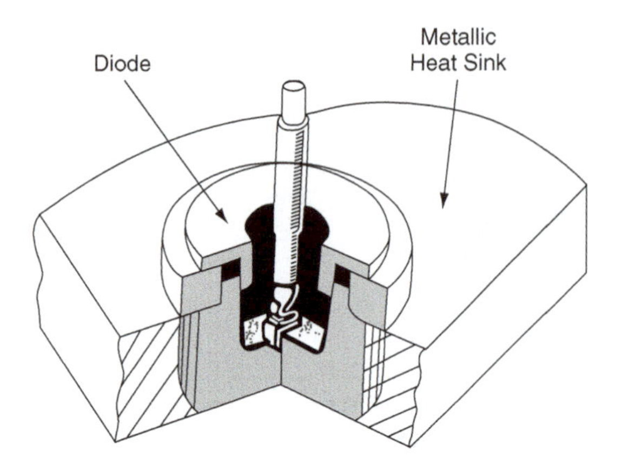

Fig. 18.12 A diode and its heat sink

Darlington pair are mounted on a metal heat sink that conducts the heat away from them.

18.13 Filter circuits

Electrical filters are used to prevent certain alternating currents from reaching parts of a circuit where they affect performance. The filter circuits shown in Figure 18.13 operate on the principle that at certain frequencies a capacitor will become a conductor of electric current.

The low-pass filter, shown in Figure 18.13(a), is designed to pass low frequencies from the input to the output. At higher frequencies the capacitor is effectively short circuited and the high frequency current bypasses the output terminals.

In the high-pass filter, shown in Figure 18.13(b), the low frequencies are effectively blocked. At a certain frequency the capacitor will become a conductor and the higher-frequency current will pass to the output terminals.

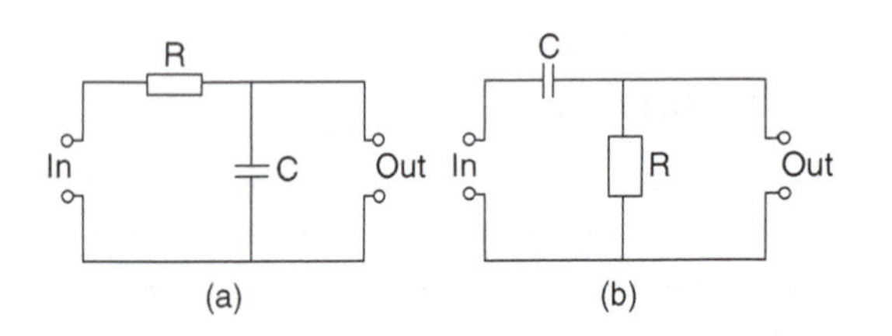

Fig. 18.13 Low-pass and high-pass filters

Voltage divider

Voltage dividers are used in circuits where a potential difference (voltage) that is a fraction of an applied emf, is required (Figure 18.14).

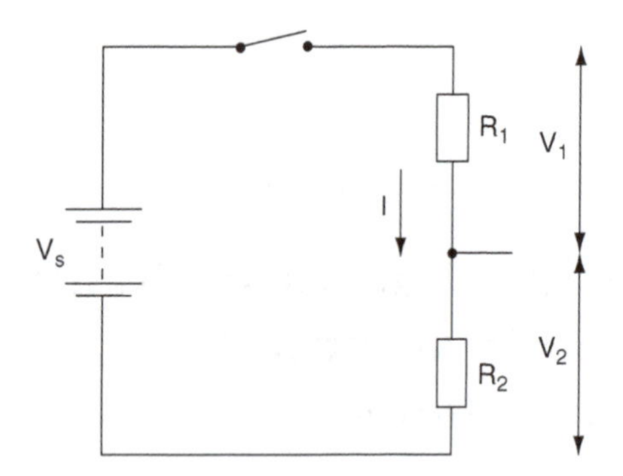

Fig. 18.14 The voltage divider

As R_1 and R_2 are in series, the same current I flows through both of them.

The resistance of the circuit $= R_1 + R_2$

The supply voltage $=$ Vs

$$\text{The current } I = \frac{Vs}{R_1 + R_2} = \frac{V_1}{R_1}$$

$$\text{From this, } V_1 = \frac{R_1 Vs}{R_1 + R_2} \qquad \textbf{(18.1)}$$

$$\text{Similarly, } V_2 = \frac{R_2 Vs}{R_1 + R_2} \qquad \textbf{(18.2)}$$

Example 18.4

Figure 18.15 shows a voltage divider circuit. Determine the value of V_1 and V_2.

Solution

Using Equation (18.1)

$$V_1 = \frac{1 \times 9}{1 + 8} = 1 \text{ volt}$$

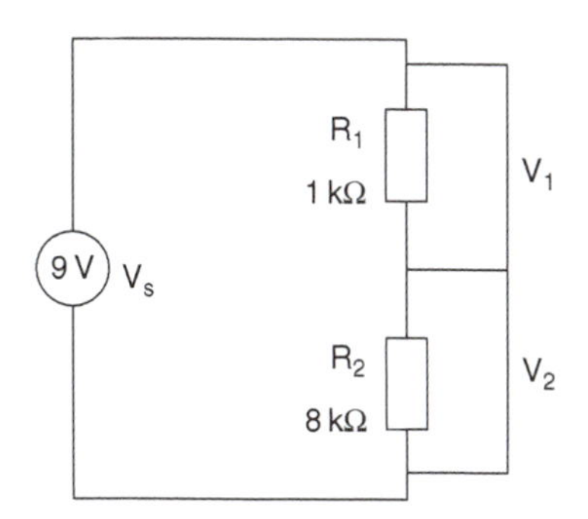

Fig. 18.15 Example – voltage divider

Using Equation (18.2)

$$V_2 = \frac{8 \times 9}{1+8} = 8\,\text{volts}$$

18.14 Integrated circuits

Large numbers of capacitors, transistors and diodes can be made on a small piece of silicon. These components are then connected to form complicated circuits. These circuits are known as integrated circuits (ICs) and the small piece of silicon is referred to as a 'chip'.

18.15 Sensors and actuators

A sensor is a device that detects some aspect of vehicle performance and converts it into an electrical signal that is fed to the control computer for use in controlling factors such as engine fuelling and ignition timing. A sensor may be classed as one of two types:

1. **Active sensors** – these produce electricity that represents the variable that they are detecting. They are often electromagnetic or, as in the case of zirconia oxygen sensors, electrochemical.
2. **Passive sensors** – they receive a supply of electricity and process it so that the control unit detects changes. In the case of the negative temperature coefficient sensor used to detect coolant temperature, the changes in resistance produce a varying sensor voltage that represents coolant temperature.

Typical examples of sensors are:

- air flow;
- exhaust gas oxygen;
- engine speed;
- crank position;
- throttle position;
- engine coolant temperature.

An actuator is a device that is controlled by commands that are outputs from the control computer, typical examples being;

- fuel injector;
- anti-lock braking modulator;
- exhaust gas recirculation valve.

The simple moving flap type of air flow sensor shown in Figure 18.16 makes use of a potential divider that interprets air flow as a voltage. The wiper of the potential divider is fixed to the spindle of the sensor. When the flap moves in response to changes in air flow, the wiper moves along the resistance strip. The voltage Vs that is transmitted to the control computer is an exact representation of air flow.

The petrol injector shown in Figure 18.17 is an electromagnetic actuator that operates in a way that is similar to the action of a solenoid. Current to the injector is supplied via a relay and the control computer switches the power transistor on, so that the injector is operated. The duration of the injection period determines the amount of fuel that is injected into the engine. The duration of the injection period (and thus the amount of fuel injected) is determined by the computer program, which makes use of the input from the air flow sensor and from other sensors for throttle position, engine speed and engine coolant temperature, for example.

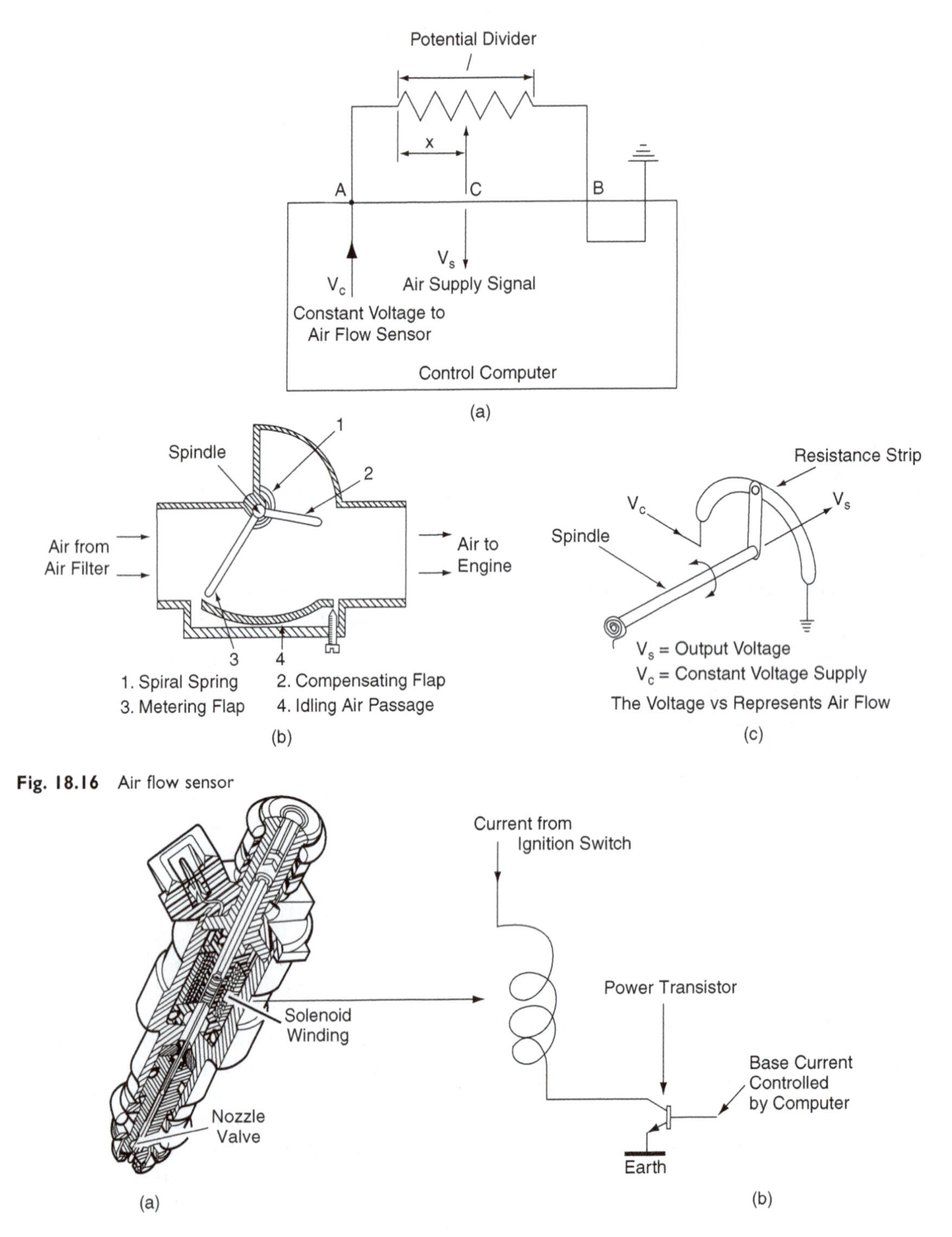

Fig. 18.16 Air flow sensor

Fig. 18.17 Petrol injector

18.16 Control unit (computer) inputs and outputs

Inputs to the control unit are often of varying voltage (analogue) form. These inputs are received at an interface that contains an analogue-to-digital (A/D) converter. The A/D converter changes the input voltage to a binary (digital) code that is used by the central processing unit (cpu) to enable the control unit to deliver output signals. The outputs are of varied forms, for example an instruction to operate the fuel injectors for a set period in each power stroke of the engine.

18.17 Logic gates

Transistor switches can be built into small circuits to make devices called logic gates. These gates operate on digital (binary) principles.

The RTL NOR gate

Figure 18.18 shows how a 'logic' gate is built up from an arrangement of resistors and a transistor (RTL = resistor transistor logic). There are three inputs: A, B and C. If one or more of these inputs is high (logic 1), the output will be low (logic 0).

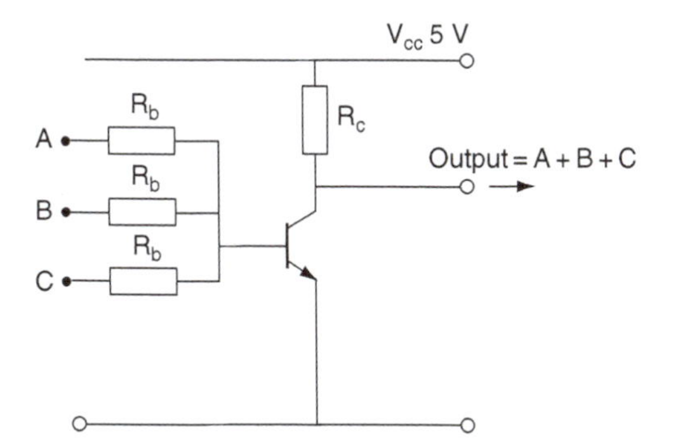

Fig. 18.18 RTL NOR gate

The output is shown as A+B+C with a line, or bar, over the top; the + sign means OR. Thus the A+B+C with the line above means 'not A or B or C' (NOR: NOT OR).

The base resistors R_b have a value that ensures that the base current, even when only one input is high (logic 1), will drive the transistor into saturation to make the output low (logic 0).

Truth tables

Logic circuits operate on the basis of Boolean logic, and terms such as NOT, NOR and NAND derive from Boolean algebra. This need not concern us here, but it is necessary to know that the input–output behaviour of logic devices is expressed in the form of a 'truth table'.

The truth table for the NOR gate is given in Figure 18.19.

In computing and control systems, a system known as TTL (transistor to transistor logic) is often used. In TTL logic 0 is a voltage between 0 and 0.8 volts. Logic 1 is a voltage between 2.0 and 5.0 volts.

In the NOR truth table, when the inputs A and B are both 0 the gate output, C, is 1. The other three input combinations each gives an output of C = 1.

A range of other commonly used logic gates and their truth tables is given in Figure 18.20.

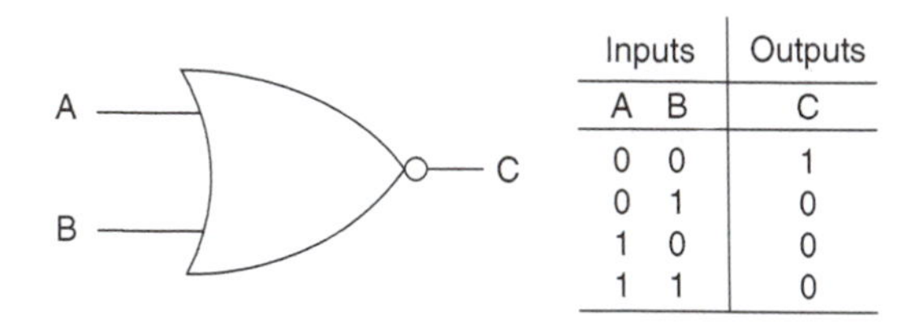

Inputs		Outputs
A	B	C
0	0	1
0	1	0
1	0	0
1	1	0

Fig. 18.19 NOR gate symbol and truth table

Type of Logic Gate	USA Symbol	UK Symbol	Truth Table
AND	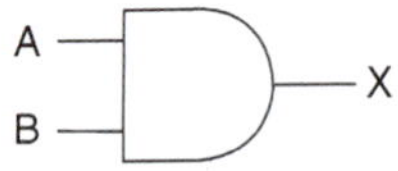 A, B — X	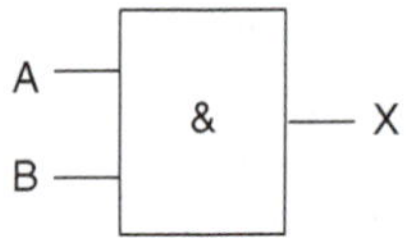 A, B — & — X	Inputs A B / Outputs X: 0 0 → 0; 0 1 → 0; 1 0 → 0; 1 1 → 1
OR	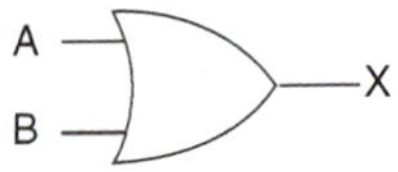 A, B — X	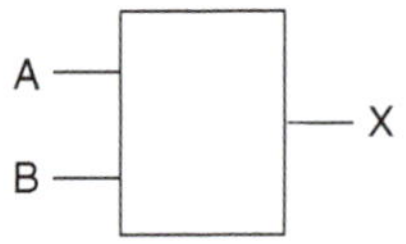 A, B — X	Inputs A B / Outputs X: 0 0 → 0; 0 1 → 1; 1 0 → 1; 1 1 → 1
(NOT AND) NAND 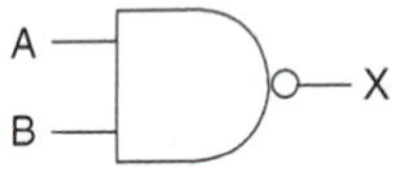 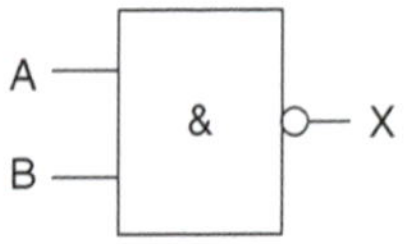	A, B — X	A, B — & — X	Inputs A B / Outputs X: 0 0 → 1; 0 1 → 1; 1 0 → 1; 1 1 → 0
NOT Inverter 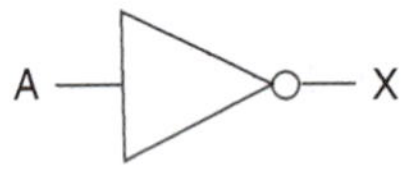 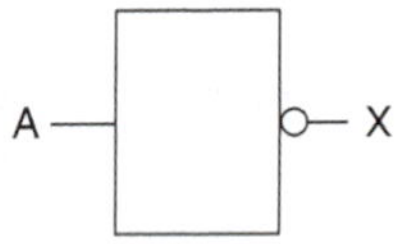	A — X	A — X	Inputs A / Outputs X: 0 → 1; 1 → 0
(NOT OR) NOR 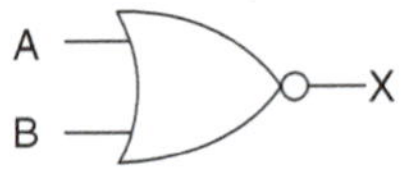 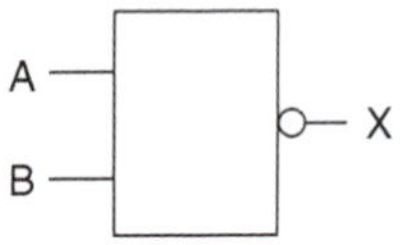	A, B — X	A, B — X	Inputs A B / Outputs X: 0 0 → 1; 0 1 → 0; 1 0 → 0; 1 1 → 0

Fig. 18.20 A table of logic gates and symbols

18.18 Bits, bytes and baud

Each 0 and 1 in a binary code is known as a **bit**, and 8 bits make one **byte**. Because large numbers of bits are used in computer processes, larger units are used. Examples of these larger units are:

a kilobyte, 1 kbyte $= 2^{10} = 1024$ bytes
a megabyte, 1 Mbyte $= 2^{20} = 1\,048\,576$ bytes

The rate at which data in the form of binary bits is moved around a computer system is known as the **baud** rate; a rate of data transfer of one bit per second is = 1 baud. The term baud derives from Jean Baudot, the French technologist. In vehicle systems such as traction control data transfer rates of 250 kbaud to Mbaud are used.

18.19 Summary of formulae

Forward voltage of silicon diode = 0.7 V (approx.)

The diode current may be calculated from

$$I = \frac{Vs - Vd}{R}$$

where R = resistance in ohms, Vs = supply voltage, Vd = diode voltage

The ratio $\frac{I_c}{I_b}$ is called the gain and it is denoted by the symbol β

$$\beta = \frac{I_c}{I_b}$$

18.20 Exercises

18.1 Are electronic components such as alternator diodes mounted on a metal heat sink
- (a) to avoid damage by vibration?
- (b) to conduct surplus electric current to earth?
- (c) to allow heat to be conducted away from the diode to prevent damage?
- (d) to improve the electrical conductivity of the circuit?

18.2 A certain transistor has a current gain of 100. Is the collector current I_c when the base current $I_b = 0.1$ mA
- (a) 10 A
- (b) 10 mA
- (c) 99 mA
- (d) 1 A

18.3 Do active sensors
- (a) process an electrical supply?
- (b) produce electricity that represents the variable they are detecting?

18.4 Do electrical filters
- (a) filter out a.c. frequencies that may interfere with the operation of an electronic system?
- (b) only work in a d.c. circuit?
- (c) only find use in radio circuits?
- (d) never find use in automotive systems?

18.5 State the reason for using an analogue-to-digital converter on computer-controlled systems.

18.6 In an OR logic gate, is
- (a) the output 1 when both inputs are 0?
- (b) the output zero when both inputs are 1?
- (c) the output 1 when any of the inputs are 1?
- (d) the output 0 when any input is 1?

18.7 Do traction control systems have a data transfer rate of
- (a) 10 bits per second?
- (b) 1000 bits per second?
- (c) 10 000 bits per second?
- (d) 250 kbaud to 1 Mbaud?

18.8 In a computer-controlled automotive system, do actuators
- (a) operate devices under the instructions of the control computer?
- (b) sense engine temperature and convert it into a voltage?
- (c) convert analogue signals to digital ones?
- (d) store fault codes?

18.9 Is the part of the computer that controls the movement of data
(a) the A/D interface?
(b) the clock?
(c) the baud rate?
(d) the speed of the vehicle?

18.10 Is a petrol injector
(a) a sensor?
(b) an actuator that operates on the principle of a solenoid?
(c) operated from the cam shaft?
(d) a variable capacitance oscillator?

18.11 Are Zener diodes used
(a) to convert a.c. to d.c.?
(b) to provide surge protection?
(c) as an element in a liquid crystal display?
(d) to block all current flow in the reverse direction?

18.12 In order to limit the current flow through an LED, is
(a) a resistor placed in parallel with the LED?
(b) the supply voltage limited?
(c) a resistor placed in series with the LED?
(d) it essential to increase the dopant level?

Answers to self-assessment questions

Chapter 1

1.1 0.50 mm

1.2 15.12 mm

1.3 Total gap = 0.35 mm. This is to be reduced to 0.15 mm. Total space to be taken up = 0.20 mm. Each thrust washer must be 0.10 mm thicker

1.4 Offset = 2 mm

1.5 X = 0.9 mm

1.6 X = 4.2 mm

1.7 Overall gear ratio = $1.531 \times 4.3 = 6.58 : 1$

1.8 (a) 3, (b) 2, (c) 144, (d) 78.55

1.9 16.15 km/l

1.10 Swept volume = CSA × L × number of cylinders

Swept volume = $55.6 \times 9 \times 4\,\text{cm}^3$

$= 2000.00\,\text{cm}^3 = 2.00$ litres

1.11 (a) 6.00, (b) 14.12, (c) 0.69

1.12 (a) 37.87, (b) 48.70, (c) 39 487 000, (d) 0.0077

1.13 £9.92 p

1.14 Total cost of labour before VAT = 6.33 × 42 = £265.86

Total cost before VAT = £265.86 + £102.58 = £368.44

1.15 Fuel used = 186 ÷ 10.36 = 17.95 litres, correct to two decimal places

1.16 Weight of fuel = 220 × 0.83 = 182.6 kg

Total weight of tank and fuel = 30 + 182.6 = 212.60 kg

1.17 (a) a^6, (b) x, (c) 5^6, (d) t^6, (e) 10

1.18 (a) 3, (b) 9, (c) 81, (d) 5

1.19 (a) 0.125, (b) 0.125, (c) 2, (d) 0.200

1.20 (a) 3, (b) 9, (c) 19, (d) 8

1.21 (a) 13, (b) 18, (c) 7

1.22 $\frac{1}{3} + \frac{1}{2} + \frac{5}{8} + \frac{1}{4} + \frac{3}{8} =$

$$\frac{8+12+15+6+9}{24} = \frac{50}{24} = 21/12 \text{ litres}$$

The lowest common denominator 2, 3, 4, 8 = 24

1.23 $(4+3+1)+(\frac{1}{2}+\frac{3}{4}+\frac{7}{8}) = 8 + \frac{4+6+7}{8} = 8 + 2\frac{1}{8} = 10$

1.24 (a) $\frac{3}{10}$, (b) $\frac{1}{8}$, (c) $\frac{3}{8}$, (d) $\frac{5}{8}$

1.25 (a) $\frac{15+10+24-5}{30} = \frac{44}{30} = 1\frac{14}{30}$

(b) $2\frac{1}{3} - 1\frac{1}{4} + \frac{5}{8} = 7/3 - 5/4 + 5/8$

$$= \frac{56-30+15}{24} = \frac{41}{24} = 117/24$$

(c) $\frac{1}{32} - \frac{1}{16} + \frac{3}{8} + 9$

$$= 9 + \frac{1-2+12}{32}$$

$= 9\,11/32$

1.26 (a) $\frac{1}{9}$

(b) $\frac{4}{5} \times \frac{5}{8} \div \frac{7}{8} \times \frac{8}{9} = \frac{4 \times 5 \times 8 \times 8}{5 \times 8 \times 7 \times 9}$

$= \frac{4 \times 8}{7 \times 9} = \frac{32}{63}$

(c) $\frac{1}{64}$

1.27 (a) $\left(\frac{1}{3} \div \frac{1}{4}\right) - \frac{1}{6} = \frac{4}{3} - \frac{1}{6} = \frac{8-1}{6} = \frac{7}{6} = 1\frac{1}{6}$

(b) $\left(\frac{1}{81} + \frac{1}{9}\right) \times \frac{12}{4} = \frac{10}{81} \times \frac{12}{4} = \frac{30}{81}$

(c) $\left(\frac{4}{3} \div \frac{1}{6}\right) + \frac{3}{8} + \frac{1}{32} = 8 + \frac{3}{8} + \frac{1}{32}$

$= 8\,13/32$

1.28 Largest first: 0.21 0.201 0.2001 0.2 0.021

1.29 Smallest first: $\frac{3}{8}$ $\frac{1}{2}$ $\frac{4}{5}$ $\frac{5}{6}$

1.30 110111_2

1.31 $(10 \times 16^2) + (2 \times 16^1) + (15 \times 16^0)$
$= 2560 + 32 + 15 = 2607$

Chapter 2

2.1 (a) Arithmetic mean = 6.22, median = 6
(b) Mode = 9
(c) 4 and 7
(d) Median = $\frac{9+11}{2} = 10$
(e) 55 marks

2.3 Angle of pie chart for CO = 155.2°
Angle of pie chart for HC + NO_x = 170.7°
Angle of pie chart for SO_2 = 34.1°

2.4 Angle of pie chart for lack of fuel = 98.9 degrees
Angle of pie chart for electrical problems = 142.1 degrees
Angle of pie chart for tyre problems = 86.5 degrees
Angle of pie chart for fuel system faults = 32.5 degrees

2.5 Profit = 45 − 39.50 = £6.50 per hour
Angles on pie chart

Technicians' pay = 76 degrees
Heating of premises = 56 degrees
Admin. and other overheads = 184 degrees
Profit = 52 degrees

2.10 Number of people working less than 30 hours = 26

Chapter 3

Exercises – Section 3.3

1. 320
2. 2
3. 523.8
4. 2800
5. 11
6. 1437.3
7. 19.8
8. 4.22×10^7
9. 47 100
10. 40

Exercises – Section 3.4

1. $3ab + 12ac + 6ab - 9ac - 5bc = 9ab + 3ac - 5bc$
2. (a) $18ab - 8ac$, (b) $14x + 21y$, (c) $6x + 21$
3. $5x - 3x + 30 + 8 - 2x = 38$

Exercises – Section 3.5

(a) $6 + 2x + 3y + xy$, (b) $x^2 + 5x + 6$,
(c) $x^2 + x - 6$, (d) $x^2 - 6x + 8$

Exercises – Section 3.6

(a) $2x(x-2)$, (b) $(a+2b)(x+2)$, (c) $ax(x+1)$

Exercises – Section 3.7

1. $x = 4$
2. $y = 3$
3. $d = 30$
4. $x = 5$
5. $x = 4$
6. $x = 16$
7. $x = 1/3$
8. $x = -9$

Exercises – Section 3.8

1. $x = \dfrac{y - c}{m}$
2. $L = \sqrt{(T/4Mn^2)}$
3. $d = \dfrac{C}{\pi}$
4. $I = \dfrac{V}{R}$
5. $L = \dfrac{4V}{\pi d^2}$

Exercises – Section 3.10

3. Graph
4. From the graph, piston velocity at $90° = 8.4\,m/s$.
5. The law is $P = 1.5\,W + 1.2$

Chapter 4

4.1 720°

4.2 $\dfrac{3000}{60} \times 2\pi = 314.2\,rad/s$

4.3 36°24′

4.4 Largest acceptable angle $31°10' + 55' = 32°5'$
Smallest acceptable angle $= 31°10' - 55' = 30°15'$

4.5 $33°55' - 31°10' = 2°45'$

4.6 $180° + 10 + 45° = 235°$

4.7 $15° + 20° = 35°$

4.8 31°

4.9 (c)

4.10 65°

4.11 135°

4.12 12°

4.13 45°

4.14 $y = \sqrt{(140^2 - 55^2)} = \sqrt{(19600 - 3025)} = \sqrt{16575} = 128.74\,mm$
Original distance from centre of gudgeon pin to centre of crank $= 140 + 55 = 195\,mm$
Distance $x = 195 - y = 195 - 128.74 = 66.26\,mm$

4.15 $\frac{1}{2}d = \sqrt{(25^2 - 20^2)} = \sqrt{(625 - 400)} = \sqrt{225} = 15.d = 2 \times 15 = 30\,mm$
Ans. 4.16. $x = \sqrt{(200^2 + 100^2)} = \sqrt{50000} = 223.6mm.$

4.17 $\frac{1}{2}d = \sqrt{(27.5^2 - 13.75^2)} = \sqrt{(756.25 - 189.06)} = \sqrt{567.19} = 23.82$
$\therefore d = 2 \times 23.82 = 47.64\,mm$

4.18 Distance moved during one revolution of the wheel and tyre $= 2\pi r = 2 \times 3.142 \times 0.25 = 1.571\,m$
Distance moved when wheel makes 1000 revolutions $= 1.571 \times 1000 = 1571\,m = 1.571\,km$

4.19 Effective area of valve opening $= \pi \times$ diameter of valve $\times$ valve lift
$= 3.142 \times 35 \times 12 = 1319.6\,mm^2$

4.20 Distance travelled by vehicle per minute = rev/min of wheel × rolling circumference
$= 600 \times \pi \times 0.3 \times 2 = 1131.1\,m/min$
Speed of vehicle in km/h $= \dfrac{60 \times 1131.1}{1000} = 67.87\,km/h$

4.21 Original area of valve opening $= \pi \times 30 \times 14 = 1320\,mm^2$
Area after opening up ports and increasing valve lift $= \pi \times 30.5 \times 15 = 1437.5\,mm^2$
Increase in area $= 1437.5 - 1320 = 117.5\,mm^2$
Percentage increase
$= \dfrac{\text{increase in area of valve opening}}{\text{original area of valve opening}}$
$= \dfrac{117.5}{1320} \times 100 = 8.9\%$

4.22 Length of arc $= \dfrac{\theta \times \text{circumference}}{360°} = \dfrac{25° \times 3.142 \times 400\,mm}{360°} = 87.3\,mm$

4.23 (a) 0.669, (b) 0.9976, (c) 0.4848, (d) 0.2588, (e) 0.3947, (e) 0.5000

4.25 23.57 m/s

4.26 (a) 17.04, (b) 14, (c) 5

4.28 (a) $x = 4.66\,m$, (b) $x = 138.93\,mm$, (c) $x = 90.57\,mm$

4.29 (a) $\theta = 53.13°$, (b) $\theta = 28.07°$, (c) $\theta = 22.62°$

4.30 (a) 26.6°, (b) 36.87°, (c) 28.07°, (d) 53.06°

4.31 (a) $x = 7.98$ cm, (b) $x = 7.46$ cm, (c) $x =$ 156.25 mm

4.32 $\theta = 73.3°$, $x = 208.8$ mm
Distance that piston has moved from TDC = 260 − 208.8 = 51.2 mm

Chapter 5

5.1 2250 N
5.2 3 N/mm
5.3 (a)
5.4 Resultant 135 N at 214°12′ to A. The equilibrant is equal in magnitude but opposite in direction.
5.5 98.8 N. Direction towards at angle of 41°41′ to force A.
5.6 (a) 7 mm, (b) 200 N
5.7 Horizontal component = 5 × cos 40° = 5 × 0.766 = 3.83 kN
5.8 Jib = 5.25 T, tie = 3.05 T
5.9 Side thrust F = 1180 N
5.10 Force A = 13 N, force B = 48 N
5.11 Force Q = 1080 N, force P = 1340 N
5.12 Pressure on piston crown = 25 bar
15.13 (c)
15.14 (b)

Chapter 6

6.1 0.00221 mm
6.2 (b) 3.33 MN/m^2, (c) 0.007 mm
6.3 49.02 MN/m^2
6.4 Angle of twist = 0.012 degrees
6.5 (a) 28.4 MN/m^2, (b) 26.7 MN/m^2
6.6 24.54 MN/m^2
6.7 fs = 595 Mpa
6.8 Maximum permissible torque = 157.2 Nm
6.9 Tensile force = 10.6 kN
6.10 164.7 GN/m^2
6.11 19.22 MN/m^2
6.12 (a) 1500 N, (b) 19.1 MN/m^2
6.13 88 kW

Chapter 7

7.1 (b) 33.3 N
7.2 Rr = 0.955 tonne; 1.8 − 0.955 = 0.845 tonne
7.3 $x = 1.527$ m
7.4 $R_2 = 3.8$ T, $R_1 = 7 - 3.8 = 3.2$ T
7.5 $R_2 = 37.5/10 = 3.75$ T; $R_1 = 7.5 - 3.75 =$ 3.75 T
7.6 F = 2667 N
7.7 T = 108 Nm
7.8 (a) 2.8, (b) 448 Nm
7.9 40 N
7.10 $R_f = 0.96$ T; Rr = 1.8 − 0.96 = 0 84 T

Chapter 8

8.1 200 kJ
8.2 7.88 kJ
Note: The answers will vary depending on the accuracy of the diagram
8.3 (a) Work done per revolution of crank = 2514 J
(b) Power = 50.28 kW
8.4 (a) Engine power = 56.6 kW
(b) Gearbox output torque = 204.7 Nm
8.5 (a) KE = 4442.5 J
(b) Work done = 4442.5 JPower = 2.96 kW
8.6 (a) 300 rev/min
(b) Crown wheel torque = 2.394 kNm
8.7 F = 88.9 kN
8.8 (a) MR = 160
(b) Effort = 16.7 N
8.9 MR = 26.8
8.10 Effort = 667 N
8.11 MA = 100; MR = 133.3; efficiency = 0.75 = 75%
8.12 3200 kJ = 3.2 MJ
8.13 Power = 2400 watts = 2.4 kW
8.14 Input power = 3 kW

Chapter 9

9.1 $\mu = 0.48$
9.2 1 Nm
9.4 $\mu = 0.57$
9.5 4.772 kN
9.6 T = 122 Nm
9.7 (a) The axial force W = 119.5 kN
(b) The axial force W = 86.5 kN
(c) Examine a workshop manual to assess the probable effect of the reduction in the axial force
9.8 2.98 kN
9.9 0.124 kW
9.10 (a) 440 Nm
(b) 135.4 kN
9.11 59%.

Chapter 10

10.1 2.78 m/s, 27.8 m/s, 15.56 m/s, 28.44 m/s
10.2 $v = 17\,\text{m/s}$
10.3 $a = -4.5 \times 10^5\,\text{m/s}^2$. The minus sign indicates retardation
10.4 s = 53.26 m
10.5 $u = 12.96\,\text{m/s}$
10.6 (a) 13.33 m/s
(b) 25 m/s
(c) 56.9 m/s
10.7 Average velocity = 37.5 m/s; *time* = 36.4 seconds; acceleration = $0.41\,\text{m/s}^2$
10.8 Distance covered in final 30 seconds = 150 − 110 = 40 m
10.9 Time taken to reduce speed 1.2 s. Distance covered during the deceleration = 9.5 m
10.10 Distance covered = velocity × time = 20 × 16.5 = 330 m
10.11 $v = 80\,\text{m/s}$
10.12 $\omega = 377\,\text{rad/s}$
10.13 $\alpha = 523.6\,\text{rad/s}^2$
10.14 Angular velocity = 50.1 rad/s
10.15 Force = 75.9 N
10.16 Force acting on the driver = 766 N

Chapter 11

11.1 Power = 58.34 kW
11.2 (a) 108 kW
(b) $0.17\,\text{m/s}^2$
11.3 2.44 kN
11.4 Max. speed = 73 km/h; max. power for acceleration = 28kW at 40 km/h
11.5 (a) $3.88\,\text{m/s}^2$
(b) 6.208 kN
11.6 Tractive resistance = 800 N
11.7 (a) 1589 N
(b) 18.89 kW
11.8 Overturning velocity = 30.2 m/s = 109 km/h
11.9 $\mu = 16.7^2$
11.10 Max. acceleration = $4.23\,\text{m/s}^2$
11.11 Dynamic load on front axle
= static load + load transfer
= 6752 + 1108
= 7860 N
Dynamic load on rear axle
= static load − load transfer
= 6001 − 1108
= 4893 N
11.12 125.6 kN

Chapter 12

12.1 444 N
12.2 79 N
12.3 (a) 421.5 N
(b) 1.708 kg
12.6 720 N
12.7 Each balance mass = 0.90/2 = 0.45 kg
12.8 Periodic time t = 1.1 s; frequency f = 0.9 Hz = 54.6 vib/min
12.9 Acceleration = $4936\,\text{m/s}^2$; force = 3.445 kN
12.11 (a) Acceleration = $1974\,\text{m/s}^2$
(b) force = 355 N

Chapter 13

13.1 $P_2 = 12\,bar + 1\,bar = 13\,bar$
$T_2 = 463.4\,K = 190.4°C$
13.2 Fuel required $m_f = 0.55/14.7 = 0.037\,g$
13.3 8.802 bar
13.4 68%
13.5 $25.87 - 1.016 = 24.85$ bar gauge, with the 11:1 compression ratio
With the 9.5:1 compression ratio $p_2 = 1.016 \times 9.5^{1.35} = 1.016 \times 20.9 = 21.22$ bar abs and $21.22 - 1.016 = 20.2$ bar gauge
13.6 At end of compression $T_2 = 2.83 \times 343 = 971\,K = 698°C$
13.7 (a) $83.3\,cm^3$
(b) 5.531 kJ
13.8 0.78 mm
13.9 242°C
13.10 1890 kJ
13.11 $T_2 = 200°C$
13.12 $T_2 = 417 - 273 = 144°C$
13.13 $T_2 = 646\,K$

Chapter 14

14.1 31.2 kW
14.2 134 kW; 89.6%
14.3 bp = 95 kW; ip = 118.8 kW
14.4 38.4 kg/h
14.5 6.7%
14.6 91 mm
14.7 28%
14.8 ip = 33 kW; ME = 76%
14.9 $5.08\,m^3/h$
14.10 BTE = 25.6%, 28.2%, 30%, 29.2%, 27.3%

Chapter 15

15.1 (a) 9.2 : 1
(b) 0.598 = 59.8%
15.2 64.8%
15.3 0.652 = 65.2%
15.4 100.6 kW
15.6 (a) 28%
(b) ASE = 0.57 = 57%
(c) Relative efficiency = 0.49 = 49%
15.7 (a) 0.594 or 59.4%
(b) 0.471 or 47.1%
15.8 (a) $177.7\,cm^3$
(b)

Chapter 17

17.1 (b) 6 Ω
17.2 (c) 2 A
17.3 (a) 8 V
17.4 (c) 1.5 Ω
17.5 (c) 8 A
17.6 (b) 6 A
17.7 (d) 4 Ω
17.8 (b) 3 A
17.9 (c) 7.2 V
17.10 (c) 250 μF
17.11 (a)
17.12 (b)
17.13 (b)
17.14 (c) Ohms
17.15 (b)
17.16 (c) 240 W
17.17 (a)
17.18 (a)
17.19 (c) 100 seconds

Chapter 18

18.1 (c)
18.2 (b)
18.3 (b)
18.4 (a)
18.6 (c)
18.7 (d)
18.8 (a)
18.9 (b)
18.10 (b)
18.11 (b)
18.12 (c)

Index

This page intentionally left blank